the moment of complexity

MARK C. TAYLOR

the moment of complexity

EMERGING NETWORK CULTURE

the
university
of chicago press
chicago and
london

The University of Chicago Press, Chicago 60637
The University of Chicago Press, Ltd., London
© 2001 by The University of Chicago
All rights reserved. Published 2001
Paperback edition 2003
Printed in the United States of America
09 08 07 06 05 04 03   2 3 4 5
ISBN: 0-226-79117-3 (CLOTH)
ISBN: 0-226-79118-1 (PAPERBACK)

Library of Congress Cataloging-in-Publication Data
Taylor, Mark C., 1945–
    The moment of complexity : emerging network
culture / Mark C. Taylor.
        p.   cm.
    Includes bibliographical references and index.
    ISBN 0-226-79117-3 (cloth : alk. paper)
    1. Complexity (Philosophy) 2. Culture—Philosophy.
I. Title.
    B105.C473 T39 2001
    117—dc21   2001001079

For Herbert A. Allen

contents

figures

acknowledgments

A work is never the creation of a solitary individual but is always the product of a "colony of writers" whose thoughts and words circulate through the author. For this book more than any I have written, others have been critical. As I have ventured into new territories, friends and colleagues have offered me help, guidance, and, most important, criticism. I would like to express my sincere appreciation to a group of colleagues at Williams College, who have been unusually generous with their time: Colin Adams, for translating Fermat's theorem into language a humanist can understand; Susan Dunn, for help translating French texts; Helga Druxes, for explaining the subtleties of *Bildung;* Stuart Johnson, for *showing* me the beauty of fractals and cellular automata; Hank Payne, for backing ventures in which he did not quite believe; Frederick Rudolph, for sharing his uncommon knowledge of the history of American education; David L. Smith, for years of conversation and support; Betty Zimmerberg, for guidance in unraveling the complexities of neuroscience; Robert Trumbull, for locating books I never would have found; Margaret Weyers, for making it all happen day in and day out; and above all to Bill Dewitt, Chip Lovett, and Bill Wootters, for their patience in explaining difficult scientific concepts and care in reviewing drafts of the manuscript. To others beyond the Purple Valley I also owe profound debts of gratitude: Richard Powers, the novelist who understands this territory unlike any other, for the most extraordinary response to a manuscript I have ever received; John W. Chandler, President Emeritus of Williams College, for thirty years of wisdom and guidance; Michael McPherson, President of Macalester College, and Morton Owen Schapiro, President of Williams College, for sharing their sophisticated knowledge of the economics of higher education;

Clayton Spencer, Harvard University, for telling me what *not* to write; the late Michael Hooker; Ruel Tyson and his colleagues at the University of North Carolina's Institute for Arts and Humanities for a semester of stimulating discussion; Dean Smith for helping me to appreciate the complexity of basketball; Edith Wyschogrod, Rice University, and Charles Winquist, Syracuse University, for words I always know I can trust; Thomas Carlson, José Márquez, and John Kim, for reversing roles and becoming my teachers; Alan Thomas, University of Chicago Press, for always understanding when others do not; Michael Rock, for his graphic imagination; my remarkable colleagues at Global Education Network—Stephen Greenberg, Jon Newcomb, Kim Wieland, Herbert Allen III, Shana Fisher, and Alexander Parker, for teaching me lessons I never expected to learn. Closer to home: Dinny Taylor, for getting me through trigonometry and always explaining what's under the hood; Aaron and Kirsten Taylor, for making me hope. Finally, words of thanks, though they hardly seem adequate, to Herbert A. Allen, President of Allen & Company, for believing when others doubted.

the moment of complexity

A few simple rules, combined in a few elegant ways, and blamm-o. The thing works. It runs. The world does move. The rules churn. The descriptions step their way through their own internal logic. The lines of code set more switches, change more states. Commands produce results.

The word made flesh.

richard powers, *plowing the dark*

Some days I will say yes, and then odd days
It seems that things say yes to me.
And stranger still, there are those times
When I become a yes

(And they are moments of the Calm).

kevin hart, "flying home"

introduction

We are living in a moment of unprecedented complexity, when things are changing faster than our ability to comprehend them. This is a time of transition betwixt and between a period that seemed more stable and secure and a time when, many people hope, equilibrium will be restored. Awash in a sea of information that seems to have no meaning and bombarded by images and sounds transmitted by new media, many people have lost a sense of direction and purpose and long for security and stability. Stability, security, and equilibrium, however, can be deceptive, for they are but momentary eddies in an endlessly complex and turbulent flux. In the world that is emerging, the condition of complexity is as irreducible as it is inescapable. While the moment of complexity inevitably generates confusion and uncertainty, today's social, economic, political, and cultural transformations are also creating possibilities for apprehending ourselves in new ways. To understand our time, we must comprehend complexity, and to comprehend complexity, we must understand what makes this moment different from every other.

What distinguishes the moment of complexity is not change as such but rather the acceleration of the rate of change. Everything moves faster and faster until speed becomes an end in itself.[1] At this point, all that is solid seems to melt away, creating a sense of vertigo that is welcomed by some as the end of false consciousness and denounced by others as catastrophic nihilism. For many people, confusion and uncertainty create a desire for simplicity that leads to a futile longing to return to basic values and foundational beliefs. In today's world, however, simplicity has become an idle dream that no longer can be realized. While it remains uncertain whether history has a purpose, it seems clear that development—personal as well as natural and historical—has a direction: things tend to move from lesser to greater complexity. The task we now face is not to reject or turn away from complexity but to learn to live with it creatively.

As always when in the midst of extensive and rapid changes, it is difficult to assume a critical perspective from which to assess the significance of what is occurring. Far from discouraging analysis, the recognition of emerging complexity invites a variety of interpretations and prognostications. For some, our era is marked by the shift from industrialism to postindustrialism; for others, by the movement from modernism to postmodernism or from a culture of production to a culture of reproduction; and for still others, by the change from market capitalism to multinational, informational, or digital capitalism. Though such characterizations of the transition now occurring represent different but related interpretive perspectives, there is widespread consensus that these changes are inseparable from the explosive development of cybernetic, information, and telematic technologies since the Second World War.

While most of these technologies were originally developed with government support and intended for military purposes, they were quickly commercialized and spread throughout society. The clearest example of this pattern of development is, of course, the World Wide Web, which was designed as part of the military defense strategy during the Cold War and now has become integral to the new economic order. As in the past, outspoken supporters and critics of new technologies confidently predict that proliferating innovations will completely change the world as we know it. Having heard such declarations many times before, it is easy to be cynical and dismiss them as misguided hype. Nonetheless, such claims are not always exaggerated; the automobile, for example, did quite literally change the face of the earth. I am convinced that the technological developments of the last half-century are creating conditions for a revolution as profound and far-reaching as the industrial revolution. Information and telematic technologies are recasting the very social, political, economic, and cultural fabric of life.

In order to assess these changes, it is important to understand information as inclusively as possible. Information is not limited to data transmitted on wireless and fiber-optic networks or broadcast on media networks. Many physical, chemical, and biological processes are also information processes. This expanded notion of information makes it necessary to reconfigure the relation between nature and culture in such a way that neither is reduced to the other but that both emerge and coevolve in intricate interrelations. As these feedback and feed-forward loops become more complex and as change accelerates, development approaches the moment of complexity, which is "the tipping point" where *more is different*. What is emerging in this moment is a new *network culture* whose structure and dynamics we are only beginning to fathom. My aim in the following pages is to develop an interpretation of network culture that will make it possible to understand what is occurring more adequately and to respond more effectively to the challenges and opportunities we face.

The course that has brought me to the moment of complexity has been long and circuitous. For more than three decades, I have been attempting to elaborate a philosophy of culture that contributes to our understanding of contemporary experience. This journey has taken unexpected twists and turns, which have led in directions I never anticipated. As I look back, I am surprised to discover the consistency and logic in the development of my thinking over the years. I have gradually come to suspect that this errant journey is not merely my own but reflects broader currents circulating through me. While beginnings are always unclear, a decisive turn in my own life and thought occurred during the late 1960s. As a young college student caught up in the turmoil of the time, it seemed clear that then, as now, we were living through a period of significant transformation. Though the nature and importance of these changes continue to be the subject of debate and reevaluation, there can be little doubt that what took place during the late 1960s continues to shape our world in more and less obvious ways. Indeed, as I will attempt to show, there is a direct line connecting the social unrest that reached the boiling point in 1968 and the world-transforming events of 1989. During the middle of the twentieth century, the ideological, political, and cultural divisions of the Cold War formed the foundation of a world system that created stability in the midst of considerable tension. The enormous complexity of the postwar era became comprehensible and manageable by dividing the world into two opposing camps, which simultaneously threatened and needed each other. When the Berlin Wall finally fell, it was much more the result of economic developments, technological change, and the spread of media and popular consumer capitalism than of a calculated military strategy. Often overlooked when considering these developments is the fact that the same 1960s counterculture that resisted the war in Vietnam and marched for civil rights also inspired the technological revolution that made a new economic order all but inevitable and that created the necessary conditions for the emergence of network culture.[2] As the aftereffects of many of these social and technological changes continue to reverberate, oppositions and antitheses that have long structured thought and guided action are rapidly unraveling.

5

Even during the heady days of 1968, social and political engagement was for me always inseparable from philosophical and theological reflection. I believed that the crisis in which we found ourselves was not merely political and economic but was, perhaps more importantly, also cultural and even spiritual. It was at this time that I began studying Hegel and Kierkegaard. Far from abstract speculations, their often opaque writings opened new perspectives from which to interpret current events. It seemed as if the struggle between "the system" and its critics, which was being discussed daily in classrooms and acted out on the streets, had been scripted a century and a half earlier in Hegel's and Kierkegaard's writings.[3] As I have continued to study these formative figures over the years, my appreciation for the importance of their insights has grown. I would now go so far as to insist that it is impossible to understand network culture adequately without recognizing the lasting significance of the philosophical, theological, and artistic contributions of major figures during the late eighteenth and early nineteenth centuries.

Among the many lessons I have learned from Hegel and Kierkegaard as well as the 1960s, two are noteworthy in this context. First, there is a religious dimension to *all* culture. In order to appreciate the far-reaching implications of religion, it is necessary to move beyond its manifest forms to examine the more subtle and complex ways in which it influences personal, social, and cultural development. Religion is often most intriguing and influential where it is least obvious. If we are to understand network culture, the eye must be trained to glimpse religion where it remains nearly invisible. Second, religion is inseparable from philosophy, literature, literary criticism, art, and architecture, as well as science, technology, capitalism, and consumerism. Multiple threads have been intricately interwoven to create the complex webs now entangling us. While these webs cannot be unraveled, their strands can be distinguished and analyzed in ways that illuminate contemporary experience.

As I was completing a book on Hegel and Kierkegaard in the late 1970s, I began reading the work of leading French post-structuralists, whose influence was then spreading rapidly from literary studies to other fields in the arts and humanities.[4] The most important of these figures for me was Jacques Derrida. When I first encountered Derrida's texts, it seemed obvious to me that he was obsessed with many of the same theological and philosophical issues with which I had been wrestling in my work on Hegel and Kierkegaard. Indeed, post-structuralism's critique of structuralism appeared to repeat Kierkegaard's "unscientific" criticism of Hegel's purportedly comprehensive system. The more deeply I immersed myself in Derrida's controversial writings, the more I became convinced that he is one of the most significant *theological* thinkers of the latter half of the twentieth century. At the time, the association of Derridean deconstruction with religion and theology was far from obvious. Most of Derrida's supporters saw no plausible connection between deconstruction and religion, and most of his opponents tended to understand deconstruction as a pernicious form of nihilism that threatened religious beliefs, moral values, and sociopolitical action. Protests to the contrary notwithstanding, I remained persuaded that there is an important connection between religion and post-structuralism. In a series of books written during the 1980s and early 1990s, I traced the relation between the tradition of Continental philosophy and literature from which post-structuralism emerged, on the one hand, and, on the other, major trajectories in nineteenth- and twentieth-century theology.[5] My goal in these books was both historical and constructive. I was attempting to appropriate deconstructive insights to develop a "constructive" theological or, more precisely, a/theological position, and to elaborate a creative philosophy of culture. Having already learned the impossibility of separating philosophy and theology from other dimensions of culture, my investigation quickly expanded to literature, architecture, and the visual arts. I became especially interested in painting and architecture. While Derrida repeatedly writes about drawing, painting, and architecture, his preoccupation with textuality leads him to overlook distinctive features of the visual arts. It is no accident that his most extensive analysis of the visual arts is devoted to the theme of blindness—*Memoirs of the Blind: The Self-Portrait and Other Ruins.* As my investigation expanded, I came to suspect that Derrida's ambivalence toward the visual arts results from his indebtedness to the aniconic Jewish religious and theological tradition.

In an effort to confirm this suspicion, I undertook what eventually proved to be a long exploration of the interplay of religion, art, and architecture throughout the twentieth century.[6] Once again I found it necessary to return to the end of the eighteenth century to understand the end of the twentieth century. Events set in motion during the crucial decade of the 1790s in Europe unexpectedly culminate in developments that occur in the 1990s. In response to the personal disruptions and social dislocations wrought during the early years of the industrial revolution, artists assumed the mantle of religious prophets. Influential members of what eventually came to be known as the avant-garde advanced utopian visions, which, though not acknowledged as such, were actually artistic versions of the kingdom of God. For those with eyes to see, these alternative visions and the controversies they provoked repeated and extended ancient theological disputes. One of the most important questions in these debates

7

was whether the utopia is imagined as immanent in, or transcendent to, natural and historical processes. While some artists established an opposition between art and world by insisting on the autonomy of the work of art, others maintained that true art must be socially productive and politically effective. Though the substance of utopian visions has varied, most have agreed that the task of the avant-garde artist is to lead the struggle to transform the world into a work of art. This sociopolitical agenda has inspired artistic programs as different as Russian Constructivism, the Bauhaus, de Stijl, and Futurism. The dreams of the avant-garde, however, foundered on the battlefields of twentieth-century Europe. And yet, this was not the end of the program. What the European avant-garde dreamed in the early decades of this century has become a reality—albeit in totally unexpected ways—in late-twentieth-century America.

When Marcel Duchamp knocked art off its pedestal by signing a urinal and placing it on display, he implicitly signaled the end of art. The end of art, however, is not its disappearance. Just as God dies when natural and historical processes are deemed sacred, so art effectively ends when it becomes indistinguishable from other forms of cultural production and reproduction. If an industrially produced object is a work of art, the question is no longer, "What *is* art?" but is, "What *is not* art?" In Andy Warhol's provocative work, the broad implications of Duchamp's readymades become unmistakable.[7] For Warhol, postwar consumer capitalism is the parodic fulfillment of the avant-garde's effort to transform the world into a work of art. Warhol's art stages what is, in effect, a realized eschatology in which all traces of a beyond are erased in a kingdom that is of this world. If art is to be socially transformative, it cannot remain limited to so-called high culture and confined to the sacred precincts of the museum but must take to the streets. On the street, the line separating art from nonart becomes endlessly obscure. Determined to annul the opposition between high and low culture, Warhol turns to the kitsch of popular culture, long disdained by most modernists, for artistic material and inspiration. Though lacking the theoretical language with which to express his insights, Warhol realizes that contemporary media and visual culture make it virtually impossible to distinguish image and reality with any degree of certainty. Furthermore, he understands that a new cultural condition is created by technological innovation and changing economic practices. Far from mere entertainment, television and related reproductive technologies transform our very sense of reality. Simultaneously interpreting and promoting the burgeoning world of media and celebrity in his studio, The Factory, and his magazine, *Interview,* Warhol calculatedly stages what Baudrillard eventually describes as "the disappearance of the real." In his endless play of images and simulacra, Warhol demonstrates the transformative effect of the interrelation of art, technology, and consumer capitalism. At one point, he goes so far as to claim that "Business art is the step that comes after Art. I started as a commercial artist, and I want to finish as a business artist. After I did the thing called 'art' or whatever it's called, I went into business art. I wanted to be an Art Businessman or a Business Artist." While obviously an ironic criticism of the flourishing gallery system, which, of course, promoted and marketed Warhol's art with remarkable success, the implications of this suggestive comment extend far beyond the art world. In a visual culture increasingly governed by electronic media, the currency of exchange is image. The so-called new economy of media and information society transforms use value into exchange value; wealth is created by buying and selling images as well as by trading information, which appears as electronic data on computer screens. "Being good at business," as Warhol concludes, "is the most fascinating kind of art."[8]

In the late 1960s, it was impossible to predict the ways in which the information and telematic technologies Warhol deemed so significant would transform society and culture by the end of the century. From my investigation of art and architecture in the 1980s, I concluded that it was necessary to extend my analysis by examining how the interplay of religious, artistic, economic, and technological processes is creating a new cultural condition. This work resulted in a series of essays, which I eventually published in *About Religion: Economies of Faith in Virtual Culture.* As my appreciation for the importance of information and telematic technologies grew, I began experimenting with new ways of teaching and writing. During the 1980s, many literary critics argued for the importance of developing new writing strategies, which, in some sense, would be performative. Most of these discussions focused on the endless play of written language but did not extend to other modes of cultural production. The more I worked on architecture and the visual arts, the more convinced I became of the importance of teaching and writing with images as well as words. In the fall of 1991, I was writing the chapter of *Disfiguring* on Warhol, which is titled "Currency," when I had the idea of using teleconferencing technology to teach a global seminar with my colleague Esa Saarinen, who was at the time a professor of philosophy at the University of Helsinki. One year later, we successfully completed the first international tele-seminar in which students met weekly for a semester to discuss the philosophical presuppositions and implications of new technologies. Hard though it is to believe, none of the students in that seminar had previously used E-mail. By the end of the course, we were all persuaded of the social significance and educational potential of these new technologies. Recognizing the novelty of what we were doing, Saarinen and I carried on an extensive E-mail correspondence about our experiment throughout the semester. At the end of the term, we reworked the E-mails for a book—*Imagologies: Media Philosophy*. Just as we sought to bring together theory and practice in the teleseminar, so we wanted to integrate form and content in the book. We were fortunate to be able to work with Marjaana Virta, who is an extraordinarily talented graphic designer, to create a highly visual book in which word and image interact in creative ways.[9]

Since the Helsinki seminar, I have continued to explore new ways to integrate technology and teaching. In one of my most successful courses, *Cyberscapes*, I have developed a media lab in which students learn to create multimedia hypertexts to probe philosophical questions and analyze cultural developments. With the growth of the web, I also created a CyberCollege for Alumni/ae in which my courses for undergraduates at Williams College were webcast synchronously and asynchronously to alumni/ae throughout the world. Students and graduates met in virtual environments to discuss issues raised in the course. To help faculty colleagues explore new technologies, I established a Center for Technology in the Arts and Humanities and secured funding to support research. Somewhat to my surprise, I discovered that these efforts were met with indifference and often overt hostility by many of my colleagues both at Williams and elsewhere. Having become discouraged by the inability or unwillingness of colleges and universities to meet the challenges of electronic technologies for higher education, I have recently pursued an alternative course. In 1999, I founded the Global Education Network with Herbert A. Allen, who is the president of the leading New York investment firm, Allen & Co. In my work with the Global Education Network (GEN), I am attempting to extend the experiments begun in the Helsinki seminar and the CyberCollege by putting the theory of network culture developed in the following pages into practice. GEN brings together educators, educational institutions, investors, and businesses to provide high quality on-line education in the liberal arts for people of all ages throughout the world.

While the Helsinki seminar was an enormous success, I was never completely satisfied with *Imagologies*. I was intrigued by the possibilities created by the interplay of words and images but was disappointed that circumstances made it necessary to develop the graphic design for the book *after* the text had been written. If visual design is not to be merely illustrative, it must be integrated into the formation of the argument *from the beginning* of the process. As I worked with my students in the media lab, I discovered new ways to formulate analyses and arguments by linking words, sounds, and images. From the interplay of the verbal, visual, and auditory, new and often surprising insights emerge. Drawing on collage and montage strategies long used by artists, I experimented with new kinds of printed and electronic works. As I began to imagine my next book—*Hiding*—I decided to collaborate with artists and designers *before* I started to write. My goal was to have the designers become effective coauthors of a multimedia work that would explore new alternatives for philosophical analysis and cultural interpretation. In Susan Sellars and Michael Rock, of the New York design firm 2 x 4, I found extraordinarily creative artists with unusual critical acumen. *Hiding* performs the argument it articulates by fashioning a work in which written text, visual images, contrasting colors, and even different textures and material surfaces work together in ways that were both planned and unexpected. I never intended *Hiding* to stand by itself but from the outset designed it to be part of a more complex work. At the same time I was writing *Hiding*, I was also working with two of my former students, José Márquez and Noah Peeters, to develop a multimedia CD-ROM on Las Vegas. *The Réal: Las Vegas, Nevada* probes the same issues explored in *Hiding* in a different medium and from a different angle.[10] Using Las Vegas as the prism for understanding the postmodern condition, *The Réal* collapses high and low in a style of cultural analysis that is more like a video game than a traditional text. These two works complement but do not complete each other: *Hiding* pushes the book as far in the direction of multimedia hypertext as possible while still remaining a book, and *The Réal* deploys multimedia to formulate arguments usually presented on the printed page. While part of one chapter of *Hiding* is devoted to Las Vegas, more interesting insights emerge through the indirect interaction of the printed and electronic works. Each text in itself as well as their interrelation is designed as a complex whole, which has a structure and follows a logic quite different from that of traditional books. The arguments within and between *Hiding* and *The Réal* are not closed, sequential, or linear but create multiple open circuits that establish constantly shifting connections. The complex text I attempted to create through the interplay of the printed and electronic works approximates the structure and logic of expanding webs that are creating network culture. In the last pages of *Hiding*, I wrote:

Neither totalizing structures that repress differences nor oppositional differences that exclude commonality are adequate in the plurality of worlds that constitute the postmodern condition. To think what post-structuralism leaves unthought is to think a nontotalizing structure that nonetheless acts as a whole. Such a structure would be neither a universal grid organizing opposites nor a dialectical system synthesizing op-

posites but a seamy web in which what comes together is held apart and what is held apart comes together. This web is neither subjective nor objective and yet is the matrix in which all subjects and objects are formed, deformed, and reformed. In the postmodern culture of simulacra, we are gradually coming to realize that complex communication webs and information networks, which function holistically but not totalistically, are the milieu in which everything arises and passes away. These webs and networks are characterized by a distinctive logic that distinguishes them from classical structures and dialectical systems. Though always eluding classificatory schemes constructed to capture them, webs and networks nevertheless display certain rules that guide their operation. The articulation of these rules defines the contours of nontotalizing structures that function as a whole. [11]

This conclusion simultaneously looks back to provide a summary of the logic of *Hiding* and *The Réal* and ahead to anticipate the argument of *The Moment of Complexity.*

The more deeply I immersed myself in information and telematic technologies, the more dissatisfied I became with the critical approaches that have governed much influential cultural analysis for more than three decades. As I struggled to find new strategies, I turned in what might seem an unlikely direction: the painting of Mark Tansey. I had been interested in Tansey's art since I used his suggestive painting *Doubting Thomas* on the cover of *Altarity* (1987). Having always found Tansey's painting intriguing, I have been fascinated by the similarity between our intellectual trajectories. Tansey's study of modernism and its critics eventually led to a serious engagement with structuralists and post-structuralists, including Lévi-Strauss, Derrida, Foucault, Baudrillard, and de Man. Throughout the 1980s and early 1990s, Tansey not only studied their works but actually used their texts in his paintings. By the mid-1990s, he felt he had exhausted these critical resources and started looking elsewhere for new insights. Always interested in the natural sciences, Tansey began tracking discussions of fractal geometry, catastrophe theory, and chaos theory. In the spring of 1996, he spent several weeks at the Santa Fe Institute, where he discovered complexity theory and entered into dialogue with some of the leading figures in this emerging field. Since I had been following the debates in complexity studies for several years, I once again was struck by the similarities between Tansey's work and my own. In an initial effort to probe the interpretive potential of complexity theory, I wrote *The Picture in Question: Mark Tansey and the Ends of Representation* (1999). In chapter 6, "Figuring Complexity," I began the investigation that comes to fruition in *The Moment of Complexity.* This approach proved particularly productive in my study of Tansey because it enabled me to begin to work out the relation of religion, philosophy, literature, and criticism to science and technology through the study of a contemporary artist.

While the problem of complexity is fast becoming a focus of investigation in a variety of fields, the Santa Fe Institute remains the most visible center of complexity studies. Through an ambitious program of research and publication, scientists and scholars at SFI are attempting to define new approaches to natural, social, and cultural processes. Some of this work is understandably controversial, but I find the firm commitment to rigorous interdisciplinary research especially promising. Lines of communication among artists, humanists, and scientists, which have been closed far too long, are finally beginning to open.

In order to avoid confusion, it is important to distinguish complexity studies from catastrophe theory as well as chaos theory. The central ideas of catastrophe theory can be traced to pioneering work done in the 1950s and 1960s by the French mathematician René Thom. Thom argues that Euclidean geometry and the axioms of classical mathematics and physics defined by Newton are ill-suited to describe a world governed by Heisenberg's uncertainty principle and quantum mechanics. In his highly influential book *Structural Stability and Morphogenesis: An Outline of a General Theory of Models,* Thom examines a broad range of natural, social, and cultural phenomena. Convinced that important changes are not the result of quantitative and continuous developments, he examines abrupt, qualitative changes that issue in sudden dislocations and disruptions. Thom identifies seven types of catastrophe, which he describes as "singularities."[12] As the analysis unfolds, it becomes clear that he is more interested in the taxonomy than in the dynamics of morphogenesis. Chaos theory, by contrast, investigates nonlinear systems. In his informative study *In the Wake of Chaos,* Stephen Kellert writes: "chaos theory is *the qualitative study of unstable aperiodic behavior in deterministic nonlinear dynamical systems.*" A dynamic system, Kellert proceeds to explain, "includes both a recipe for producing such a mathematical description of the instantaneous state of a physical system and a rule for transforming the current state description into a description for some future, or perhaps past, time."[13] The systems that are most interesting to chaos theorists are those in which extreme sensitivity to initial conditions creates effects that are disproportionate to their causes. While the inability to identify all relevant initial conditions makes it impossible to predict behavior accurately, nonlinear dynamical systems are not indeterminate but follow definable rules.

Complexity theory shares catastrophe theory's preoccupation with discontinuous change and chaos theory's concern with the dynamics of nonlinear systems but explores the activity of systems "far from equilibrium" or "at the edge of chaos." According to complexity theorists, all significant change takes place *between* too much and too little order. When there is too much order, systems are frozen and cannot change, and when there is too little order, systems disintegrate and can no longer function. Far from equilibrium, systems change in surprising but not necessarily random ways. Most catastrophe, chaos, and complexity theorists agree about the importance of not restricting investigation to natural phenomena and encourage the extension of research to social, political, economic, and cultural systems. Though chaos theory has long interested artists, writers, and critics, humanists only recently have begun to recognize the importance of complexity theory. To appreciate the far-reaching implications of these issues, it is necessary to understand complexity as comprehensively as possible. *The Moment of Complexity* is not, however, a primer for nonspecialists; rather, my aim is to synthesize ideas currently being debated in complexity studies and related fields in a way that extends and expands the philosophy of culture I have been developing for more than three decades.

I begin with a chapter entitled "From Grid to Network" in which I approach the moment of complexity in terms of the work of three architects, Mies van der Rohe, Robert Venturi, and Frank Gehry. Throughout the first half of the century, the pristine forms and Euclidean geometry of classical modern and International Style architecture rendered transparent the operational logic of industrial society. By midcentury, Robert Venturi had become convinced that the austere minimalism and insistent simplicity of modern architecture were ill-suited for a world increasingly governed by new media and communications technologies. In his books *Complexity and Contradiction in Architecture* (1966) and *Learning from Las Vegas* (studio 1968, book 1972), Venturi calls for an architecture that privileges image over structure, and ornament over form. While criticizing the simplicity of modernism, the complexity Venturi extols is a matter of façade rather than structure and thus remains superficial. In Gehry's recent work, by contrast, form itself becomes complex. With the development of sophisticated software design programs such as Catia, Gehry is able to create forms and structures that display the architecture of contemporary network culture. In an effort to situate these developments historically, I argue that the transition from modern industrial society to network culture begins in the late 1960s and reaches the tipping point with the collapse of the Berlin Wall on November 9, 1989.

Having established the framework of the analysis, I proceed in chapter 2, "Critical Emergency," to examine the inadequacies of the approaches that have dominated cultural criticism for the past thirty years. The three tendencies I consider are social constructivism (Michel Foucault), deconstruction (Jacques Derrida), and the theory of simulation (Jean Baudrillard). To explain the contributions and limitations of these perspectives, I show how they remain caught in the structural oppositions of the Cold War era while at the same time they anticipate more complex structures of psychological, social, and cultural organization. Rather than simply rejecting these critical positions, I turn their interpretive strategies back on themselves to detect new lines of analysis, which might overcome the current theoretical impasse.

After placing the moment of complexity in the context of events occurring in the last half of the twentieth century, I next, in chapter 3, trace its roots to cultural developments at the end of the eighteenth century. In his systematic development of Kant's philosophy, Hegel explicates the mechanical logic that characterizes the Newtonian world and formulates a logic of organisms that is still indirectly influential in important areas of theoretical biology. Hegel's system, however, is of more than historical interest; if his work is reread through certain aspects of information theory, new theoretical insights begin to emerge. To show how this might be possible, I compare Hegel's system to the autopoietic systems analyzed by theorists like Francisco Valera, Humberto Manturana, and Niklas Luhmann. Drawing on Douglas Hoftstadter's imaginative reading of the paintings of René Magritte and helpful account of Gödel's theorem, I show how such systems can be understood as open rather than closed. As a way of joining these issues with problems considered in the previous chapter, I explain how Hoftstadter's account of the paradoxes of self-reflexivity can clarify the often obscure accounts of the aporia of self-reflexivity presented by leading post-structuralists.

In chapter 4, "Noise In Formation," I consider Claude Shannon's communication theory and Norbert Weiner's analysis of cybernetic systems. My purpose at this juncture is to determine how the interaction between information and noise leads to the emergence of increasingly complex structures. Taking Michel Serres's imaginative interpretation of the relation between literary criticism and information theory as a point of departure, I argue that the relationship between information and noise can clarify recent philosophical and critical debates about interplay between system and structure, on the one hand, and, on the other, otherness and difference. The work of Ilya Prigogine on dissipative structures provides a way of understanding the systematic implications of this intersection of communication and critical theory. According to Prigogine, disorder does not merely destroy order, structure, and organization, but is also a condition of their formation and transformation. New dynamic states, which emerge in conditions far from equilibrium, can temporarily check entropic processes.

Complex structures that resist the seemingly inexorable flow of time can be described as emerging self-organizing systems. In chapter 5, "Emerging Complexity," I draw on the work of Henri Atlan, Stuart Kauffman, John Holland, John Casti, Murray Gell-Mann, and others to formulate an account of complex adaptive systems. Returning to issues raised in chapter 1, I use the paintings of Chuck Close to complicate the relationship between grids and networks. In Close's remarkable work, vital forms emerge from complex gridlike structures in a manner that recalls the uncanny operation of cellular automata. This formation of figure from structure provides a suggestive example of the process of emergence described in John Holland's *Emergence: From Chaos to Order.* After examining how these processes operate in biological organisms and systems such as insect colonies, I extend the analysis of complex self-organizing systems to media and information networks.

Complex adaptive systems, which always emerge at the edge of chaos far from equilibrium, are not static but are in a state of continual evolution. In chapter 6, "Evolving Complexity," I discuss the ways in which these systems change over time. I situate my argument at this point by discussing the often overlooked dependence of Darwin's theory of evolution on Adam Smith's understanding of market capitalism. With this background in mind, I turn to contemporary revisions of Darwin in the work of biologists like Richard Dawkins, Stuart Kauffman, and Brian Goodwin, as well as the heated debate between Stephen Jay Gould and Daniel Dennett. The aim of this chapter is to show how the process of evolution can be understood as a complex adaptive process, which moves episodically toward greater complexity.

Chapter 7, "Screening Information," marks the theoretical culmination of the book. Weaving together strands of the argument developed in preceding chapters, I attempt to show how complex adaptive systems can help us to understand the interplay of self and world in contemporary network culture. Drawing on the multiple meanings of "screen," I propose the closely related notions of *nodular subjectivity* and the *technological unconscious* to interpret experience in emerging webs and global networks. This account of subjectivity requires a rethinking of traditional binaries and oppositions like nature/culture, subject/object, mind/body, mind/brain, organism/machine, interiority/exteriority, and even life/nonlife. When extended to cultural processes, the notion of complex adaptive networks illuminates the ways in which symbols, concepts, myths, and theories transform noise into information, which can be organized in meaningful patterns to inform thinking and guide action. Effective patterns of meaning require a new architecture of complexity, which simultaneously embodies and articulates the logic of networking. I conclude that we are gradually discovering that we are, in effect, *incarnations* of worldwide webs and global networks whose complexity is fraught with danger as well as opportunity.

In the coda, "The Currency of Education," I describe the way in which I am trying to put the theory developed in *The Moment of Complexity* into practice in the Global Education Network. As I have noted, this venture has grown out of my experiments with technology in my teaching and writing. I strongly believe emerging network culture is going to thoroughly transform higher education. As I have tried to persuade colleagues of their responsibility in shaping this new educational environment, I have been very discouraged by their profound suspicion of business and technology. In an effort to understand their resistance to change, I go back to the beginning of the modern university as Kant defined it in *The Conflict of the Faculties* (1798). Kant's model of the university has survived with remarkable consistency from the University of Berlin, founded in 1810, to the present day. Indeed, as I attempt to show, Kant's vision of the university and his division of the curriculum underlies many current criticisms of the bureaucratization and corporatization of the university. The arguments of contemporary critics, who fashion themselves radical, actually reflect an understanding of the university that is over two hundred years old. Once we recognize the philosophical underpinnings of the modern university, the reasons for the resistance to change on the part of people, who otherwise seem so forward-looking, becomes more understandable, though no more defensible. While insisting that the unholy alliance of technology, business, and education marks the end of higher education as we have known it, these critics rarely—if ever—offer any constructive alternative. Though it is impossible to predict these changes with confidence, it is beginning to seem clear that the university of the twenty-first century will be an intricate network or network of networks which is structured like a complex adaptive system. Within this expanding web, individuals and organizations which have never worked together will have to learn to cooperate.

Optimism has never come easily to me. But after three decades of teaching, my students have taught me the imperative of hope. It is not enough for critical practice to have as its primary aim the production of texts that fewer and fewer people read. I can no longer look into the eyes of my remarkable students and tell them that all they have to look forward to is the endless struggle to undo systems and structures that cannot be undone. They deserve more than being told repeatedly that nothing can be fundamentally changed. The world I am at-

17

tempting to comprehend in *The Moment of Complexity: Emerging Network Culture* is the world of my children and students more than my own. If this book helps them to understand *their* world and negotiate *their* future, it will have served its purpose.

SEPTIMUS: Geometry, Hobbes assures in the *Leviathan,* is
the only science God has been pleased to bestow on mankind.
LADY CROOM: And what does he mean by it?
SEPTIMUS: Mr. Hobbes or God?
LADY CROOM: I am sure I do not know what either means by it.
THOMASINA: Oh, pooh to Mr. Hobbes! Mountains are not
pyramids and trees are not cones. God must love gunnery and
architecture if Euclid is his only geometry. There is another
geometry which I am engaged in discovering by trial and error,
am I not, Septimus?

tom stoppard

from grid to network

COLLAPSING WALLS

At pivotal moments throughout history, technological innovation
triggers massive social and cultural transformation. Apparently
unrelated developments, which had been gradually unfolding for
years, suddenly converge to create changes that are as disruptive as
they are creative. We are currently living in a moment of extraordinary complexity when systems and structures that have long organized life are changing at an unprecedented rate. Such rapid
and pervasive change creates the need to develop new ways of understanding the world and of interpreting our experience.

While moments of radical transformation can never be defined with precision, the collapse of the Berlin Wall on November 9, 1989, signaled a decisive shift from an industrial to an information society. With the ostensible triumph of multinational, informational, or digital capitalism, walls, which once seemed secure, become permeable screens that allow diverse flows to become global. What is emerging from the flux of these flows is a new *network culture*. To understand the distinctive logic and dynamics of network culture, it is necessary to consider how it grows out of and breaks with previous historical moments.

Though 1989 marks a decisive juncture in the formation of network culture, this moment of transition had actually been emerging for nearly half a century. During the postwar years, modern industrial organizations, which had been so functional and productive during the first half of the century, began to change in ways that were not immediately obvious. These changes were, in large measure, brought on by new information and communications technologies. By the late 1960s, the proliferation of multiple media networks had created a world some were beginning to label "postmodern." Developments set in motion in the 1960s reached a turning point in the 1990s. As electronic information and telematic technologies became more sophisticated, their social, political, economic, and cultural impact became more significant.

In order to understand the scope and importance of these changes, I would like to begin somewhat indirectly by considering the work of three architects: Mies van der Rohe, Robert Venturi, and Frank Gehry. Architecture might seem to be an unlikely angle of entry for understanding the complex dynamics of network culture, but, as we will discover, architectural practices both reflect and shape broader social and cultural currents. After all, societies and cultures, as well as computers and software, have architectures as distinctive as any building. By examining the architecture of Mies, Venturi, and Gehry, it becomes possible to trace the movement from industrial society, through media culture, to network culture. This trajectory suggests that the moment of complexity can be understood in terms of the shift from a world structured by grids to a world organized like networks. Since the contrasting figures of the grid and the network return repeatedly in the following pages, we must try to determine what they are and how they differ.

What, then, is a grid, and what is a network? This question, which generates many more questions, is deceptively simple. We might begin to appreciate its complexity by conducting what Kierkegaard once described as a "thought experiment."

Imagine a grid.

What is its structure?

What is its function?

Are all grids the same or are they different?

If they are the same, why?

If they are different, how?

Do grids change or remain the same over time?

What is the relation of parts to whole and whole to parts in grids?

When did grids first emerge?

Where did grids first emerge?

Who invented the grid?

Where can grids be observed today?

What is a grid today?

What is not a grid today?

What is the function of grids today?

What is the architecture of grids today?

Are grids simple or complex?

Imagine a network.

What is its structure?

What is its function?

Are all networks the same or are they different?

If they are the same, why?

If they are different, how?

Do networks change or remain the same over time?

What is the relation of parts to whole and whole to parts in networks?

When did networks first emerge?

Where did networks first emerge?

Who invented the network?

Where can networks be observed today?

What is a network today?

What is not a network today?

What is the function of networks today?

What is the architecture of networks today?

Are networks simple or complex?

What is the relationship between grids and networks?

Picture two buildings: Mies van der Rohe's Seagram Building in New York City (1954–58) and Frank Gehry's Guggenheim Museum Bilbao (1998) (figs. 1 and 2).

What is their structure?
What is their function?
What is the relation of form to function?
How are these buildings similar?
How are they different?
What is the relation between interior and exterior?
What is the relation between structure and surface?
What is the relation between building and environment?
What do these buildings teach us about architecture?
What do they disclose about the architecture of the society and culture in
    which they emerged?
Is either building a grid?
Is either building a network?

Is either building both a grid and a network?

During the 1970s and 1980s, rapid technological change combined with the privatization and deregulation of major industries to create the conditions for the emergence of a new political and economic order. The *New York Times* foreign affairs columnist Thomas L. Friedman argues that in the 1990s a new international system emerged, which has replaced the Cold War system that had governed the world for half a century. Globalization, he explains,

has its own defining technologies: computerization, miniaturization, digitization, satellite communications, fiber optics, and the Internet. And these technologies helped to create the defining perspective of globalization. If the defining perspective of the Cold War was "division," the defining perspective of globalization is "integration." The symbol of the Cold War system was the wall, which divided everyone. The symbol of globalization is the World Wide Web, which unites everyone. The defining document of the Cold War was "The Treaty." The defining document of the globalization system is "The Deal."[1]

These processes of globalization are now creating a new network culture whose complex logic and dynamics we are only beginning to understand.[2]

The contrast between grids and networks clarifies the transition from the Cold War system to network culture. The Cold War system was designed to maintain stability by simplifying complex relations and situations in terms of a grid with clear and precise oppositions: East/West, left/right, communism/capitalism, etc. This is a world in which walls seem to provide security. Walls and grids, however, offer no protection from spreading webs; as webs grow, walls collapse and everything begins to change. A new economy displaces the old and a "new world order" appears on the horizon. In this situation, the structural oppositions, which had long informed thinking and guided policy, unravel and the political balance of power disappears. Whereas walls divide and seclude in an effort to impose order and control, webs link and relate, entangling everyone in multiple, mutating, and mutually defining connections in which nobody is really in control. As connections proliferate, change accelerates, bringing everything to the edge of chaos. This is the moment of complexity.

THE EDGE OF CHAOS

Complexity is both a marginal and an emergent phenomenon. Never fixed or secure, the mobile site of complexity is always momentary and the marginal moment of emergence is inevitably complex. Far from a *nunc stans,* the emergent moment, which repeatedly constitutes and reconstitutes the flux of time, harbors a momentum that keeps everything in motion. It is significant that the word "moment" derives from the Latin *momentum,* which means movement as well as momentum. Though often represented as a simple point, the moment is inherently complex. Its boundaries cannot be firmly established, for they are always shifting in ways that make the moment fluid. This is the intermediate domain that complexity theory attempts to understand.

As I have noted, though the dynamics of chaos and complexity share certain characteristics, they also differ in important ways.[3] Chaos theory was actually developed as a corrective to the closed and linear systems of Newtonian physics. Rather than the absence of order, chaos is a condition in which order cannot be ascertained because of the insufficiency of information. While Newtonian physics imagines an abstract world governed by definable laws that are completely determinative, in the "real world" things are never transparent because adequate information necessary to establish laws and understand their operation is always unattainable. Though there are many reasons for this situation, two are noteworthy in this context. First, finite systems are not closed but are open and thus incomplete; and, second, some systems involve recursive relations, and cannot be understood in terms of linear models of causality. In many open, recursive systems, it is impossible to measure initial conditions with enough precision to determine causal relations accurately beyond a very limited period of time. Unpredictability, therefore, is unavoidable. Unlike linear systems, in which causes and effects are proportional, in recursive systems, complex feedback and feed-forward loops generate causes that can have disproportionate effects.

In contrast to chaos theory, complexity theory is less concerned with establishing the inescapability of "determinate chaos" than with what John Casti aptly labels "the science of surprise."[4] Falling *between* order and chaos, the moment of complexity is the point at which self-organizing systems emerge to create new patterns of coherence and structures of relation. Having grown out of investigations in the biological sciences, the insights of complexity theory can be used to illuminate social and cultural dynamics. In his wide-ranging study *At Home in the Universe: The Search for the Laws of Self-Organization and Complexity,* theoretical biologist Stuart Kauffman, who is one of the leading figures in complexity studies, writes:

For what can the teeming molecules that hustled themselves into self-reproducing metabolisms, the cells coordinating their behaviors to form multicelled organisms, the ecosystems and even economic and political systems have in common? The wonderful possibility . . . is that on many fronts, life evolves toward a regime that is poised be-tween order and chaos. The evocative phrase that points to this working hypothesis is this: life exists at the edge of chaos. Borrowing a metaphor from physics, life may exist near a kind of phase transition. Water exists in three phases: solid ice, liquid water, and gaseous steam. It now begins to appear that similar ideas might apply to complex adapting systems. For example, we will see that the genomic networks that control development from zygote to adult can exist in three major regimes: a frozen ordered regime, a gaseous chaotic regime, and a kind of liquid regime located in the region between order and chaos. It is a lovely hypothesis, with considerable supporting data, that genomic systems lie in the ordered regime near the phase transition to chaos. Were such systems too deeply into the frozen ordered regime, they would be too rigid to coordinate the complex sequences of genetic activities necessary for development. Were they too far into the gaseous chaotic regime, they would not be orderly enough. Networks in the regime near the edge of chaos—the compromise between order and surprise—appear best able to coordinate complex activities and best able to evolve as well.[5]

While the networks with which Kauffman is primarily concerned are biological, his analysis can be extended to social and cultural dimensions of experience. Poised between too much and too little order, the moment of complexity is the medium in which network culture is emerging. With these insights in mind, let us return to grids and networks.

GRID WORK

The grid is the figure of modernism. In the foreword to his 1924 manifesto, *The City of To-Morrow and Its Planning,* Le Corbusier suggests why grids and their geometry are so important for modern art and architecture as well as modern life:

Geometry is the means, created by ourselves, whereby we perceive the external world and express the world within us.

Geometry is the foundation.

It is also the material basis on which we build those symbols which represent to us perfection and the divine.

It brings with it the noble joys of mathematics.

Machinery is the result of geometry. The age in which we live is therefore essen-tially a geometrical one; all its ideas are so orientated in the direction of geometry.

25

Modern art and thought—after a century of analysis—are now seeking beyond what is merely accidental; geometry leads them to mathematical forms, a more and more generalized attitude.[6]

Consisting of ideal forms and perfectly precise lines, the foundational geometry Le Corbusier worships at this point in his career is Euclidean.[7] Far from a mere aesthetic artifice, straight lines and right angles, he believes, characterize human existence. Indeed, people are distinguished from animals by their ability to follow a straight-and-narrow line:

Man walks in a straight line because he has a goal and knows where he is going; he has made up his mind to reach some particular place and he goes straight to it.

The pack-donkey meanders along, meditates a little in his scatter-brained and distracted fashion, he zigzags in order to avoid the larger stones, or to ease the climb, or to gain a little shade; he takes the line of least resistance.

But man governs his feelings by his reason; he keeps his feelings and his instincts in check, subordinating them to the aim he has in view. He rules the brute creation by intelligence. His intelligence formulates laws which are the product of experience. His experience is born of work; man works in order that he may not perish. In order that production may be possible, a line of conduct is essential, the laws of experience must be obeyed. Man must consider the result in advance.[8]

"The Pack-Donkey's Way," Le Corbusier proceeds to argue, "is responsible for the plan of every continental city," which must be destroyed to make way for "the city of the future." In contrast to the accidental growth and arbitrary structure of cities in the past, the modern city "lives by the straight line, inevitably; for the construction of buildings, sewers, and tunnels demands the straight line; it is the proper thing for the heart of a city. The curve is ruinous, difficult and dangerous; it is a paralyzing thing."[9] His drawings of the contemporary city underscore Le Corbusier's idealization of the grid (fig. 3).

Le Corbusier's argument as well as his architecture is structured by a series of binary opposites, which he does not always explicitly articulate:

THE PACK-DONKEY'S WAY	MAN'S WAY
Premodern	Modern
Aimless	Directed
Heedless	Disciplined
Distracted	Concentrated

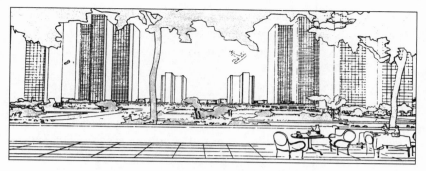

FIGURE 3 Le Corbusier, A Contemporary City. (© 2001 Artists Rights Society [ARS], New York/ADAGP, Paris/FLC.)

Messy	Exact
Accidental	Essential
Paralysis	Circulation
Disorder	Order
Matter	Idea
Body	Mind
Feeling	Reason
Curve	Straight line
Errant	Proper

These opposites obviously are not of equal value; in each case, Le Corbusier privileges the latter over the former term. Primitives and infants might be excused for being aimless, heedless, distracted, and messy, but if moderns display these traits, they violate the very essence of "human nature." So understood, Le Corbusier's binaries are axiological as well as temporal and historical. In the course of moving from premodernity to modernity as well as from infancy to maturity, human beings progress from desire to discipline. Straight lines and the grids they form both represent and impose the strict discipline through which the rule of reason is secured. As feelings and emotions are controlled, order is wrought from disorder. What Le Corbusier describes as the "march towards order" imposes disciplinary practices necessary for the efficient functioning of industrial society.[10]

Philosophy, art, and life intersect in the architecture of the grid. Not only buildings but the city as a whole become "a machine to live in." Grids graph Cartesian space, which is supposed to be completely rational, maximally efficient, and perfectly transparent. Anticipating Joseph Schumpeter's principle of "creative destruction," Le Corbusier argues that the space of modernity is created by destroying everything that is both natural and premodern:

WE MUST BUILD ON A CLEAR SITE. The city of today is dying because it is not constructed geometrically. To build on a clear site is to replace the "accidental" layout of the ground, the only one that exists today, by the formal layout. Otherwise nothing can save us. And the consequence of geometrical plans is Repetition and Mass-production.[11]

Inasmuch as the structure of this "formal layout" is the grid, the logic of the grid appears to be isomorphic with the serial logic of mass production.

The broader implications of Le Corbusier's argument for modern industrial society can be understood when his argument is extended from architecture to the assembly line. In 1913, Henry Ford created an automated assembly line in Dearborn, Michigan, which made it possible for the first time to produce homogeneous products by rationalizing, standardizing, and regulating management, the production process, and labor. Two years before cars started rolling off Ford's assembly line, Frederick Winslow Taylor published his influential *Principles of Scientific Management*. In a way unforgettably satirized by Charlie Chaplin in *Modern Times*, machine, workers, and managers are coordinated to operate at maximum efficiency. Taylor summarizes his conclusions:

To repeat then throughout all of these illustrations, it will be seen that the useful results have hinged mainly upon (1) the substitution of a science for the individual judgment of the workman; (2) the scientific selection and development of the workman, after each man has been studied, taught, and trained, and one may say experimented with, instead of allowing the workmen to select themselves and develop in a haphazard way; and (3) the intimate cooperation of the management with the workmen, so that they together do the work in accordance with the scientific laws which have been developed, instead of leaving the solution of each problem in the hands of the individual workman. In applying these new principles, in place of the old individual effort of each workman, both sides share almost equally in the daily performance of each task, the management doing that part of the work for which they are best fitted, and the workmen the balance.[12]

As described by Taylor, the labor process can be divided into discrete movements, which can be regulated precisely. So understood, the logic of the assembly line is obviously mechanical, serial, and linear. Separate individuals, entities, and events are joined in a predictable chain where effects are proportional to causes. The repetition of the process, as Le Corbusier stresses, results in the uniformity of the product. The revolutionary innovation of this mechanical logic is, in part, made possible by the identification of rationalization with simplification. Extending the principles of analytic reason to management and labor, the assembly line breaks down complicated processes into simple elements, which are then ordered through a series of discrete yet related activities. In addition to establishing a division of labor, the assembly line separates managers from workers: while managers plan and direct, workers implement and execute. The management/labor relation mirrors the mind/body distinction. As we have seen, these binaries tend to be hierarchical; in this case, management controls labor as the mind rules the body.

It is important to stress that the principles of scientific management structure time as well as space. In ways Newton and deistic theologians never could have anticipated, the modern mechanistic world is a clockwork universe. The operation of the assembly line depends on the repeated execution of tasks carried out with precise timing. In 1914, Ford initiated an eight-hour workday, for which workers were paid five dollars. Ford and his engineers realized that efficient production required the imposition of mechanical regularity on natural rhythms. The workers' day, like the assembly line, was divided into equal but separate parts. Within the all-encompassing logic of industrialism, work, leisure, and rest are designed to promote efficiency and thus increase profitable production. Workers not only have to produce but also must consume more and more products. Indeed, one of the first products of mass production must be mass consumption. For industrial society to thrive, a sufficient number of people must have the time, money, and desire to consume what they do not necessarily need. The intricate relation between production and consumption underscores one of the inescapable contradictions of capitalism. Profitable production demands, on the one hand, the rational control of emotions and desires and, on the other, the cultivation of the desire to consume, which often is unreasonable.

One way to mediate this tension is to regulate time by reconfiguring space. Mechanical engineering, in other words, cannot work without an equally calculated social engineering. Toward this end, social engineers develop strategies to separate work and home in a way that secures different domains for different activities. From Ford's early efforts to organize the lives of his workers and Pullman's town planning, to postwar American advertising and suburbanization, industrialists have realized that for capitalism to thrive, it must become a "total

29

way of life." This totality is implicitly structured by a series of binary opposites, which, we have discovered, are never equivalent:

Factory	Home
Work	Leisure
Reason	Desire
Man	Woman
Husband	Wife

Since home, leisure, desire, and women are ordered to serve factory, work, reason, and men, industrial rhythms pervade all of life. Within this total way of life, nothing seems to escape the logic of mass production.

As we have come to suspect, the figure of this all-encompassing logic is the grid. The assembly line extended beyond the factory floor to create supposedly rational urban and suburban grids where workers spend what they earn and relax and rest so they can work efficiently another day. Within this economy, spending and play cannot become excessive but must be carefully monitored and regulated. The straight lines and right angles of streets and avenues as well as modern houses and buildings channel desires in ways that allow controlled moments of release necessary to keep the wheels of industry turning. This is not to imply that modern architecture and urban design originate with the assembly line. From the beginning of modernity, the grid functions as an instrument for rationalizing and thus controlling nature. Long before Ford and Taylor, Enlightenment thinkers saw in the grid a figure of universal reason and human equality, which, when effectively deployed, could level social hierarchies. In France, for example, commitment to the principles of Cartesian rationalism was expressed in the establishment of an all-inclusive grid of administrative departments throughout the country. The goal of this organization was to create governmental efficiency and political democracy. Importing the principles of the European Enlightenment, Jefferson imposed the grid on America, to create rational systems of regulation and control, which, he believed, would allow representative democracy and political equality to flourish.

There is, of course, a darker side to this idealized vision. The very structures that make possible democratic representation and egalitarian administration also create technologies of surveillance, control, and even repression. The invasive eye of reason can turn back on credulous citizens to destroy the freedom it is supposed to promote. This prospect becomes even more troubling with the recognition of the Eurocentric view of reason during the Enlightenment. All too often rationalization and colonization seem to be inseparable. When the ideal of universality is put into practice uncritically, it can quickly lead to a uniformity that excludes or represses everything and everyone deemed different.

By the beginning of this century, important transformations of basic Enlightenment principles were taking place. First and most important, with the spread of industrial capitalism, the notion of rationality changed from a universal capacity of the human mind to a strategy of calculation devised to maximize economic profit. Reasonable activity came to be associated with economic benefit. This redefinition of reason led to a second crucial revision of the Enlightenment worldview. The spread of reason does not necessarily lead to universal equality but tends to establish and reinforce social hierarchies and economic inequalities. These social and economic differences are exacerbated if the ideal of universality degenerates into the crude uniformity of standardized consumer products, which are mechanically produced and sold for substantial profits. Disequilibrium keeps this system running: social and economic inequalities generate desires, which strengthen the very consumer economy that enriches a few at the expense of many. Third and finally, by the early twentieth century, it became abundantly clear to people other than philosophers, writers, and artists that the problem of representation is an aesthetic as well as a political and economic issue. Inasmuch as art, politics, and economics mutually condition each other, the modernist dictum "Make it new!" intersects with the industrial requirement of "planned obsolescence," to extend the principle of "creative destruction." While appearing to escape the utilitarianism of industrialism, the founding principle of the avant-garde actually reinforces the system it is designed to criticize. A consideration of the interplay of the mechanical logic of industrialization and modernist aesthetic principles and artistic practices illuminates the complexity of these relations.

The modernist preoccupation with making it new represents a concerted effort to break with the past by living fully in the present. This present, Rosalind Krauss points out, is associated with the grid: "By discovering the grid, cubism, de Stijl, Mondrian, Malevich . . . land in a place that was out of reach of everything that had gone before. Which is to say, they landed in the present and everything else was declared to be past." The modernist present figured by the grid is not merely temporal but involves a specific mode of experience embodied in the work of art. In keeping with a tradition dating back to Kant's *Critique of Judgment* (1790), modernist art is regarded as autonomous or self-referential. Fine or high art serves no practical purpose but is created and enjoyed *for its own sake*. Referring to nothing beyond itself, the modernist work of art provides the

31

occasion for the experience of *simple* presence in the immediacy of the present moment. "The grid," Krauss concludes, "declares the space of art to be at once autonomous and autotelic."[13]

Few architects have been more devoted to the logic of the grid than Mies van der Rohe. By leveling several blocks in Chicago to construct the Illinois Institute of Technology (IIT), he gave concrete expression to the principle of creative destruction. Both the site as a whole and the individual buildings conform to the austere structure of the grid (fig. 4). While IIT is the most comprehensive project Mies ever realized, the significance of his use of the grid can be better understood by considering his influential Seagram Building in New York City. Set back and elevated above surrounding Manhattan streets, Mies's structure rises as a monument to the elegance of simplicity. Unadorned by useless ornament, the building seems to turn in on itself to suggest "a world apart"; indeed, this pristine form appears poised to lift off and leave the weight of the world behind. This moment of apparent separation freezes the flux of the world and human experience until eternity appears to enter time to arrest its corrosive flow. Insofar as Mies's architecture captures this moment, its attraction is spiritual or even religious as well as aesthetic.

But separation is always incomplete, for we remain entangled with that from which we struggle to escape. Mies maintains that architecture does not express timeless values but always remains bound to its time. In a "personal statement" drafted in 1964, he confessed:

It was my growing conviction that there could be no architecture of our time without the prior acceptance of these new scientific and technical developments. I never lost that conviction. Today, as for a long time past, I believe that architecture has little or nothing to do with the invention of interesting forms or with personal inclinations.

True architecture is always objective and is the expression of the inner structure of our time, from which it stems.[14]

If understood in terms of the inner structure of Mies's time, the grid work of the Seagram Building appears to embody the analytic simplicity and rational organization of modern industrial society. For Mies, however, the task of creating order out of chaos is never complete. Always dissatisfied with the present, he clung to utopian aspirations. Mies remained committed to the tradition of the modern avant-garde and thus continued to believe that architecture must be socially effective as well as aesthetically satisfying. The challenge for the architect is to transform the world into a work of art. Though a recurrent theme in Euro-

FIGURE 4 Mies van der Rohe, Illinois Institute of Technology. (By permission of University Archives, Paul V. Galvin Library, Illinois Institute of Technology, Chicago. Reproduction without permission prohibited.)

pean avant-garde art and architecture, this task was nowhere taken up with more vigor and consistency than at the Bauhaus. By bringing together artists, architects, and designers with the engineers of mass production and financiers of modern industrialism, the Bauhaus became a research and development studio

33

for twentieth-century industrial society. In the Illinois Institute of Technology, Mies deliberately recreated the Bauhaus in the heartland of America. Following Le Corbusier, he never gave up his conviction that artistic and historical progress is measured by the "march towards order." Architecture, Mies repeatedly insists, is driven by the need to "create order out of the desperate confusion of our time."[15]

SUPERFICIAL COMPLEXITY

The confusions of postwar America were, of course, quite different from the confusions of war-torn Europe. Having suffered destruction that was not creative, Europe had to build the new from the ashes of the old. In America, by contrast, two world wars increased the country's industrial strength and economic power. When modern architecture was imported from Europe, it lost its utopian edge and quickly became the preferred style of corporate capitalism. In 1932, six years before Mies's arrival in the United States, this change was signaled by the Museum of Modern Art's exhibition on the "new architecture." Organized by Henry-Russell Hitchcock and Philip Johnson, who eventually would become Mies's partner in the Seagram Building, the MOMA show introduced the American public to the revolution wrought by Europe's leading modern architects. In the catalog accompanying the exhibition, which bore the significant title *The International Style: Architecture since 1922,* Hitchcock and Johnson describe the new architecture in familiar terms: "absence of ornament," "functionalism," "principle of order," and, most important, "the formal simplification of complexity." "The finest buildings since 1800," they conclude, "were those least ornamented."[16] There is, however, an important difference between Johnson's and Hitchcock's reading of modern architecture and the self-understanding of architects like Le Corbusier and Mies. In what seems to be a passing aside, the curators of the MOMA exhibition note: "The fact that there is so little detail increases the decorative effect of what there is. Its ordering is one of the chief means by which consistency is achieved in the parts of a design."[17] Ornament, then, is not, as Adolf Loos insists, a crime but is a means by which the consistency of design is enhanced. Hitchcock and Johnson underscore this reassessment of the nature and function of ornament by labeling the modernist principle of universality an International Style. Instead of privileging function, this version of modernism uses formalism to achieve *stylistic* effect. The far-reaching implications of this development were not realized for over three decades.

In 1972, Robert Venturi and his colleagues Denise Scott Brown and Steven Izenour published the book credited with beginning postmodern architecture:

Learning from Las Vegas. In the years following its appearance, the influence of this pivotal work extended far beyond the field of architecture. Venturi's attack on modern architecture and appropriation of popular culture quickly became associated with what Jean-François Lyotard describes as "the postmodern condition." Though rarely acknowledged, Venturi laid the groundwork for his criticism of modernism in *Complexity and Contradiction in Architecture,* published in 1966. For Venturi, modernism's preoccupation with "the formal simplification of complexity" is ill-suited to a world that is becoming increasingly complex. Passing beyond the logic of grids, Venturi envisioned an architecture that anticipates the moment of complexity, which eventually defines network culture.

Venturi's best-known comment on Mies and the architecture he represents is his revision of the Miesian maxim "Less is more" to read "Less is a bore." Rejecting the rigidity of the grid, Venturi argues that "in Modern architecture we have operated too long under the restrictions of unbending rectangular forms supposed to have grown out of the technical requirements of the frame and the mass-produced curtain wall."[18] The paradigm of this formalist tendency, according to Venturi, is Mies's Seagram Building. Mies and Johnson, he maintains, "reject all contradictions . . . in favor of an expression of a rectilinear frame." Indeed, "Mies allows nothing to get in the way of the consistency of his order, of the point, line, and plane of his always complete pavilions."[19] For Venturi, the critical issue is not the technical requirements imposed by the materials and processes of mass production but Mies's preference for an aesthetic that represents values at odds with the complexities and contradictions of the contemporary world. Far from developing an architecture that is "the expression of the inner structure of our time," classically modern architecture actually represents devotion to puritanical moralism that is "dissatisfied with *existing* conditions":[20]

Orthodox modern architects have tended to recognize complexity insufficiently or inconsistently. In their attempt to break with tradition, they idealize the primitive and elementary at the expense of the diverse and the sophisticated. As participants in a revolutionary movement, they acclaimed the newness of modern functions, ignoring their complications. In their role as reformers, they puritanically advocated the separation and exclusion of elements, rather than the inclusion of various requirements and their juxtapositions.[21]

In contrast to the moralistic rejection of contemporary reality for the sake of abstract forms and ideals, Venturi proposes a "more tolerant approach" that accepts "the existing landscape."[22]

The existing conditions Venturi seeks to accommodate differ in important ways from the conditions to which Mies and fellow modernists attempted to adapt. Venturi identifies some of the crucial differences between the modern and the postmodern condition by posing a series of polar opposites:

Form	Image
Pure architecture	Mixed media
Instant city	Process city
Nineteenth-century	Twentieth-century
industrial vision	communication technology
The easy image	The difficult image
The easy whole	The difficult whole[23]

The most prescient observation in this list is Venturi's recognition that the nineteenth-century industrial vision has given way to twentieth-century communication technology. In ways not fully evident at the time, this transition transformed the existing landscape. While Venturi usually is understood as urging architects to reflect contemporary culture by replacing pure forms and structures with images and signs drawn from popular media and culture, his point is considerably more subtle. Formalism, he argues, is actually ornamental: "Modern architects have substituted one set of symbols (Cubist-industrial-process) for another (Romantic-historical-eclecticism) but without being aware of it."[24] What modernists declared to be the nonornamental foundation of the decorative façade is every bit as ornamental as ostensibly superficial decoration. With the recognition that style is all-encompassing, it becomes clear that there is no escape from the endless play of signs and images. Seemingly abstract forms and structures, in other words, are actually ornamental images reflecting aesthetic values. As we shall see in the next chapter, the implications of this insight extend far beyond the realm of architecture. Venturi is suggesting that in the world created by communication technology, there is nothing outside signs and images.

In the late 1960s and early 1970s, it was impossible to anticipate the social, economic, and cultural transformations communications technologies eventually would bring about. While recognizing the importance of the shift from an industrial to a postindustrial society, Venturi's architectural revision remains bound to and by industrialism as it is embodied in the automobile.[25] Unlike his predecessors, who were interested in the efficient functioning of rational machines, Venturi revels in the "messy" roadside culture automobiles spawn. The place where the changes in the postwar American landscape are most prominently displayed is, according to Venturi, the Strip in Las Vegas:

The emerging order of the Strip is a complex order. It is not the easy, rigid order of the urban renewal project or the fashionable "total design" of the megastructure. It is, on the contrary, a manifestation of an opposite direction in architectural theory. . . . The Strip includes; it includes at all levels, from the mixture of seemingly incongruous land uses to the mixture of seemingly incongruous media plus a system of neo-Organic or neo-Wrightian restaurant motifs in Walnut Formica. It is not an order dominated by the expert and made easy for the eye. The moving eye in the moving body must work to pick out and interpret a variety of changing, juxtaposed orders. . . . It is the unity that "maintains, but only just maintains, a control over the clashing elements which compose it. Chaos is very near; its nearness, but avoidance, gives . . . force."[26]

The complexity of the Strip, then, emerges *at the edge of chaos.* This is not a world where chance is eliminated and control assured. To the contrary, aleatory associations and unexpected juxtapositions create "difficult" wholes that cannot be comprehended by the neat-and-clean distinctions of a logic of noncontradiction based on the exclusive principle of either-or. The Strip displays the synthetic or even dialectical logic of both-and. For Venturi, modern architecture's simplifications and rationalizations distort the ambiguities, paradoxes, and complexities of life in the latter half of the twentieth century:

The movement from a view of life as essentially simple and orderly to a view of life as complex and ironic is what every individual passes through in becoming mature. But certain epochs encourage this development; in them the paradoxical or dramatic outlook colors the whole intellectual scene. . . . Amid simplicity and order rationalism is born, but rationalism proves inadequate in any period of upheaval. Then equilibrium must be created out of opposites. Such inner peace as men gain must represent a tension among contradictions. . . . A feeling for paradox allows seemingly dissimilar things to exist side by side, their very incongruity suggesting a kind of truth.[27]

Though posed in terms of the difference between the primitive and the modern or the infantile and the mature, Venturi uses the distinction between simplicity and complexity to differentiate high modernism from the postmodern architecture needed in today's world. Grids, which might have worked in industrial society, are obsolete in network culture.

Venturi's efforts to capture this shift in his architecture, however, are less successful than his theoretical arguments. He rejects modernism's aversion to ornamentation but does not explore alternative architectural forms and structures. When Venturi translates his theories into practice, the complexities he probes prove to be superficial. The application of juxtaposed signs and images does not fundamentally change architectural form. Consider, for example, his famous sketch of a "decorated shed" (fig. 5). As this drawing makes clear, the decorative façade leaves the gridlike structure of the building intact. What is not obvious from the sketch is that Venturi directly appropriates the ornamental surface

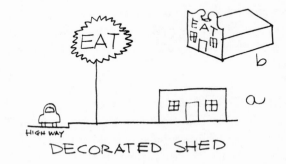

FIGURE 5 Robert Venturi, Decorated Shed. (From *Learning from Las Vegas*, by Robert Venturi, Denise Scott Brown, and Steven Izenour [Cambridge, Mass.: MIT Press, 1988].)

from a roadside sign. In a chapter entitled "Symbol in Space before Form in Space: Las Vegas as a Communication System," he notes: "The sign for the Motel Monticello, silhouette of an enormous Chippendale highboy, is visible on the highway before the motel itself. This architecture of styles and signs is antispatial; it is an architecture of communication over space; communication dominates space as an element in the architecture and in the landscape."[28] This comment suggests the inspiration for one of Venturi's most influential buildings—the Vanna Venturi House (1963–65), where the shape of the façade is a literal representation of the motel sign (fig. 6). The influence of the Chippendale highboy does not end with this project. When Philip Johnson forsakes modernist formalism for postmodern decoration, he appropriates Venturi's appropriation of the highway sign. Johnson's AT&T building (1979–84) is obviously a comment on the Seagram Building on which he had worked with Mies twenty-five years earlier (fig. 7). By replacing the flat roof with a highboy design, Johnson adds the excessive supplement Mies had insisted on erasing. Inasmuch as Venturi maintains that the emerging landscape is being created by twentieth-century communication technology, it is significant that Johnson's signature building was done for the AT&T Corporation. Yet for Johnson, as for Venturi, surface

FIGURE 6 (top) Robert Venturi, The Vanna Venturi House. (Photograph by Rollin R. La France.)

FIGURE 7 (left) Phillip Johnson, AT&T Building.

differences do not signal profound changes; the formal structure of the AT&T building is virtually the same as the Seagram Building. The failure of classical postmodern architects to develop alternative *forms* and *structures* reflects a society that is no longer industrial but not yet postindustrial. By creating an architecture around the automobile, Venturi, Johnson, and others remain stuck along the industrial highway and do not venture into the world created by the information superhighway. *Their complexities are not complex enough.* In network culture, not only surfaces but structures that once seemed simple become irreducibly complex.

NET WORK

What postmodern architects anticipate, Frank Gehry realizes. Consider the differences between Johnson's AT&T building and Gehry's Guggenheim Museum Bilbao (figs. 8–11). In place of a decorated shed or an ornate rectilinear high-rise structure, Gehry develops a complex horizontal network of dispersed yet interrelated forms. Ever suspicious of simplicity, Gehry revels in a riot of forms that appear to morph even while standing still. From his earliest work, he has never found beauty in pristine forms and rigid structures. To the contrary, Gehry has always been drawn to raw materials and rough edges. The works he creates—be they buildings, cardboard chairs, or fish lamps—are neither autonomous nor complete but are deliberately unfinished or even broken. When Euclidean geometries are fractured, surfaces left unfinished, and forms rendered incomplete, structures are opened to allow complexity to emerge in surprising ways. Having found the mechanistic logic of modernism inadequate, Gehry seeks an alternative logic that approximates the logic of networking. This is not to imply that Gehry simply negates modernism and the world it represents; to do so would be a thoroughly modernist gesture. It is as if the complicated lines of Gehry's buildings echo the lines Tom Stoppard puts in Thomasina's precocious mouth: "Mountains are not pyramids and trees are not cones. God must love gunnery and architecture if Euclid is his only geometry. There is another geometry which I am engaged in discovering by trial and error."[29] Thomasina, whose nickname, Thom, is the name of the inventor of catastrophe theory, deliberately echoes the creator of fractal geometry.[30] In the opening lines of *The Fractal Geometry of Nature,* Benoit Mandelbrot observes:

Clouds are not spheres, mountains are not cones, coastlines are not circles, and bark is not smooth nor does light travel in a straight line.

More generally, I claim that many patterns of Nature are so irregular and frag-
mented that, compared with Euclid—a term used in this book to denote all of stan-
dard geometry—Nature exhibits not only a different degree but an altogether differ-
ent level of complexity.31

While Gehry's forms are not precisely fractal, the new geometry he, like
Thomasina, seeks, involves an "altogether different level of complexity."

In probing new frontiers of complexity, as I have noted, Gehry does not simply negate modernism and the world it represents. Instead of repeating modernism's gesture by destroying its grids, Gehry subtly folds the mechanical logic of
industrialism into his work in ways that paradoxically negate and preserve its
traces. This strategy is evident in his design process as well as in the finished
building. Consider, for example, how the computerized rendering of the
Guggenheim Museum Bilbao translates his drawing into a schematic that can actually be constructed. For years, engineers told Gehry that forms he drew and
models he created could not be built. It was not until new computers and software programs were created that Gehry and his associates could build what they
had long imagined. When Gehry's office started using the software program
Catia, his work took a sudden, and for many unexpected, leap forward, which
would not have been possible without the convergence of art and technology.
The new software not only enabled Gehry to transform his models into programs
but also significantly influenced the structures he designed. What is intriguing
about the plans generated by Catia is the way in which they torque grids to create
complex structures embodying another geometry with a different logic. The grid
does not merely disappear but morphs into forms that are dynamic rather than
rigid, organic rather than mechanical, complex rather than simple.

In the completed structure, the interplay of grid and network extends from
building to its surroundings. By weaving the museum into the industrial fabric
of the city and vice versa, Gehry places it in the midst of a transportation network formed by intersecting roads, railroad tracks, and the Nervión River. When
viewed from above, the museum appears to be a complex network of complex
forms (figs. 8 and 9). Traces of modernist structures are incorporated only to be
overwhelmed by shimmering surfaces and fluid shapes. Though the building's
signature titanium surface is a grid, its structure fades in the constant play of
fleeting images (fig. 12). Conversely, vital forms emerge from a fading grid as if
they were faces of a Chuck Close painting or pixel figures displayed on the screens
and scrims of contemporary media and information culture. With the moving
images on these mobile surfaces, Gehry seems to achieve the impossible: he si-

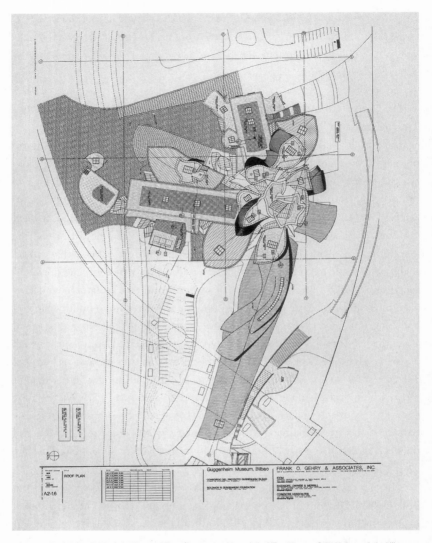

FIGURE 8 Frank Gehry, Guggenheim Museum Bilbao. (Provenance: Guggenheim Bilbao Museoa. © FMGB Guggenheim Bilbao Museoa.)

multaneously sets forms in motion and gives movement form. Far from a static structure, Gehry's building is a *complex* ongoing event.

This sense of complexity and mobility carries over from outside to inside. Instead of removing structure from context by raising it on a pedestal as Mies had done, Gehry actually sinks his building into the urban environment. De-

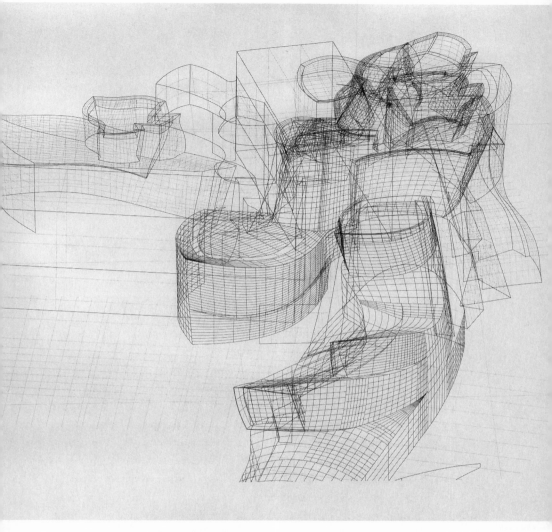

FIGURE 9 Frank Gehry, Guggenheim Museum Bilbao. (Provenance: Guggenheim Bilbao Museoa. © FMGB Guggenheim Bilbao Museoa.)

scending the sloping stairs, one enters a space that is futuristic yet strangely familiar. The exterior steel and glass as well as the rich beige limestone are repeated inside, thereby creating an interplay of interiority and exteriority that simultaneously dissolves and maintains the walls (fig. 11). The resulting forms range from the rectilinear to the curvilinear as well as from the small and intimate to

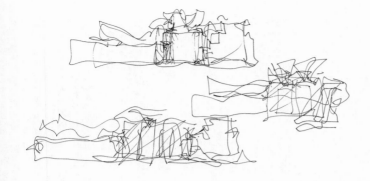

FIGURE 10

Frank Gehry, Guggen-
heim Museum Bilbao.
(Provenance: Guggen-
heim Bilbao Museoa. ©
FMGB Guggenheim
Bilbao Museoa.)

the vast and overwhelming. Though clearly distinct, these forms remain intricately interconnected; indeed, it is precisely their complicated relations that articulate their distinguishing differences. The complex structure of the museum is not quite in equilibrium and thus keeps everyone who roams through it slightly off balance. Since the forms are irregular and their relation unpredictable, the possibility of surprise lurks along every curve. Wandering through the museum's circulatory circuits, its structure seems to change repeatedly in unexpected ways. Instead of preprogrammed or permanent, the order of this structure is emergent and transient. It is as if the flow of the space follows the swirling eddies and turbulent whirlpools of the Nervion rushing nearby.

In the Guggenheim Museum Bilbao, *form itself becomes complex.* The difference between Venturi's superficial complexity and Gehry's radical complexity is symptomatic of changes in social conditions that have occurred since the late 1960s. While automobile culture has not, of course, disappeared, its importance has been eclipsed by the extraordinary growth of the economies of information, telematics, media, and virtuality.[32] What Gehry begins in Bilbao, he pushes to the tipping point in Times Square. The site is significant: as the node where information, finance, and entertainment networks converge, Times Square forms a lens through which the complexities of network culture can be glimpsed. Throughout the 1990s, this area of New York City was transformed from the seedy to the spectacular. Upscale hotels and expensive restaurants displaced peep shows and porno shops. No longer mechanically puffing smoke, signs have gone electronic and now display everything from high fashion to high finance. The studios of ABC, MTV, and ESPN confuse inside and outside until the street becomes a set for latter-day flâneurs to be actors and actresses in a show that has become global. On Broadway between 42d and 43d Streets, a stunning high-rise building is nearing completion. This is the new home of the global news and entertainment conglomerate Condé Nast, whose print and electronic publications simultaneously reflect and promote the crisscrossing circuits of contemporary culture: *Vogue, The New Yorker, Architectural Digest, Traveler Magazine, Swoon, Epicurious, Investors Business Daily, Technical Analysis of Stocks and Commodities,* and even *Wired.* In addition to offices for Condé Nast's multiple ventures, the building also houses the Nasdaq marketing center and ticker board from which fluctuating stock prices are telecast and webcast worldwide in real time. In Times Square, Nasdaq becomes entertainment competing with ABC, MTV, ESPN, and Disney for viewers and ratings. Surrounding Nasdaq's corner offices there is perhaps the most spectacular sign ever built: a 120-foot-tall $37 million cylindrical video screen displaying stock information and advertisements in 16.7 million colors. The 18 million light-emitting diodes (L.E.D.s) create a density of pixels that allows the transmission of colors and images of unprecedented intensity and veracity. In contrast to the traditional stock ticker projected on the Morgan Stanley Building at the other end of Times Square, the Nasdaq sign displays the vividly colored logos rather than the printed abbreviations of the companies driving the new economy. Here it becomes undeniable that "business art *is,*" as Warhol insists, "the step that comes after Art."

Tucked away near the Nasdaq sign, Gehry puts the business of art to work to create an extraordinary environment. In the Condé Nast cafeteria, Gehry

FIGURE 11 Frank Gehry, Guggenheim Museum Bilbao. (© FMGB Guggenheim Bilbao Museoa. Photographer, Erika Barahona Ede. All rights reserved. Partial or total reproduction prohibited.)

FIGURE 12 Frank Gehry, Guggenheim Museum Bilbao. (© FMGB Guggenheim Bilbao Museoa. Photographer, Erika Barahona Ede.

turns Bilbao in on itself to produce strange loops that surprise as much as they delight. Most important, the shimmering surfaces of Bilbao are folded inward and tinted blue to form the ceiling and walls. The result is a simulated sea or river, which is certainly not natural but not quite artificial. When transposed from outside to inside, opaque forms become transparent and translucent. Rippled surfaces of undulated glass shapes reflect the reflections of the titanium panels. Along a mirrored corridor, reflections of reflections of reflections create figures that flow, torque, morph, and liquefy only to reform and return to circulation. In this virtually aqueous environment, the surging currents of network culture pulsate through mind and body. As forms swirl and images flicker, everything drifts far from equilibrium and rapidly approaches the edge of chaos, where it becomes clear that this moment of complexity is where the action is.

two

Palladio faced a fork in the road, and he took the
wrong turn. He should have recognized that there's
chaos; he should have gone ahead and done what Borro-
mini did. He would have been a pioneer . . . to be in the
middle of Venice, so close to Palladio—and so much architecture today refers to Palladio . . . to be talking about
disorder, another kind of order is a bit irrelevant. . . .
Western culture just thinks of our kind of order . . . of
symmetry, classicism, and the idea of central focus.
But the whole world can't be built on axes alone.

frank gehry

critical emergency

POLITICS AND CRITICISM

When we begin to try to understand the intricate dynamics of the
moment of complexity, it quickly becomes apparent that we are
facing a critical emergency. The theoretical resources informing
social and cultural analysis for more than three decades have been
exhausted, and alternative interpretive strategies have yet to be defined. Innovative perspectives once bristling with new insight by
now have been repeated and routinized until they yield arguments
that are utterly predictable and familiar. For more than thirty
years, there has been a widespread consensus that the most pressing critical challenge is to find ways to resist systems and structures that totalize by repressing otherness and by reducing differ-

ence to same. Whether understood socially, psychologically, politically, economically, or culturally, so-called hegemonic structures, it is argued, must be subverted or overturned by soliciting the return of repressed otherness and difference in a variety of guises. For many years, writers and critics who were preoccupied with the theoretical problem of alterity were institutionally marginalized. Not merely a traditional generational conflict, the highly touted culture wars grew out of contrasting visions of the constitution of knowledge and of the legitimacy of different cultural traditions. In the heat of debate, exaggerated claims on all sides led to misunderstandings and distortions, which have tended to obscure more than clarify. In recent years, however, a significant shift has occurred: once marginal critical practices have become widely institutionalized, and with this change a split between theory and practice has developed. As critics who had built careers on the resistance to centralized authority assume positions of institutional responsibility, they frequently repeat precisely the kind of exclusionary gesture they once suffered. Accordingly, critics of the Western canon canonize noncanonical modes of analysis, thereby relegating other approaches to the periphery. Institutional position and critical approach are implicated in such a way that each conditions the other. As the marginal becomes institutionally central, the theoretical concern with difference and otherness is gradually transformed into a preoccupation with the same. Though critics repeatedly claim to recover *difference,* their arguments always come down to the *same:* systems and structures inevitably totalize by excluding difference and repressing otherness. Since the point is always the same, difference in effect collapses into identity in such a way that this undeniably influential critical trajectory negates itself and turns into its own opposite. At this moment, theory, as it recently has been understood, reaches a dead end.

There is profound irony in this turn of events, for it illustrates the pivotal insight of the thinker who, many critics insist, represents everything that must be resisted—Hegel. Whether labeled deconstructive or post-structuralist, criticism in recent years has tended to depict Hegelianism as both theoretically inadequate and practically destructive. Hegel, many insist, philosophically anticipates and indirectly promotes social and political processes of totalization that have ravaged the twentieth century. But this point of view fails to do justice to the subtleties of Hegel's system and obscures the contribution his thought can make to understanding emerging network culture. If the critical strategies designed to resist Hegel are turned back on themselves and read against the grain, it becomes apparent that various philosophies of difference harbor complexities and entail relational structures to which they remain blind. The nature of these complexities and the implications of these alternative structures can be understood by rethinking Hegel's system through complexity theory and vice versa.[1] This strategy points to a way out of the impasse we now face. To appreciate what is at stake in this critical turn, it is necessary to consider the historical genesis of the interpretive trajectories that have reached closure.

In matters of culture as well as politics, 1968 was a pivotal year. From the jungles of Vietnam and streets of Prague to universities in the United States and throughout Europe, social unrest seemed to many observers to portend radical change. While it is easy to dismiss the euphoria of the 1960s in retrospect, it is a mistake to underestimate the ongoing significance of the global dislocations that occurred during this pivotal decade. The events of 1968 set in motion processes of change that eventually culminated in the collapse of the Berlin Wall in 1989. If the fall of the Wall signals the emergence of network culture, then an adequate understanding of what now is occurring must begin with an examination of certain aspects of what took place in 1968. Global events unleashed in 1968 marked a decisive turning point in world politics and the international order. The ideology of the Cold War divided the world into two opposed camps. While creating conditions of extreme tension, exacerbated by the nuclear arms race, this system nonetheless functioned as a stabilizing structure. By drawing seemingly clear distinctions between East and West, Left and Right, right and wrong, good and evil, the complexities of the postwar world were simplified and a semblance of order bestowed on conditions that seemed to be at the edge of chaos. Since this binary system proved to be an effective means of exercising control and establishing a certain equilibrium, it proved advantageous to both sides. The United States became the leader of the so-called free world and gained control of regions that provided the opportunity for expanding markets, while the Soviet Union secured a military buffer zone protecting it from the West and allowing it to control Eastern Europe.

Though the social and cultural differences between these contending systems were obviously significant, political and economic factors eventually proved decisive. When conceived in binary terms, the centralized economic planning and top-down hierarchical political control of socialism and communism are often set over against the decentralized markets and bottom-up political power of democratic capitalism. Such stark contrasts however, are misleading, for oppositions are never so simple. On the one hand, continuing devotion to Keynesian economics and centralized programs of social welfare guided American policy, and, on the other hand, the imposition of Soviet economic principles and policies was never as complete or monothilic as both sides of the ideological divide would have had us believe. Furthermore, the contending ideologies of the Cold War era actually share commitments to fundamental Enlightenment ideas and ideals. Soviet communism and American capitalism are, in effect, alternative secularized eschatologies, which emerged in the wake of the critiques of religion formulated during the eighteenth century. By rewriting biblical history as secular metanarratives of human emancipation, these ideologies offered contrasting strategies for personal and social salvation.

When conflict broke out in the United States and Europe in 1968, the trigger was America's expanding war in Vietnam. International opposition to American political and economic policies was an indirect reflection of their success. Having rebuilt war-torn Europe, the United States both created new markets for its consumer goods and helped to establish strong national economies, which quickly became its competitors. As stability and security returned to Europe, resistance to U.S. economic and cultural power spread. The war in Vietnam provided the occasion for the expression of long-simmering resentments. In attempting to understand the implications of these developments for seemingly unrelated issues in critical theory, it is important to note that for many critics, the differences between the contending world systems were less important than their similarities. Far from a decentralized structure promoting distributed authority and local control, America's military-industrial complex appeared to many to be a repressive system, which imposed uniformity by attempting to turn everybody and everything to its own ends. While the criticisms of Americanism often grew out of commitments to different versions of socialism, this political and cultural conflict cannot be understood in traditional terms of Left and Right. In the 1960s, a "New Left" emerged, which found many of the policies and practices of what they dubbed the "Old Left" as repressive as those of the Right. For the New Left, the Old Left had been ineffective in resisting capitalism and had failed to provide an adequate standard of living for people living in countries under different forms of communism. For many social critics, the contending ideologies were mirror images of each other and were equally unsatisfactory. Whether capitalist or socialist, systems seeking global power inevitably seemed to become repressive. What united otherwise diverse ideological factions was a common resistance to the centralized organization and hierarchical authority of "the System." All such social and political movements of resistance were, in some sense, antisystematic. While some people became devoted libertines preoccupied with individual subjectivity, others sought more immediate forms of community through "authentic" political engagement and social action. Eventually, all of these efforts failed because they lacked constructive programs of social engagement and political reform. In hindsight, it is clear that criticism alone is not sufficient; what is required is the articulation and formation of alternative systems and novel networks that can function creatively.

POWER AND STRUCTURE

Though crucial events in 1968 took place throughout the world, what occurred in Paris in May effectively caught the drift of global change. There are many reasons Paris was the center of heated conflict. The prolonged Vietnam peace talks held in the city not only focused world attention on what many saw as American imperialism but also served as a vivid reminder of the long tradition of French colonialism. In local rather than global politics, the unprecedented alliance between students and workers throughout France created an unusually volatile situation. When the universities erupted in May, unrest quickly spread beyond the walls of universities to create a pervasive sense of urgency and widespread fear that things were spinning out of control. Within the precincts of the university, sociopolitical conflict was translated into intellectual disputes and vice versa. Among the many strands of intellectual controversy at this time, the most important for our purposes was the debate between structuralists and their critics. For many postwar French intellectuals, structuralism had provided an interpretive method that created the possibility of overcoming the limitations of traditional humanistic approaches to personality, society, and culture. The chaos and confusion created by World War II generated a pervasive longing for order and meaning. Instead of following existentialists like Sartre and Camus by attempting to plumb the depths of despair and absurdity in a world gone mad, structuralists sought to discover reason in history by uncovering forms and patterns that are permanent and universal rather than transient and arbitrary. In contrast to Sartre, who associated individual subjectivity with a freedom that precedes any trace of essence, Lévi-Strauss identified universal structures operating in self, society, and culture, structures which are *essentially* rational and as such completely comprehensible. Rather than disinterested speculation, structuralism's search for "scientificity" expressed a profound desire for things to make sense. In the opening chapter of *The Savage Mind,* entitled, significantly, "The Science of the Concrete," Lévi-Strauss echoed Le Corbusier and Mies's call for order: "Any classification is superior to chaos and even a classification at the level of sensible properties is a step towards rational ordering."[2] This search for "rational ordering" in the architecture of mind and society eventually issued in the articulation of conceptual structures that reflect concrete social, political, and economic realities. It is not surprising that in the polarized world of the Cold War, the structures Lévi-Strauss identified are always binary opposites. While it is debatable whether these structures are invented or discovered, it is indisputable that they display a systematic logic consistent with the machinations of modern industrialism. Lévi-Strauss was preoccupied with binary structures that are hierarchical, static, and stable. The task of critical analysis, he insisted, is twofold: one must first break down the complexities of the world and human experience into their constitutive parts, and then proceed to determine the simple principles, procedures, and rules by which these elements are combined to form social and cultural phenomena. The results of this mechanical process are as standardized as products rolling off an assembly line. Lévi-Strauss's approach is explicitly reductive and his logic is strictly linear. In this interpretive system, effects are always proportionate to their causes.

Given our consideration of the aesthetics of modernism, it should not be surprising that Lévi-Strauss associates the architecture of structuralism with the logic of the grid:

Now, on the theoretical as well as the practical plane, the existence of differentiating features is of much greater importance than their content. Once in evidence, they form a system, which can be employed as a grid used to decipher a text, whose original unintelligibility gives it the appearance of an uninterrupted flow. The grid makes it possible to introduce divisions and contrasts, in other words the formal conditions necessary for a significant message to be conveyed.[3]

When the grid is understood in this way, it functions like a universal code that is hardwired in the mind. This code is the foundational origin of self, society, and culture. While in his early work Lévi-Strauss resists the temptation to reduce psychological, social, and cultural structures to natural processes, in his later work he tends to naturalize mind, society, and culture. He signals this change in critical perspective by recasting his understanding of the code from linguistics, with morphemes and phonemes as combinatorial units, to the gridlike periodic table of basic elements:

By drawing up an inventory of all the customs that had been observed, all those imagined in myths, those evoked in the games of children and adults, and the dreams of healthy and sick individuals and psychopathological behaviors, one would be able to draw up a periodic table like that of the chemical elements, in which all real or merely possible customs would appear grouped in families, and in which we would simply need to recognize those which societies have in fact adopted.[4]

The motive for this change is Lévi-Strauss's desire to achieve what he regards as greater scientific rigor and thus to insure the scientific legitimacy of his work. Attempting to justify his effort to fuse nature and culture, he draws on both the physical and biological sciences. To revise his approach in a way that moves beyond structural linguistics, Lévi-Strauss first considers certain aspects of René Thom's catastrophe theory, developed in his *Structural Stability and Morphogenesis.* As I have noted, Thom identifies seven types of catastrophe, which he describes as "singularities." In topology, singularities are "phenomena that occur when the points of one surface are projected onto another as the surfaces are topologically distorted," and in calculus, the term designates "a point on a graphic curve where the direction or quality of curvature changes."[5] Thom's use of "singularity" appropriates both of these meanings to specify sites of discontinuous phase transition. After lengthy mathematical analysis, he concludes that all catastrophic changes can be classified under seven headings: fold, cusp, swallowtail, butterfly, parabolic, hyperbolic, and umbilic.[6] While Thom is preoccupied with the complex dynamics of morphogenesis as well as formal classification, Lévi-Strauss uses Thom's typology of catastrophes to substantiate his claims for the isomorphism of natural, social, and cultural structures. Nature, like culture, he argues, is organized by a limited inventory of forms. Lévi-Strauss is more interested in Thom's account of the stability of these forms than in his analysis of the ways in which they emerge and change.

Lévi-Strauss sees even stronger support for his efforts to naturalize cultural forms in the revolutionary developments in genetic biology. After Watson and Crick elucidated the structure of DNA in 1953, the temptation to reduce cultural processes to an apparently natural substrate became overwhelming. Lévi-Strauss goes so far as to call for "a positive collaboration between geneticists and ethnologists."[7] This collaboration however, is not between equals, because the physical and biological worlds, he believes, are more fundamental than culture. The laws of the physical world that are most relevant for Lévi-Strauss's argument are genetic. From a structuralist perspective, genetic and linguistic codes appear to be perfectly isomorphic: while the genotype functions as something like an elemental universal grammar (i.e., language or *la langue*), the phenotype operates like particular speech events (i.e., *la parole*). The four nucleotide bases (adenine [A], guanine [G], cytosine [C], and thymine [T]), which are analogous to elements or morphemes, are combined through chemical reactions that approximate syntactic rules of grammar. Just as Lévi-Strauss is more interested in the typology of catastrophe than the dynamics of morphogenesis, so he lends greater weight to genetic than to evolutionary biology. By naturalizing structures and structuring nature, he creates what is, in effect, an objective idealism in which temporal events and historical development become epiphenomenal. In this scheme, time and history are but the pale shadows of the eternal structures and codes that constitute the abiding substance and substrate of all reality.

With this turn in Lévi-Strauss's thinking, the reasons for the eventual criticism of and resistance to structuralism become clear. Though claiming to institute interpretive practices that overcome the limitations of traditional humanism, many critics in the late 1960s were convinced that his structuralism was unavoidably conservative. By naturalizing structuralism, Lévi-Strauss not only ignored the temporality of experience but also denied the possibility of *fundamental* change. In the "Overture" to *The Raw and the Cooked: Introduction to the Science of Mythology,* he explains:

Throughout, my intention remains unchanged. Starting from ethnographic experience, I have always aimed at drawing up an inventory of mental patterns, to reduce apparently arbitrary data to some kind of order, and to attain a level at which a kind of necessity becomes apparent, underlying the illusions of liberty. In Les Structures, *behind what seemed to be the superficial contingency and incoherent diversity of the laws governing marriage, I discerned a small number of simple principles, thanks to which a very complex mass of customs and practices, at first absurd (and generally held to be so), could be reduced to a meaningful system.*[8]

From this point of view, order, it seems, can be established only if freedom and change are denied, and meaning can be secured only if diversity is reduced to uniformity, and complexity is reduced to simplicity.

GENERATIONAL GAPS

By the 1960s, memories of the chaos wrought by wars on European soil were fading and many considered the price Lévi-Strauss paid to secure order and stability to be too high. Far from a revolutionary methodology, structuralism came to be associated with institutional repression, which had to be overturned in the name of freedom, history, and difference. Though the objections to the structuralist agenda were many, I will concentrate on three critical trajectories, which have been particularly important during the past three decades: social constructivism, deconstruction, and the theory of simulation. These alternative interpretive strategies are associated with three individuals: Michel Foucault, Jacques Derrida, and Jean Baudrillard. When these influential figures are approached through their shared suspicions of structuralism, important similarities in their positions, which usually are overlooked or denied, become apparent. Moreover, the limitations of their critical responses to structuralism point toward a way out of the critical impasse that continues to make it difficult to develop an adequate understanding of network culture. To insist that Foucault, Derrida, and Baudrillard are suspicious of structuralism is not to claim that they dismiss it out of hand. As savvy critics, these writers do not attack the positions they oppose from without but attempt to subvert them in ways that preserve what they nonetheless displace.

The intricate imbrication of structuralism and its critique is evident throughout Foucault's work. In the introduction to *The Archaeology of Knowledge,* he attempts to explain his complicated relation to structuralism:

My aim is to uncover the principles and consequences of an autochthonous transformation that is taking place in the field of historical knowledge. It may well be that this transformation, the problems that it raises, the tools that it uses, the concepts that emerge from it, and the results that it obtains are not entirely foreign to what is called structural analysis. But this kind of analysis is not specifically used.[9]

Foucault's efforts to distinguish his archaeological approach from structuralism proved so unsuccessful that a year later he was forced to conclude his foreword to the English translation of *The Order of Things* with a searing attack on his interpreters: "In France, certain half-witted 'commentators' persist in labeling me a 'structuralist.' I have been unable to get it into their tiny minds that I have used none of the methods, concepts, or key terms that characterize structural analysis."[10] Foucault, needless to say, protests too much; while admitting that there is an "order of things," he insists that this order is neither natural nor essential and thus cannot be preordained or unchanging. Whatever order is at work in the world is historically contingent and therefore to a certain extent arbitrary. By supplementing analyses of synchronous structures with a detailed examination of their diachronic emergence, Foucault lends temporality and historicity a weight denied them by Lévi-Strauss and his followers. Structuralism can in a certain sense be understood as an archaeological investigation devoted to stripping away layer after layer of superficial differences to uncover the foundational structures constituting the common origin of diverse natural, social, and cultural phenomena. Foucault is convinced that this search for origins is futile: "In setting itself the task of restoring the domain of the original, modern thought immediately encounters the recession of the origin."[11] The recession of the origin insinuates a *radical* temporality in the midst of every structure designed to repress it. In an exceedingly complex but very important passage, Foucault writes:

A task is thereby set for thought: that of contesting the origin of things, but of contesting it in order to give it a foundation, by rediscovering the mode upon which the possibility of time is constituted—that origin without origin or beginning, on the basis of which everything is able to come into being. Such a task implies the calling into question of everything that pertains to time, everything that has formed within it, everything that resides within its mobile element, in such a way as to make visible that rent, devoid of chronology and history, from which time issued. Time would then be suspended within that thought, which nevertheless cannot escape from it since it is never contemporaneous with the origin; but this suspension would not have the power to revolve the reciprocal relation of origin and thought; and as it pivoted upon itself, the origin, becoming what thought has yet to think, and always afresh, would be forever promised in an imminence always nearer, yet never accomplished. In that case, the origin is that which is returning, the repetition towards which thought is moving, the return of that which has always already begun, the proximity of a light that has been shining since the beginning of time.[12]

Instead of foundational, this "origin without origin or beginning" faults every foundation believed to be permanent or secure. This "rent" or tear is "devoid of chronology and history" because it is the trace of the withdrawal that releases the emergence of chronology and history. Not present but always near, the receding origin repeatedly approaches as the eternal return of what never arrives. Such an ever-elusive "origin" cannot be captured or contained by any structure; rather, its absence constitutes the condition of the possibility of structures and systems. The space opened by this strange origin fascinates Foucault.

It should now be obvious why Foucault insists that his archaeological investigations should not be labeled structuralist. The structures, patterns, and systems through which knowledge is ordered are not universal but are *historical* a prioris, which emerge in *specific* circumstances. With the recognition that interpretive grids are temporally contingent and historically relative, it becomes evident that things might always have been otherwise. The grids that form the codes of language, perception, and practice are neither hardwired nor natural but are both *instituted* and *constructed*. This is not to imply, of course, that Foucault attributes these codes to creative subjects or authors; to the contrary, socio-cultural construction occurs in the absence of any constructive subject. "Consequently," Foucault concludes, "no one is responsible for an emergence; no one can glory in it, since it always occurs in the interstice."[13] This critical emergence disrupts and destabilizes the structures Lévi-Strauss struggles to secure. The interstice is the "gap," which is the "middle region" that "disturbs what was previously considered immobile; . . . fragments what was thought unified; . . . [and] shows the heterogeneity of what was imagined consistent with itself."[14] Foucault explains: "This middle region, then, insofar as it makes manifest the modes of being of order, can be posited as the most fundamental of all: anterior to words, perceptions, and gestures, which are taken to be more or less exact, more or less happy, expressions of it."[15] Since the middle region is "anterior to words," it cannot be adequately comprehended by linguistic codes or rules of reason. This does not mean that the intermediate gap is irrational; neither rational nor irrational, it is the condition of the possibility of both reason and unreason. The codes of language and reason—as well as all other codes—are *constituted, instituted, constructed* and, therefore, presuppose something that is not coded. Always implicated in something that can never be ordered, order is always incomplete and codes are inevitably partial.

In an effort to explain the far-reaching consequences of his archaeological inquiries, Foucault turns to Nietzsche's genealogy of culture. Nietzsche develops an analysis of what he describes as "'effective' history"—*wirkliche Historie.*"History becomes effective," Foucault explains, "to the degree that it introduces discontinuity into our very being." This discontinuity results from the ceaseless play of instincts, impulses, and desires. Within this field of contending forces, culture—and, by extension, knowledge—is never disinterested and its forms and structures are constantly changing. Cultural forms as well as the knowledge they make possible are *inventions* designed to regulate individual behavior and social processes. For as long as they maintain their vitality, cultural systems operate far from equilibrium and are always on the brink of dissolving into chaos.

In his later work, Foucault combines his archaeology of knowledge with Nietzsche's genealogy of culture to develop a sophisticated analysis of the production of knowledge and construction of social and cultural codes. The central category in this analysis is "discursive practices." For Foucault, "discursive practices are not purely and simply ways of producing discourse. They are embodied in technical processes, in institutions, in patterns for general behavior, in forms for transmission and diffusion, and in pedagogical forms which, at once, impose and maintain them."[16] By extending the analysis from the physical and psychological aspects of experience to technological processes and institutional practices, Foucault enriches and complicates his analysis. Like psychological, social, and cultural codes, these discursive practices are neither natural nor produced by creative subjects; to the contrary, subjects as well as objects are constructed by discursive practices and disciplinary structures that are both restrictive and productive. Since the constitution of the knowing subject and known object occurs in a field of fluctuating powers, subjects, objects, and their interrelation are always changing and thus ever incomplete.

These insights suggest the reasons for the confusions about Foucault's relation to structuralism as well as his repeated efforts to distance himself from it. On the one hand, Foucault's discursive formations function something like Lévi-Strauss's organizing structures. There is an order of things, which arises through structuring processes that are formally regulative. But, on the other hand, Foucault denies that these forms or structures are universal or atemporal. Insisting on a radical diachronicity at the heart of all synchronicity, Foucault argues that the order of things emerges from a recessive origin that can never be comprehended. The danger he sees in structuralism is an essentialism that makes history irrelevant and change impossible. To counter this essentialism, he develops an account of discursive practices, which, he believes, exposes the constructed character of *all* reality. In the hands of his less sophisticated followers, this constructivism often becomes as reductive as the essentialism it is designed to correct.

In recent years, what has come to be known as "social constructivism" or "cultural constructivism" has provoked heated debate and has led to a widening of the gap between C. P. Snow's "two cultures." Concerned about what they describe as postmodern relativism and skepticism, a growing number of scientists have begun to voice severe criticisms of certain trends in the arts and humanities. The excessive rhetoric of these exchanges often leads to misunderstandings on both sides and makes responsible engagement difficult if not impossible. In their calculatedly provocative book *Higher Superstition: The Academic Left and Its Quarrels with Science,* Paul Gross and Norman Levitt attack what they regard as the pernicious perspectivism promoted by critical theorists and promulgated in the humanities:

Science, arguably the dominant mode of thought in the contemporary world, has thus come under the scrutiny of Foucault, Derrida, and their followers. In the case of Foucault, skepticism is expressed in the form of doubts about the human importance of scientific truth, rather than on the possibility of achieving it. Nonetheless, the basic

idea, that a mode of discourse is inevitably a code of power relations among the peo-
ple who use it, has contributed importantly to the notion that science is simply a cul-
tural construct which, in both form and content, and independently of any individ-
ual scientist's wish, is deeply inscribed with assumptions about domination, mastery,
and authority.[17]

For Gross and Levitt, this line of analysis is the problematic legacy of the social
and political activism of the late 1960s and early 1970s. The danger they and
their colleagues see in sociocultural constructivism and deconstruction is a relativism they believe undercuts the authority of science and renders true knowledge impossible. Such criticisms are not limited to marginal figures in the sciences; indeed, several widely respected scientists have recently entered the fray.
Oxford biologist Richard Dawkins, often described by some as "one of the most
influential scientists of our time," dismissively declares that "the meaningless
wordplays of modish francophone *savants,* splendidly exposed in Alan Sokal and
Jean Bricomont's *Intellectual Impostures* (1998), seem to have no other function
than to impress the gullible. They don't even want to be understood."[18] On this
side of the Atlantic, Edward O. Wilson echoes Dawkins's criticism in his much-
discussed *Consilience: The Unity of Knowledge:*

All movements tend to extremes, which is approximately where we are today. The ex-
uberant self-realization that ran from romanticism to modernism has given rise to
philosophical postmodernism (often called post-structuralism, especially in its more
political and sociological expressions). Postmodernism is the polar antithesis of the
Enlightenment. The difference between the two extremes can be expressed roughly as
follows: Enlightenment thinkers believe we can know everything, and radical post-
modernists believe we can know nothing.

   The philosophical postmodernists, a rebel crew milling beneath the black flag of
anarchy, challenge the very foundations of science and traditional philosophy. Reality,
they propose, is a state constructed by the mind, not perceived by it. In the most ex-
travagant forms of this constructivism, there is no "real" reality, no objective truths
external to mental activity, only prevailing versions disseminated by ruling social
groups. Nor can ethics be firmly grounded, given that each society creates its own
codes for the benefit of the same oppressive forces.[19]

While admitting concern about the epistemological skepticism and moral relativism he believes inherent in postmodernism and post-structuralism, Wilson
initially seems more open to responsible dialogue with humanists than many

other scientists. In the opening pages of his book, he actually claims that "the greatest enterprise of the mind has always been and always will be the attempted linkage of the sciences and humanities." But first impressions are deceptive; a few pages later, he admits: "I believe the enterprises of culture will eventually fall out into science, by which I mean the natural sciences, and the humanities, particularly the creative arts." Feigning respect for the arts and humanities, Wilson actually attempts to seize their territory:

The cutting edge of science is reductionism, the breaking apart of nature into its natural constituents. . . . Critics of science sometimes portray reductionism as an obsessional disorder declining toward a terminal stage one writer recently dubbed "reductive megalomania." That characterization is an actionable misdiagnosis. Practicing scientists, whose business it is to make verifiable discoveries, view reductionism in an entirely different way: It is the search strategy employed to find points of entry into otherwise impenetrably complex systems. Complexity is what interests scientists in the end, not simplicity. Reductionism is the way to understand it. The love of complexity without reductionism makes art; the love of complexity with reductionism makes science.[20]

A complexity that can be reduced to simplicity, however, is no complexity at all. Wilson's characterization of the opposition between science and art does not hold up under careful examination. When insisting on the superiority of scientific simplicity over artistic complexity, Wilson implicitly expresses *aesthetic* values, which, as we have seen, inform the work of modernists like Le Corbusier and Mies van der Rohe. All too often scientists do not accord artists and humanists the same respect they demand from them. While condemning humanists for not studying science seriously, many scientists have no reservations about attacking important writers whose work they do not understand and often appear not to have read. There is no doubt that the work of many of Foucault and Derrida's epigones is misleading, but nothing is gained by perpetuating culture wars through opposing positions which are, in fact, misguided mirror images of each other. For many scientists, constructivists make the object investigated as well as the aim of the investigation disappear in a relativistic perspectivism, which, though supposedly not the creation of a subject, nonetheless appears to be subjective; for the constructivists, by contrast, scientific research often seems to involve an essentialism that leads to repressive reductionism. Instead of reducing nature to culture, or culture to nature, what is needed is a way of understanding the *complex* dynamics that render them mutually constitutive. If there is to be any hope of moving beyond the current impasse, humanists and scientists must find ways to talk with and learn from each other. To see where the opening for such a dialogue might be found, we must turn to the two remaining trajectories that have informed critical debate for the past thirty years.

DIGITAL DIVIDE

Derrida's deconstructive critique of structuralism bears a striking resemblance to Foucault's genealogical analysis. Though often obscured by tensions between teacher and student, Foucault's constructivism and Derrida's deconstruction are, in effect, different versions of the same criticism of systems and structures. In his early essay "Cogito and the History of Madness," originally delivered as a lecture at the College Philosophique on March 4, 1963, and published four years later in *Writing and Difference,* Derrida frames his analysis of *Madness and Civilization* in terms of Kierkegaard and Hegel. He signals the general direction of his argument in the epigram, drawn from Kierkegaard: "The Instant of Decision is Madness."[21] By citing this text, Derrida suggests that his reading of *Madness and Civilization* effectively extends Kierkegaard's critical perspective. The gist of Derrida's argument is that Foucault remains essentially Hegelian. Since Derrida, among others, regards Hegelianism as protostructuralism and structuralism as latter-day Hegelianism, he concludes that in spite of his vehement denials, Foucault is actually a structuralist.

While agreeing with Foucault's insistence on the inseparability of madness and civilization, Derrida argues that the madness Foucault probes is not mad enough. Foucault's aim, Derrida explains, is "to write a history of madness *itself. Itself.* Of madness itself. That is, by letting madness speak for itself. Foucault wanted madness to be the *subject* of his book in every sense of the word: its theme and its first-person narrator, its author, madness speaking about itself. Foucault wanted to write a history of madness *itself,* that is madness speaking on the basis of its own experience and under its own authority, and not a history of madness described from within the language of reason."[22] According to Derrida, there are three problems with this project. First, Foucault proceeds as if he knows what madness is. If, however, he can identify "madness *itself*" and describe it as the theme or subject of his book, then madness remains "within the language of reason," and, thus, is not really mad. Second, Foucault's assumption that madness has a history discloses a commitment to the ideals of the Western metaphysical tradition: "The attempt to write the history of the decision, division, difference runs the risk of construing the division as an event or a structure subsequent to the unity of an original presence, thereby confirming metaphysics in its fundamental operation."[23] Finally, a corollary of the previous point: by approaching madness as reason's other, Foucault interprets it in terms of negativity. Negativity is, of course, the fundamental principle of the Hegelian dialectic. When history is interpreted as a process of negation and double negation, nothing escapes reason's grasp. From this point of view, Derrida argues, "the revolution against reason can be made only within it, in accordance with a Hegelian law to which I myself was very sensitive in Foucault's book, despite the absence of any precise reference to Hegel." Far from exposing the faults of reason by subverting its comprehensive gestures, Derrida argues that Foucault reinscribes precisely the Cartesianism he claims to overturn: "In this sense, I would be tempted to consider Foucault's book a powerful gesture of protection and internment. A Cartesian gesture for the twentieth century. A reappropriation of negativity."[24]

For Derrida, there is a direct line of descent from Descartes's geometric rationalism and Hegel's dialectical logic to Lévi-Strauss's structural grids. The problem with all of these perspectives is what Derrida describes as "structuralist totalitarianism." In his most devastating critique of Foucault, he writes:

Structuralist totalitarianism here would be responsible for an internment of the Cogito similar to the violences of the classical age. I am not saying that Foucault's book is totalitarian, for at least at its outset it poses the question of the origin of historicity in general, thereby freeing itself of historicism; I am saying, however, that by virtue of the construction of his project he sometimes runs the risk of being totalitarian.[25]

As we have seen, the violence of structuralism results from its persistent effort to reduce difference and repress otherness. By reading madness as the negative of reason, Foucault incorporates it in a dialectic that reconciles opposites in a binary structure that is thoroughly rational. The madness whose history can be narrated is not, in the final analysis, mad at all. Reason is not thrown into crisis until it encounters—but, of course, it cannot really encounter—what Derrida regards as a different or an other madness: "This crisis of reason in which reason is madder than madness—for reason is non-meaning and oblivion—and in which madness is more rational than reason, for it is closer to the wellspring of sense, however silent or murmuring—this crisis has always already begun and is interminable."[26] The madder madness, which is necessarily implicated in reason, is the madness of Kierkegaard's instant of decision. This "bifurcation point" opens every system "to an undecidable resource that sets the system in motion."[27]

Derrida's sustained criticism of *Madness and Civilization* actually misrepresents Foucault's genealogical analysis. What Derrida seeks in his appropriation of Kierkegaard's instant of decision is what Foucault pursues in his "recessive origin." Over thirty years later, Derrida returns to questions addressed in his critique of Foucault in a work, which is particularly important for our purposes, entitled *Archive Fever: A Freudian Impression.* Derrida's argument underscores his lifelong obsession with what he sees as the totalizing propensities of *all* structures and systems. Though he has written an astonishing number of texts on a remarkable range of subjects, Derrida's analysis always comes down to the *same* point: structures and systems, which appear to be closed and complete, are inevitably ruptured because they cannot incorporate what they nevertheless presuppose. Whether in the guise of madness or something other, there is always the trace of an unassimilable remainder, which is the condition of the possibility of systems and structures as well as the impossibility of their integrity and stability. What is particularly interesting about *Archive Fever* in this context is Derrida's effort to push his recurrent analysis of systems into the digital domain. As will become apparent, Derrida mistakenly believes that digital networks involve the same logic and thus are subject to the same limitations as the dialectical systems that Foucault reinscribes and the binary structures that Lévi-Strauss analyzes.

Like Foucault, Derrida has long been preoccupied with the question of the archive and, by extension, the problem of memory. *Archive Fever* actually updates the argument of another essay in *Writing and Difference,* "Freud and the Scene of Writing," in which Derrida reinterprets the unconscious as a writing machine operating like an extensive postal system. Since the state of technology in a given period influences the way in which the psyche is understood, Derrida wonders how Freud's analysis might have been different had he known about computers and E-mail. Beginning on an etymological note, Derrida points out that "archive" derives from *arkhe,* which "names at once the *commencement* and the *commandment.* The word apparently coordinates two principles in one: the principle according to which nature or history, *there* where things *commence*—physical, historical, or ontological principle—but also the principle according to the law, *there* where authority, social order are exercised, *in this place* from which *order* is given—nomological principle."[28] By focusing on *arkhe* as commencement and commandment, Derrida returns to the issue of origin, which he explores in his analysis of Freud. Through a consideration of the interplay of eros and thanatos in *Beyond the Pleasure Principle,* Derrida identifies what he describes as "a non-origin, which is originary." Though never acknowledged, this nonoriginary origin actually corresponds to Foucault's "origin without origin or beginning." As an origin that is not originary or is without beginning, the archive is irreducibly paradoxical: "The concept of the archive shelters in itself, of course, this memory of the name *arkhe*. But it also *shelters* itself from this memory which it shelters: which comes down to saying that it forgets it."[29] The *arkhe,* in other words, is remembered *as forgotten.* Rather than the fading of a memory once present, such forgetting is a *primal* oblivion of what has never been known. This reading of the *arkhe* is actually an elaboration of Derrida's understanding of Kierkegaard's instant of decision. As commandment, *arkhe* institutes the "nomological principle" through which "*order* is given." As such, the *arkhe* falls within Foucault's "*middle region,*" which is anterior to words. Whether conceived as Foucault's institutional moment or Derrida's decisive instant, the activity through which the principles of order and rules of reason are constituted cannot be ordered or coded. In this moment, beyond yet within reason and order, Derridean deconstruction and Foucauldian constructivism become indistinguishable.

While the technologies of archivization change, Derrida argues, the dynamics of the process remain constant. From Freud's mystic writing pad to the most sophisticated techniques of data storage and processing, archivization "not only incites forgetfulness, amnesia, the annihilation of memory, as *mneme* or *anamnesis,* but also commands the radical effacement, in truth the eradication, of that which can never be reduced to *mneme* or to *anamnesis.*"[30] This "annihilation of memory" leads Derrida to accept Freud's association of the technologies of archivization with the death instinct. The machinations of techno-science, he maintains, entail an inescapably destructive violence. Faced with this threat, Derrida deploys his familiar deconstructive strategy. Since digital networks and information technologies totalize by reducing everything to binary code, they can be subverted by exposing the way in which they inevitably entail that which cannot be coded. Archivization presupposes a "space" or "place" of inscription that is always already "there" prior to the process of recording. This antecedent site is "an *internal* substrate, surface, or space" which is "a *prosthesis on the inside.*" For Derrida, "*there is no archive without a place of consignation, without a technique of repetition, and without a certain exteriority. No archive without outside.*"[31] This "outside" that is "inside" displaces the archive as if from within. The prosthesis on the inside is a *digital divide,* which disrupts the closure of the code. Since every code—be it linguistic, genetic, or digital—presupposes something that cannot be coded, the machinations of techno-science are inevitably incomplete and unavoidably faulty.

It is obvious that Derrida sees digital technology and network culture as extremely dangerous. Expanding telecommunications networks and proliferating worldwide webs, he believes, carry the threat of a hegemony more thorough than the military-industrial complex. Furthermore, the extraordinary rate of technological development lends urgency to the need to respond to these changes. Through his deconstructive analyses, Derrida attempts to disrupt digital technologies and the systems they produce by turning the digital divide into a rupture that can never be overcome. However, his critique is, in the final analysis, ineffective: *deconstruction changes nothing.* While exposing systems and structures as incomplete and perhaps repressive, deconstruction inevitably leaves them in place. This is not merely because deconstruction involves theoretical analyses instead of practical action but also because of the specific conclusions reached by the theoretical critique. Instead of showing how totalizing structures can actually be changed, deconstruction demonstrates that the tendency to totalize can never be overcome and, thus, that repressive structures are inescapable. For Derrida and his followers, all we can do is to join in the Sisyphean struggle to undo what cannot be undone.

This is not, however, the only conclusion to be drawn from deconstructive analysis. By turning Derrida's criticism back on itself, it becomes possible to discern creative possibilities he overlooks. Upon closer examination, it is clear that Derrida commits the very error of totalization for which he criticizes others. In his effort to preserve psychological, political, social, and cultural differences, he ignores important distinctions between and among different kinds of systems and structures. As we will see in later chapters, it is simply wrong to insist that *all* systems and structures necessarily totalize and inevitably repress. What Derrida cannot imagine is *a nontotalizing system or structure that nonetheless acts as a whole.* Important work now being done in complexity studies suggests that such systems and structures are not merely theoretically conceivable but are actually at work in natural, social, and cultural networks.

DECODING THE REAL

While Foucault and Derrida are acutely aware of the significance of digital technologies, they do not consider the growing impact of the media and popular culture in any detail. This oversight is at least in part the result of their failure to appreciate the formative influence of visual culture in an increasingly multimedia world. Baudrillard was one of the first theorists to realize the ways in which informatic and telematic media change experience and reconstitute the world.

Though rarely considered in relation to Foucault and Derrida, Baudrillard actually extends many of their insights into areas they do not explore.

For Baudrillard, the events of May 1968 marked a decisive turning point in Western culture. In an essay entitled "The End of Production," published in 1976, he writes:

May '68: The Illusion of Production. *The first shockwaves of this transition from production to pure and simple reproduction took place in May '68. They struck the universities first, and the faculty of human sciences first of all, because that was where it became most evident (even without a clear "political" consciousness) that* we were no longer productive, *only reproductive (and that lecturers, science and culture were themselves only relays in the general reproduction of the system).*[32]

The shift from a productive to a reproductive society marks the transition from a manufacturing to an information economy governed by new media. With remarkable prescience, Baudrillard foresaw the ways in which the events of the late 1960s would reverberate throughout later decades. "These shockwaves," he writes, "are still being felt. They cannot but reach the very limits of the system, as soon as entire sectors of society topple from the rank of *productive forces* to the pure and simple status of *reproductive forces.* Although this process was first felt in the cultural sectors of science, justice, and the family—the so-called 'superstructural' sectors—it is clear today that it is progressively affecting the entire so-called 'infrastructural' sector."[33] When production becomes reproduction, the line separating superstructure from infrastructure is hopelessly obscured.[34] Baudrillard realizes that the changes wrought by technology during the last three decades concretely realize many of the developments anticipated in twentieth-century critical theory.

For Baudrillard, the critical turning point in theory as well as practice is what he describes as "the structural revolution of value" in which meaning and value lose their referential moorings and are set free in an endless "structural play":

Referential value is annihilated, giving the structural play of value the upper hand. *The structural dimension becomes autonomous by excluding the referential dimension, and is instituted upon the death of reference. The systems of reference for production, signification, affect, substance and history, all this equivalence of a "real" content, loading the sign with the burden of "utility," with gravity—its form of representative equivalence—all this is over with. Now the other stage of value has the*

upper hand, a total relativity, general commutation, combination, and simulation—
simulation in the sense that, from now on, signs are exchanged against each other
rather than against the real. . . . The emancipation of the sign: remove this "archaic"
obligation to designate something and it finally becomes free, indifferent and totally
indeterminate, in the structural or combinatory play that succeeds the previous rule
of determinate equivalence.[35]

In the structural revolution of value anticipated by Saussure and developed by
Lévi-Strauss, the meaning of signs is a function of their systematic relations and
not their ostensible referents. This line of analysis does not simply abolish reference but refigures it in terms of the interplay of signifiers rather than the association between signs and referents of a different order. Instead of referring to
something that is not a sign, signs are always signs of other signs. For Baudrillard, this play of signs marks the end of meaning and the disappearance of the
real. Since he associates meaning with reference to that which is not already a
sign, he insists that the "liquidation of referentials" is also the "liquidation of
meaning."[36] It is important to stress that the real "dies" and meaning is "annihilated" *only* if they are assumed to entail something different from and antecedent to structures of signification.

   The collapse of the "real" marks the emergence of the "hyperreal" in
which "the simulation principle" takes precedence over "the reality principle."
In his best-known text, Baudrillard argues that "the precession of the simulacrum" both obliterates every referent and subverts the very possibility of originality:

Abstraction today is no longer that of the map, the double, the mirror or the concept.
Simulation is no longer that of a territory, a referential being or substance. It is the
generation by models of a real without origin or reality: a hyperreal. The territory no
longer precedes the map, nor survives it. Henceforth, it is the map that precedes the
territory—PRECESSION OF SIMULACRA—it is the map that engenders the ter-
ritory and if we were to revive the fable today, it would be the territory whose shreds
are slowly rotting across the map.[37]

In more familiar philosophical terms, experience is never immediate or direct
but is always mediated by organizing patterns and structures. While Baudrillard
wavers on the question of whether hyperreality involves a novel sociocultural
condition or is the recognition that "the real . . . never existed," he is convinced
that the structures of mediation are neither a priori categories of the mind nor

hardwired patterns of the brain but are inseparably bound to the changing technologies of production and reproduction in different societies.[38]

Nowhere is the structural revolution of value, which issues in the "collapse" of the opposition between superstructure and infrastructure, more evident than in the economic sphere. "Hyperreality" and what Baudrillard describes as "hypercapitalism" mirror each other in such a way that each is a condition of the possibility of the other. With the movement from an economy of production to reproduction, exchange value displaces use value. Baudrillard contends that this transition is strictly parallel to the change from a referential to a structural understanding of value and meaning. While use value designates needs that are supposed to be natural, exchange value is a function of a proliferating play of signs. By the early 1970s, something like a structural revolution of value transforms actual economic relations. Two developments were decisive in bringing about significant changes: first, the movement away from the gold standard, and second, the emergence of electronic currencies and other financial instruments. The end of the gold standard, Baudrillard maintains, is the "annihilation of the referent," which marks the end of fixed value—and creates the conditions for the speculative economy of the 1990s. As the economy becomes ever more wired, new financial instruments proliferate. Freed from grounding referents, capital becomes a matter of "pure circulation," which proliferates as its speed increases. With these developments, "everything within production and the economy becomes commutable, reversible and exchangeable according to the same indeterminate specularity as we find in politics, fashion or the media."[39]

When economic relations are mediated by blips of light transmitted through fiber-optic networks and displayed on computer screens, the economy and media become virtually indistinguishable. If everything within the economy is "commutable, reversible, and exchangeable" with the media, traditional reductive explanations of the relation between economics and the media—and, by extension, the economy and culture—must give way to interpretations of complex recursive relations, which are mutually determinative. Though Baudrillard discerns similar processes at work in hypercapitalism and the media, he thinks the destructive force of hyperreality is more evident in today's visual culture than in the subtleties of the global economy. Expressing a latter-day version of the ancient fears of iconoclasts, he warns anyone willing to listen about "the murderous capacity of images, murderers of the real."[40] Never re-presenting the "real," images annihilate what they claim to represent: *the image is the death of the thing.* As inescapably destructive, images harbor "evil demons," which are manifested in their "non-signifying fury."[41] Negating every vestige of transcendence, this fury unleashes an "immanent logic" that seems to turn back on itself in such a way that it consumes every vestige of a beyond. In this consumer economy, media becomes the only reality there is.

The immanent logic of simulacra is not limited to culture but extends to what once was regarded as the "natural" world in the form of the genetic code. The structural law of value, Baudrillard argues, entails a "metaphysics of the code." This "binary divinity" is the most recent expression of a god whose line-age can be traced at least to Genesis:

After the metaphysics of being and appearance, after energy and determinacy, the metaphysics of indeterminacy and the code. Cybernetic control, generation through models, differential modulation, feedback, question/answer, etc.: this is the new op-erational *configuration (industrial simulacra being* mere operations*). Digitality is its metaphysical principle (Leibniz's God), and DNA is its prophet. In fact, it is in the genetic code that the "genesis of simulacra" today finds its completed form. At the limits of an ever more forceful extermination of references and finalities, of a loss of semblances and designators, we find the digital, programmatic sign, which has a purely* tactical *value, at the intersection of other signals ("bits" of information/tests) and which has the structure of a micro-molecular code of command and control.[42]*

Whether generating currency, media, or biological organisms, programs, for Baudrillard, pre-scribe the course of development in advance in such a way that what appears to constitute change is actually the unfolding of a code implicit from the outset. The "anticipation and immanence of the code," therefore, seem to make actual change as impossible as Lévi-Strauss's universal structures. Dis-playing the same immanent logic at work in speculative finance and media net-works, genetic programs inevitably fold back on themselves to form closed struc-tures that incorporate differences and erase all "interstitial spaces." This closure results in what Baudrillard describes as the "terrorism of the code." When noth-ing escapes binary regulation, society and culture become *obscene.* Obscenity, for Baudrillard, is not pornographic display but is a condition of perfect trans-parency in which "all secrets, spaces, and scenes [are] abolished in a single di-mension of information."[43] In a world of invisible data banks, genetic engineer-ing, and nanotechnology, inside and outside are thoroughly confounded in such a way that the private is public and every secret is told. Baudrillard longs for what he realizes is impossible: the decoding of the real. To decode, in this con-text, does not mean to render transparent but *to free the real from the codes that seem to destroy it.*

69

Baudrillard's argument complements the analyses of Foucault and Derrida by extending the critique of structuralism to contemporary technologies of reproduction. Digital technologies become repressive when they absorb everything once believed to be real. With the death of the real, resources for resistance seem to disappear and change becomes an idle dream. The code, however, is not as seamless as it often appears. While insisting that the real is annihilated, Baudrillard admits that its traces remain to haunt the structures that destroy it. For Baudrillard, as for Foucault and Derrida, systems and structures that seem to be complete inevitably include what they are designed to exclude. Lingering vestiges of the real appear in a most unlikely place—the world of fashion. Fashion stages "the enchanting spectacle of the code," which "consumes the world and the real in advance."[44] The consumption fashion promotes creates the consumer economy that fuels hypercapitalism. When viewed through the lens of fashion, the structural revolution of value appears to be less an innovative linguistic or philosophical theory than an economic necessity. Hypercapitalism can thrive only in a world where exchange value displaces use value in a play of signs whose worth grows as the speed of trading increases:

The acceleration of the simple play of signifiers in fashion becomes striking to the point of enchanting us—the enchantment and vertigo of the loss of every system of reference. In this sense, it is the completed form of the political economy, the cycle wherein the linearity of the commodity comes to be abolished. There is no longer any determinacy internal to the signs of fashion, hence they become free to commute and permutate without limit. At the term of this unprecedented enfranchisement, they obey, as if logically, a mad and meticulous recurrence. This applies to fashion as regards clothes, the body and objects—the sphere of "light" signs. In the sphere of "heavy" signs—politics, morals, economics, science, culture, sexuality—the principle of commutation nowhere plays with the same abandon. We could classify these diverse domains according to a decreasing order of "simulation," but it remains the case that every sphere tends, unequally simultaneously, to merge with models of simulation, of differential and indifferent play, the structural play of value. In this sense, we could say that they are all haunted by fashion, since this can be understood as both the most superficial play and as the most profound social form—the inexorable investment of every domain of the code.[45]

Fashion, which "haunts" every domain of the code, is itself haunted by specters it can never escape. Baudrillard goes so far as to claim that "the enjoyment of fashion is . . . the enjoyment of a spectral and cyclical world of bygone forms endlessly revived as effective signs."[46] In this way, fashion discloses the spectral machinations of a society that is no longer productive but only reproductive. Never original, fashion is always *retro;* creation is a recombinant process in which the old is resurrected and recycled to appear as new and the new is always haunted by the old. Requiring ever greater and faster expenditure, hypercapitalism unleashes a furious spectacle of "the abolition of forms," which is simultaneously creative and destructive. Engendering an endless desire for the new, fashion promotes the excessive expenditure required to expand markets. Though preoccupied with the present, fashion remains bound to the past. The specters haunting fashion are not merely out-of-date styles that return but are uncanny traces of a more profound past that is neither precisely absent nor present. This haunting *revenant* issues in an "eclipse of presence," which engenders a sense of loss and lack, which, for Baudrillard, can never be overcome. Inescapable specters are lingering traces of a process of interminable mourning whose impossible aim is to recover what has never been possessed.[47]

While Baudrillard's account of "the spectral and cyclical world of bygone signs" leads to a detailed diagnosis of what he regards as the ills besetting emerging information and media culture, he can prescribe no cure but only counsel the endless expression of regret. His inability to move beyond this impasse is the result of his continuing commitment to structuralist strategies of analysis. All of Baudrillard's arguments rest upon a series of familiar binary oppositions:

Map	Territory
Code	Substance
Information	Matter
Image	Reality
Virtuality	Actuality
Simulacra	Real

In contemporary culture, Baudrillard argues, all such polarities are eroding as opposites "implode." This implosion does not involve the collapse of these structural differences; rather, the former pole absorbs or consumes the latter, while itself remaining more or less unchanged. Once unleashed, the processes of simulation cannot be stopped. With no prospect of productive activity, all we can do is to mourn interminably the death of the real.

But is our condition really so hopeless? The structure of binary opposition that Baudrillard borrows from structuralism reflects a world that has been disappearing for half a century. The transition from modern industrial to contemporary network culture does not involve a shift from one to the other pole of a stable polar structure. The technologies of production and reproduction in network culture are creating strange loops that are transforming rather than destroying differences and oppositions that long seemed secure. In a world where screens displace walls, neither map nor territory, code nor substance, information nor matter, image nor reality, virtuality nor actuality, simulacra nor the real is what it had seemed to be when it was the opposite of its presumed other. Something else, something different, something *new* is emerging.

For more than three decades, Foucault's constructivism, Derrida's deconstruction, and Baudrillard's account of simulacra have combined to form rich resources from which to assess various social and cultural developments. However, as we move into a new era, the limitations of these lines of analysis are becoming obvious. The creative possibilities of network culture cannot be understood unless new interpretive trajectories are fashioned. By now it should be clear that this conclusion does not imply a simple rejection of positions that now seem partial. Rather, a careful reconsideration of constructivism, deconstruction, and the theory of simulation discloses previously overlooked insights for understanding emerging network culture. Any adequate interpretive framework must make it possible to move beyond the struggle to undo what cannot be undone as well as the interminable mourning of what can never be changed.

The Grand Tortue grasped the loop between his feet and, with a few simple manipulations, created a complex string, which he proffered wordlessly to the monk. At that moment, the monk was enlightened.

douglas hofstadter

strange loops

In 1968, Foucault published an early version of an essay on the paintings of René Magritte, which eventually appeared as his celebrated book *This Is Not a Pipe* (1973). Magritte fascinated Foucault because of the creative way his work extends the Surrealist investigation of the interplay between reason and madness by presenting provocative explorations of the complex relations between words and things. While many Surrealists were interested in both the visual and verbal arts, Magritte was one of the few who effectively combined words and images to create original works of art. At first glance, this use of words and images in the same work seems to violate the principle of the autonomy of different arts, which critics like Clement Greenberg consider the cardinal principle of modernism. Further reflection, however, suggests that the positions of Magritte and Greenberg are much closer than they initially appear, for they share a profound interest in the resources and paradoxes of self-reflexivity and self-referentiality. Though rarely acknowledged, Greenberg formulates the basic tenets of his position by appropriating the conclusions of Kant's critical philosophy. "The essence of Modernism," Greenberg claims, "lies in the use of characteristic methods of a discipline to criticize the discipline itself, not in order to subvert it but in order to entrench it more firmly in its area of competence. Kant used logic to establish the limits of logic, and while he withdrew much from its old jurisdiction, logic was left all the more secure in what there remained of it." Just as Kant turns thought back on itself to articulate its limits, so each art, Greenberg argues, must define its limits by becoming self-reflexive. In this version of modernism, the autonomy of the work of art is twofold. First, modern art breaks with the regime of representation by referring to nothing beyond itself; art, in other words, is always about art. Second, each art is separate from every other art. "The task of self-criticism," for Greenberg, is "to eliminate from the specific effects of each art any and every effect that might conceivably be borrowed from or by the medium of any other art."[1] From this point of view, words and images should not be permitted to "contaminate" each other. Self-reflexivity both defines the limits and forms the substance of critical art.

While obviously rejecting Greenberg's ideal of autonomy as it bears on different arts, Magritte's work is nonetheless obsessed with the problem of self-reflexivity and, by extension, self-referentiality. Unlike Greenberg, for whom self-reflexivity serves to purify each art of all alien traces, Magritte probes paradoxes of self-reflexivity, which, he believes, infect all systems of signification. The problem of the relationship between words and images is a recurrent theme in much Surrealist art. Apollinaire, for example, was particularly interested in the graphic potential of words in works he labels "Calligrams." In a poem like "Fumées," he anticipates the "concrete poetry" of the 1960s by arranging the words to suggest the image of a pipe. Words and image are woven together to form a self-reflexive loop: words form an image, which represents that to which the words refer. In a series of works that includes *The Treachery of Images* (1952–53), *Force of Habit* (1960), *The Two Mysteries* (1966), and, most important, *This Is Not a Pipe* (1926), Magritte pushes Apollinaire's Calligrams beyond the bounds of representation to create something like a visual version of the Liar's Paradox (fig. 13).[2] To differentiate Magritte's painterly paradoxes from Apollinaire's verbal-visual puns, Foucault draws a distinction between resemblance and similitude:

Resemblance has a "model," an original element that orders and hierarchizes the increasingly less faithful copies that can be struck from it. Resemblance presupposes a primary reference that prescribes and classes. The similar develops in series that have neither beginning nor end, that can be followed in one direction as easily as in another, that obey no hierarchy but propagate themselves from small differences among small differences. Resemblance serves representation, which rules over it; similitude serves repetition, which ranges across it. Resemblance predicates itself upon a model it must return to and reveal; similitude circulates the simulacrum as an indefinite and reversible relation of the similar to the similar.[3]

Simultaneously anticipating and echoing Baudrillard, Foucault argues that resemblance gives way to similitude when meaning is relational rather than referential. If meaning is constituted by signs that point to nothing beyond themselves, then the structure of signification must be self-reflexive:

Henceforth similitude is restored to itself—unfolding from itself and folding back upon itself. It is no longer the finger pointing out from the canvas in order to refer to something else. It inaugurates a play of transferences that run, proliferate, propagate, and correspond within the layout of the painting, affirming and representing nothing. Thus in Magritte's art we find infinite games of purified similitude that never overflow the painting.[4]

While the play of signs does not "overflow the painting," the work of art is neither a closed structure nor a complete whole. Folding back on itself, the painting creates what Douglas Hofstadter describes as "strange loops." Strange loops are self-reflexive circuits, which, though appearing to be circular, remain paradoxically open. Within these loops, meaning becomes as "undecidable" as a Magritte painting.

Hofstadter explores the implications of strange loops in a provocative section of his highly acclaimed *Gödel, Escher, Bach: An Eternal Golden Braid* entitled "Magritte's Semantic Illusions," where he examines *The Two Mysteries,* in which Magritte complicates the problem of reference by including a painting of his earlier painting *This Is Not a Pipe* within the frame of a new work (fig. 14). As word and image are doubled and redoubled, the paradox of the painting deepens in a way that simultaneously provokes and eludes reflection. "The only way not to be sucked in," Hofstadter explains, "is to see both pipes merely as colored smudges on a surface a few inches in front of your nose. Then, and only then, do you appreciate the full meaning of the written message *Ceci n'est pas une*

FIGURE 13 René Magritte, *This Is Not a Pipe*. (© 2001 C. Herscovici, Brussels/Artists Rights Society [ARS], New York.)

pipe—but ironically, at the very instant everything turns to smudges, the writing too turns to smudges, thereby losing its meaning! In other words, at that instant, the verbal message of the painting self-destructs in a most Gödelian way."[5] This self-destruction of the verbal message is, of course, precisely what makes the painting both so provocative and productive. It is as if Magritte's work were declaring: This painting is a lie. Hofstadter realizes that the paradoxical implications of such true lies extend beyond the framework of art. In one of the imaginative dialogues woven throughout his book, a character Hofstadter dubs "Crab" uses another Magritte painting to help explain to his friend Achilles how Gödel's theorem illuminates the baffling ability of "Tobacco Mosaic Viruses and certain other bizarre biological structures . . . to self-assemble spontaneously."[6] In *The Fair Captive*, Magritte once again depicts a picture within a picture by presenting a painting of the horizon where sea and sky meet, which is leaning on an easel placed on the beach portrayed on the canvas. Standing beside the easel is a tuba engulfed by flames. To demonstrate the complex dynamics of "spontaneous self-assembly," Crab suggests that Achilles use a television camera to project the image of the painting on a screen. Crab complicates the situation by asking: "What happens . . . if you point the camera at the flames on the TV screen?"

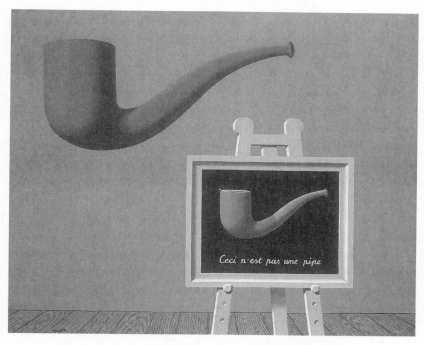

FIGURE 14 René Magritte, *Two Mysteries*. (© 2001 C. Herscovici, Brussels/Artists Rights Society [ARS], New York.)

Achilles: . . . But I don't understand what's on the screen now—not at all! It seems to be a strange long corridor. Yet I'm certainly not pointing the camera down any corridor. I'm merely pointing it at an ordinary TV screen.

Crab: Look more carefully, Achilles. Do you really see a corridor?

Achilles: Ahhh, now I see. It's a set of nested copies of the TV screen itself, getting smaller and smaller and smaller . . . Of course! The image of the flames HAD to go away, because it came from my pointing the camera at the PAINTING. When I pointed the camera at the SCREEN, then the screen itself appears, with whatever is on the screen at the time—which is the screen itself, with whatever is on the screen at the time—which is the screen itself, with—. . . I get a beautiful spiraling corridor! Each screen is rotated inside its framing screen, so that the littler they get, the more rotated they are, with respect to the outermost screen. The idea of having a TV screen "engulf itself" is weird.[7]

The images generated in these recursive screens appear to be fractals set in motion. In effect, these screens within screens create the similitude of the dynamics

77

generating what Baudrillard describes as the culture of simulacra. Far from merely destructive, the weird process of self-engulfing generates forms and structures that *do* overflow the frame. Emerging patterns are not static but repeatedly morph in unexpected ways. These changes result from the unavoidable *temporal delay* involved in the self-reflexive relation of images. As the camera turns on itself to picture itself picturing itself, a temporal gap opens. In this momentary space of this delay, patterns appear to organize themselves spontaneously. "This delay reverberates through the whole system, like a visual echo. And if things are set up so the echo doesn't die away, then you can get pulsating patterns."[8] As the visual echo resounds throughout the system, its effects become disproportionate to their seemingly slight momentary causes.

The pulsations in the loops of these screens display the rhythms of network culture. In a world where signs are signs of signs, and images are images of images, all reality is in some sense *screened.* The strange loops of information and media networks create complex self-organizing systems whose structure does not conform to the intrinsically stable systems that have governed thought and guided practice for more than three centuries. As we begin to understand the difference between simple and complex systems, the significance of the moment of complexity for emerging network culture comes into sharper focus.

MACHINES AND DEATH

The identification of the principles governing intrinsically stable systems can be traced to a discovery that occurred over three centuries ago. On April 28, 1686, Newton presented the conclusions of his *Principia* to the Royal Society of London. While his formulation of the basic laws of motion and definition of pivotal concepts like mass, acceleration, and inertia marked a milestone in the history of science, it was his identification of the universal law of gravitation that has had the most abiding impact both within and beyond the world of science. The third book of the *Principia,* in which Newton discusses universal gravitation, bears the significant title "The System of the World." In the course of describing this system, he not only transforms the understanding of the natural world but also claims to define the logic of stable systems wherever they occur. For Newton, there is actually only one system of the world, and it is governed by immutable universal laws that make natural processes potentially transparent and predictable. At the distance of more than three centuries, it is difficult to appreciate the enthusiasm with which many greeted Newton's discovery. In the words of Alexander Pope's memorable couplet:

Nature, and Nature's laws lay hid in night:
God said, Let Newton be! and all was light.[9]

The light of nature and its seemingly universal laws shines in a variety of ways through the Enlightenment down to our own day. The conviction that the order of the natural world can be rationally comprehended makes it possible to believe that disorder and chaos are merely apparent and can finally be explained. There is, however, a high price to be paid for the transparency and predictability that Newton and his followers declare. "Classical science," as Ilya Prigogine points out, "revealed to men a dead, passive nature, a nature that behaves as an automaton which, once programmed, continues to follow the rules inscribed in the program."[10] Newton's world is in effect a machine ruled by abstract laws, which are imposed upon it. While Prigogine's language suggests the model of cybernetics or computer technology, the more common image for the Newtonian universe is the clock.[11] The clock is a mechanical device made up of separate parts organized by a design which is not implicit in the parts as such. Whether this design presupposes an external designer remained a matter of dispute throughout the Enlightenment. What is indisputable in the mechanistic world of the early industrial revolution is that the order of the whole is neither intrinsic to nor emerges from the various parts. The laws that control Newton's world can be expressed in mathematical formulas, which, like ideal Platonic forms, are independent of historical circumstances and temporal variations. In this ideal realm, "natural" motion is constant and is always directed in a straight line; irregular motion, curves, and angles that are not right are nothing more than deviations. Such systems are intrinsically stable.

From these relatively simple principles, it is possible to derive the fundamental characteristics of intrinsically stable systems. While Newton was primarily concerned with systems in the physical universe, his analysis has been widely appropriated to interpret all aspects of society and culture. Intrinsically stable systems are:

1. *Closed:* Once established, they are not open to outside influences or sources of energy and information.
2. *Deterministic:* The laws of the system function universally and cannot be broken. Effects are proportionate to and can be accurately predicted from their causes.
3. *Reversible:* The laws governing the system apply to both temporal directions. As a result of this reversibility, time appears to be inconsequential.

4. *Operate at or near equilibrium:* The closure of the structure and reversibility of its governing laws incline the system to the state of inertia. Forces and counterforces as well as actions and reactions tend to balance each other. Change occurs *between* equilibrium states.

5. Finally, in intrinsically stable systems, parts are independent of each other and thus are *externally* related. The *whole,* therefore, *is the sum of its parts.* Since the independent parts are not fundamentally changed by their place in the whole, the whole can be reduced to the parts that comprise it.

As we will see in more detail in the following chapters, though intended to be universally descriptive, the mathematical formulas Newton derives cannot explain the distinctive features of complex systems, which operate far from equilibrium.

When viewed as a whole, the Newtonian world is, in Deleuze and Guattari's apt terms, "a *machinic assemblage* of bodies, of actions and passions, an intermingling of bodies reacting to one another."[12] Insofar as human beings are part of such a world, they too are machines controlled by the laws that apply to all other physical bodies. From La Mettrie's *L'Homme machine* (1747) to Andy Warhol's machinic art, some philosophers and artists have found the mechanistic universe strangely liberating. Warhol actually goes so far as to suggest only half in jest that he wanted to be a machine: "I think everybody should be a machine. I think everybody should be like everybody."[13] Most people, however, find a thoroughly mechanized universe frightfully dehumanizing. As early as the end of the eighteenth century, the spread of modern industrialism had created a pervasive feeling of alienation and nostalgia for what was believed to be a lost unity. In a world governed by the competitive laws of an industrial economy, new class divisions emerged, which created conflicts among as well as within individuals. Friedrich Schiller captured the sense of isolation and fragmentation shared by many people in his influential *Letters on the Aesthetic Education of Man* (1795):

Eternally chained to only one single little fragment of the whole, man himself grew to be only a fragment; with the monotonous noise of the wheel he drives everlastingly in his ears, he never develops the harmony of his being, and instead of imprinting humanity upon his nature he becomes merely the imprint of his occupation, of his science.[14]

Such lifeless machines harbor the specter of death, which, for many critics, haunts the industrialized world.

It was left for Schiller's contemporary, Hegel, to lay bare the abstract logic of the machine and to suggest how its deadly impact might be overcome in a way that would revitalize modern life. In his *Science of Logic,* Hegel describes the fundamental principles at work in machines:

This is what constitutes the character of mechanism, *namely, that whatever relation obtains between the things combined, this relation is one* extraneous *to them that does not concern their nature at all, and even if it is accompanied by a semblance of unity it remains nothing more than* composition, mixture, aggregation *and the like. . . . A* mechanical style of thinking, a mechanical memory, habit, a mechanical way of acting, *signify that the peculiar pervasion and presence of spirit is* [sic] *lacking in what spirit apprehends or does.*[15]

For Hegel, the definitive characteristic of mechanistic systems is a trait we have already discovered in intrinsically stable systems: self-subsistent parts are joined by the imposition of an order that establishes *external* relations. Since differences remain indifferent to each other even when bound together, a composition, mixture, or aggregate is merely a "formal totality," which involves nothing more than "a semblance of unity." Neither mechanical combination nor machinic assemblage, according to Hegel, exhibits a genuine unification or real integration. In such aggregates, rules of thought and principles of organization remain abstract even when they are imposed on determinate entities. The combination of separate parts to form composite "wholes" requires mechanical processes of calculation. In retrospect, this description of mechanical processes seems to anticipate important aspects of the operations of today's computers. In a remarkably prescient observation, Hegel writes: "Calculation being so much an external and therefore mechanical business, it has been possible to construct machines that perform arithmetical operations with complete accuracy. A knowledge of just this one fact about the nature of calculation is sufficient for an appraisal of the idea of making calculation the principle means for educating the mind and stretching it on the rack in order to perfect it as a machine."[16] From Hegel's point of view, any so-called knowledge resulting from mechanical calculations is not only specious but can actually be dangerous. When such "knowledge" becomes power, its force is repressive:

The knack of this kind of wisdom is as quickly learned as it is easy to practice; once familiar, the repetition of it becomes as insufferable as the repetition of a conjuring trick already seen through. The instrument of this monotonous formalism is no more difficult to handle than a painter's palette having only two colors, say red and green, the one for coloring the surface when a historical scene is wanted, the other for landscapes. . . . What results from this method of labeling all that is in heaven and earth with the few determinations of the general schema, and pigeon-holing everything in this way, is nothing less than a "report as clear as noonday" on the universe as an organism, viz. a synoptic table like a skeleton with scraps of paper stuck all over it, or like the rows of closed and labeled boxes in a grocer's stall.[17]

When Hegel's metaphor of painting is combined with his reference to calculating machines, his account of the dangers of monotonous formalism anticipates Baudrillard's warnings about the "terrorism of the code." Just as a palette limited to two colors cannot capture the vital hues of the world, so abstract algorithms and digital codes cannot process living organisms but inevitably reduce life to "a disordered heap of dead bones."[18] The "rows of closed and labeled boxes" constitute something like a universal grid, which is supposed to apply to everything. The mechanical imposition of this grid on reality, Hegel insists, is an act of violence that destroys the world and ultimately negates itself.

Hegel's analysis of the logic of the machine and, by extension, his account of mechanistic systems has had a significant impact on twentieth-century interpretations of science and technology. No philosopher has been more influential in exposing perceived dangers of what has come to be known as "techno-science" than Martin Heidegger. Though Heidegger never directly acknowledges a debt to Hegel on this issue, his understanding of science and technology is consistent with Hegel's argument in *Science of Logic*. While for Hegel mechanical calculation is "the antithesis of the Notion," for Heidegger thinking in the strict sense is the opposite of calculation. Indirectly extending Hegel's remarks on nineteenth-century calculating machines to twentieth-century computers, Heidegger writes:

Do we then have a right to the opinion that the thinking entry into the essential source of identity could be achieved in a day? Precisely because this entry requires a break, it must take its time, the time of thinking, which is different from the time of calculation that pulls our thinking in all directions. Today, the computer calculates thousands of relationships in one second. Despite their technical uses, they are inessential.[19]

In his seminal essay "The Question Concerning Technology," Heidegger defines this "inessentiality" as the "essence of technology." To the calculating mind, the entire world appears to be a "standing-reserve" *(Bestand)* waiting to be exploited for human purposes. In contrast to thinking, which involves a passivity that lets things be what they are "in themselves," calculation is "a challenging-forth into ordering."[20] Protests to the contrary notwithstanding, the order science claims to discover is, according to Heidegger, imposed rather than revealed. In a manner not unlike latter-day constructivists, the calculating subject creates the world in his or her own image through a process called "enframing":

> Yet when destining reigns in the mode of Enframing, it is the supreme danger. This danger attests itself to us in two ways. As soon as what is unconcealed no longer concerns man, even as object, but does so, rather, exclusively as standing-reserve, and man in the midst of objectlessness is nothing but the orderer of the standing-reserve, then he comes to the very brink of a precipitous fall; that is, he comes to the point where he himself will have to be taken as standing-reserve. Meanwhile man, precisely as the one so threatened, exalts himself to the posture of lord of the earth. In this way the impression comes to prevail that everything man encounters exists only insofar as it is his construct. This illusion gives rise in turn to one final delusion: It seems as though man everywhere and always encounters only himself.[21]

Science and technology are, for Heidegger, the culmination of a process that begins with Descartes's turn to the subject. By reducing objective truth to subjective certainty, Descartes set in motion developments that eventually led to "the will to mastery," which has resulted in twentieth-century techno-science.[22] Just as sociocultural constructivism leads to a form of subjective idealism that negates objectivity by consuming the natural world, so the will to mastery issues in a "subjective egoism," which is ultimately destructive. Through this egoism, Heidegger argues, "subjectivity only gains in power. In the planetary imperialism of technologically organized man, the subjectivism of man attains its acme, from which point it will descend to the level of organized uniformity and there firmly establish itself. This uniformity becomes the surest instrument of total, i.e., technological, rule over the earth."[23] With these developments, calculation seems to be omnipotent; the instrumentality of the machine appears to transform the entire universe into a *means* to humankind's own ends. But in this moment of apparent triumph, the will to mastery turns on itself and becomes self-destructive. On the "brink of a precipitous fall," man himself becomes a standing-reserve subject to domination, exploitation, and eventually death.

This unexpected reversal of the will to mastery repeats Hegel's dialectical analysis of utility. The calculations of techno-science involve the machinations of instrumental reason through which everything becomes useful as the means to human ends. "The useful," Hegel argues, "is the object insofar as self-consciousness penetrates it and has in it the *certainty* of its *individual self,* its enjoyment (its *being-for-self*); self-consciousness sees right into the object, and this insight contains the *true* essence of the object (which is to be something that is penetrated, or to be *for an 'other'*)."[24] Inasmuch as apparent being-for-self is really being for an "other," the self-conscious subject seems to enjoy *itself* in and through ostensible otherness. But appearances again are deceptive; when the calculating subject, in Heidegger's terms, "everywhere and always encounters only himself," he destroys the very other he needs to be himself. The resources necessary to sustain life (i.e., the standing-reserve) eventually are exhausted in a "fury of destruction" that becomes all-consuming.[25]

Within Hegel's dialectical scheme, the logic of mechanistic systems, which governs thought and guides action throughout the Enlightenment, leads to a "reign of terror" whose unavoidable outcome is death. Far from a world of light and reason, Newton's world is, for Hegel, plagued by a darkness that makes life impossible. As contemporary physicist and cosmologist Lee Smolin points out, "the movement from the Newtonian world to the modern one is a transition from a universe in which life is impossible to one in which life has a place."[26] It is precisely this transition that Hegel attempts to negotiate in his dialectical analysis of mechanical systems. Life, he concludes, presupposes an alternative logic, which creates a different kind of system. In formulating his account of this new kind of system, Hegel returns to insights Kant initially advanced in the Third Critique.

BEAUTY OF ORGANISMS

The movement from the eighteenth to the nineteenth century is characterized by a shift from mechanistic to organic images and metaphors for understanding the world. The personal and social fragmentation and alienation wrought by the industrial revolution could only be overcome, many artists, poets, and philosophers believed, by forming organic communities to nourish creative activity. In this context, the nineteenth century can be understood to begin effectively with the publication of Kant's analysis of aesthetics in the *Critique of Judgment* in 1790.[27] Intended to mediate the oppositions and resolve the contradictions within and between the *Critique of Pure Reason* (1781) and the *Critique of Practical Reason* (1788), the Third Critique marks a major change in the intellectual landscape. Kant develops an analysis of the difference between mechanisms and organisms by recasting the ancient distinction between efficient and final causality in terms of the problem of teleology or purposiveness. In mechanisms, cause and effect as well as means and end are *externally* related. Drawing on the well-worn image of God as a transcendent clockmaker, Kant illustrates his understanding of mechanism with the example of a watch. In a watch, he explains, "one part is certainly present for the sake of another, but it does not owe its presence to the agency of that other. For this reason, also, the producing cause of the watch and its form is [*sic*] not contained in the nature of this material but lies outside the watch in a being that can act according to ideas of a whole which its causality makes possible."[28] The structure of the watch conforms to the logic of the machine as Hegel defines it. Its cause is imposed from without by a designer who remains external to his creation. Furthermore, different parts of the watch are not integrally related but are held together by an *extrinsic* design. In contrast to a machine, an organism, like a beautiful work of art, is a "*self-organized being.*" Rather than imposed from without, order in the organism *emerges* from within through a complex interplay of parts, which, in the final analysis, constitutes the activity of the whole. According to the principle of "*intrinsic finality,*" "*an organized natural product is one in which every part is reciprocally both end and means.*" "The parts of the things," Kant explains,

combine of themselves into the unity of a whole by being reciprocally cause and effect of their form. For this is the only way in which it is possible that the idea of the whole may conversely, or reciprocally, determine in its turn the form and combination of all the parts, not as cause . . . but as the epistemological basis upon which the systematic unity of the form and combination of all the manifold contained in the given matter becomes cognizable for the person estimating it.[29]

Since means and end are reciprocally related and mutually constitutive, the parts of the organism do not point beyond themselves to an external telos but constitute their own end or purpose. In this way, the organism displays "inner teleology" or "purposiveness without purpose" (i.e., purposiveness without a purpose other than itself). In contrast to machinic assemblages, organisms are integrated wholes, which display "systematic unity." This unity presupposes a systematic structure that differs from mechanical systems in important ways. Hegel devotes his entire philosophical inquiry to formulating a comprehensive investigation of the complex self-organizing structure Kant identifies.

In his remarkable book, *The Logic of Life: A History of Heredity,* Nobel laureate François Jacob writes:

In the second half of the eighteenth century and the beginning of the nineteenth, the very nature of empirical knowledge was gradually transformed. Analysis and comparison tended to operate, not only on the component parts of objects, but also on the relationships between these components. It was within living bodies themselves that the very cause of their existence had to be found. It was the interaction of the parts that gave meaning to the whole. Living bodies then became three-dimensional entities in which the structures were arranged in depth, according to an order prescribed by the working of the total organism. The surface properties of a living being were controlled by the inside, what is visible by what is hidden. Form, attributes and behavior all became expressions of organization. By its organization the living could be distinguished from the non-living. Through organization organs and functions joined together. Organization assembled the parts of the organism into a whole, enabled it to cope with the demands of life and imposed forms throughout the living world. Organization became, as it were, a structure of a higher order to which all perceptible properties of organisms were referred. Thus with the start of the nineteenth century, a new science was to appear. Its aim was no longer to classify organisms, but to study the processes of life; its object of investigation was no longer visible structure, but organization.[30]

If relations are constitutive of organisms and their components, taxonomic ordering in terms of external visual attributes cannot disclose the logic of life. To understand life, it is necessary to analyze the structure of the complex relations that create and sustain it. These relations are not, as structuralists would have us believe, merely synchronic but are also diachronic. Organisms are functional wholes that develop over time. "In the nineteenth century," Jacob continues, "the problem of genesis is posed in a new way: the component elements of objects are not merely juxtaposed, but linked by a *network* of relationships." This network is an evolving web that weaves together all past and present forms of life to create a seamless whole whose operational logic is not mechanical.

Though Jacob is interested in biology rather than philosophy, his comment on Lamarck's theory of "transformism" might well be describing Hegel's system:

It develops by a continuous dialectical movement in which opposites become interpenetrated and quality is produced from quantity. The finality of the objects making up the world lies in their necessity, which itself can no longer be dissociated from their contingency. Instead of a chronological list of independent events, history becomes the movement in time by which the universe has become what it is, a process of development, a change from the most elementary to the most complex; in short, an "evolution" born of the internal sequence of transformations.[31]

For Hegel, the structure of history as well as nature is identical to the structure of life. Kant anticipates the notion of reason that forms the structure of both subjective and objective reality in Hegel's system. "One of Kant's great services to philosophy," Hegel avers, "consists in the distinction he made between relative or *external,* and *internal* purposiveness; in the latter, he has opened up the Notion of life."[32] When appropriating Kant's idea of inner teleology, Hegel transforms it from a "regulative idea," which serves as a heuristic rule, to a constitutive ontological principle. While "external purposiveness" characterizes mechanisms, "internal purposiveness" defines organisms:

Now purposiveness shows itself in the first instance as a higher being *in general, as an* intelligence *that* externally *determines the multiplicity of objects by* a unity that exists in and for itself, *so that the different determinatenesses of the objects become* essential through this relation. *In mechanism they become so through the* mere form of necessity, *their* content *being indifferent; for they are supposed to remain external, and it is only understanding as such that is supposed to find satisfaction in cognizing its own connective principle, abstract identity. In teleology, on the contrary, the content becomes important, for teleology presupposes a Notion, something* ab-solutely determined *and therefore self-determining, and so has made a distinction between the* relation *of the differences and their reciprocal determinedness, that is the* form, *and the* unity that is determined in and for itself *and therefore* a content.[33]

Within such a teleological structure, differences are neither independent nor indifferent, because reciprocal determinations render them mutually constitutive. The differences that form identity can only emerge in a unified whole that is in some sense more than the sum of its parts. Parts and whole exist in and through each other in such a way that each brings forth and sustains the other. Parts create the whole, which, in turn, creates the parts. This "unity that is reflected into itself" forms the organizational structure that Jacob identifies with life. "Life as *process,*" Hegel argues, "is the division of itself into forms and at the same time the dissolution of these persistent differences; and the dissolution of this division is just as much a division or a forming of members." So understood, life is "the self-developing whole that dissolves its development and in this movement simply preserves itself."[34] In less convoluted terms, life is nothing apart from its embodiment in individual organisms, and individual organisms cannot exist apart from the global process of life.

This reciprocal relation between parts and whole creates what Kant identifies as "self-organized being." In recent years, biochemists and theoretical biologists have begun to explore the structure Kant identifies and Hegel explicates in investigations of what they describe as autocatalytic sets and autopoietic systems. These new models of systems build upon postwar research in information theory and cybernetics. While the theories are framed in terms of different kinds of machines, the systems and structures they describe bear more similarity to the logic of organisms than to the logic of machines as Hegel and nineteenth-century romantics define it. Stuart Kauffman traces a direct line from Kant's philosophical reflections to important areas in contemporary biological research:

"Autocatalytic sets," mused Brian [Goodwin], "are absolutely natural models of functional integration. They are functional wholes." Of course, I agreed with him. Some years before, two Chilean scientists, Huberto Maturana . . . and Francisco Valera, had formulated an image of what they called autopoiesis. Autopoietic systems are those with the power to generate themselves. The image is older than Maturana . . . and Valera. . . . Kant, writing in the late eighteenth century, thought of organisms as autopoietic wholes, in which each part existed both for and by means of the whole, while the whole existed for and by means of the parts.[35]

During the Second World War, interest in cybernetics led to growing preoccupation with the problem of self-reflexivity.[36] One of the specific issues that provoked questions about reflexivity, or, more precisely, self-reflexivity, was the question of the relationship between the observer and the system observed in scientific inquiry. In a paper entitled "On Self-Organizing Systems and Their Environments," first presented in 1960, Heinz von Foerster, who at the time was conducting research at the University of Illinois's Biological Computer Laboratory, describes the paradox of "observing systems." On the one hand, if the observer remains outside the system, it is not clear whether knowledge of its inner workings is possible; on the other hand, if the observer is part of the system, self-observation leads to an infinite regress that makes complete knowledge of the system impossible. To illustrate this point, von Foerster asked Gordon Pask to draw a cartoon of "a man in a bowler hat, in whose head is pictured another man in a bowler hat, in whose head is yet another man in a bowler hat."[37] When pondering this cartoon, it is difficult not to be reminded both of Magritte's painting, *Décalomanie* (1966), which poses the paradoxes of reflexivity by depicting empty heads with bowler hats, and of Hofstadter's nested screens on which images mirror images, which mirror images. Von Foerster eventually became convinced that Maturana and Varela had found a way to resolve the paradoxes of reflexivity. Maturana's far-reaching theory grew out of insights first recorded in a paper entitled "What the Frog's Eye Tells the Frog's Brain." By implanting electrodes in frogs' brains, Maturana and his colleagues were able to detect the way in which sensory receptors process the data the brain receives. They conclude that the frog's eye "speaks to the brain in a language already highly organized and interpreted instead of transmitting some more or less accurate copy of the distribution of light upon the receptors."[38] The sensory receptors are species specific and function in a manner not unlike Kant's a priori forms of intuition and categories of understanding. Though inevitably triggering a response, Maturana believes that the world as such, like the thing-in-itself, remains inaccessible to all observers—be they frogs or humans.

There are, of course, important differences between frogs and humans. While frogs sense their environment and might even have a rudimentary form of consciousness, humans are both conscious and self-conscious. With self-consciousness, the reflexive relation between subject and object, or observer and observed, bends back on itself and becomes self-reflexive. In *Autopoiesis and Cognition: The Realization of the Living,* Maturana and Varela underscore the epistemological implications of self-reflexivity:

A living system capable of being an observer can interact with those of its own descriptive states, which are linguistic descriptions of itself. By doing so it generates the domain of self-linguistic descriptions within which it is an observer of itself as an observer, a process which can be necessarily repeated in an endless manner. . . . The observer as an observer necessarily always remains in a descriptive domain, that is, in a relative cognitive domain. No description of an absolute reality is possible. Such a description would require an interaction with the absolute to be described, but the representation which would arise from such an interaction would necessarily be determined by the autopoietic organization of the observer, not by the deforming agent; hence, the cognitive reality that it would generate would unavoidably be relative to the knower.[39]

Since the observer can never transcend the perspective from which he or she observes, the data of experience are always already processed. The effort to describe the framework of interpretation leads to an infinite regress, creating a self-reflexive structure from which there seems to be no exit. In anticipation of arguments to be developed in later chapters, it is important to stress that neither the impossibility of describing "absolute reality" nor the unavoidable self-reflexivity of self-consciousness necessarily leads to a relativism that ends in solipsism. Autopoietic systems can be members of complex networks comprised of other autopoietic systems. Furthermore, autopoietic systems are not as closed as they appear and, thus, a certain degree of interactivity remains possible.

Maturana and Varela realize that the implications of their analysis are not confined to epistemology but extend to ontology. Accordingly, they extrapolate the notion of autopoiesis discovered in their investigation of the biological dimensions of cognition to create a model that describes the structure of living organisms. Underscoring the importance of the principle of self-organization, which Jacob associates with the emergence of the biological sciences, Maturana and Varela write: "That living systems are machines cannot be shown by pointing to their components. Rather, one must show their organization in a manner such that the way in which all their peculiar properties arise, becomes obvious."[40] The organizational structure that sustains life is autopoietic:

An autopoietic machine is organized (defined as a unity) as a network of processes of production (transformation and destruction) of components that produces the components, which: (i) through their interactions and transformations continuously regenerate and realize the network of processes (relations) that produced them; and (ii) constitute it (the machine) as a concrete unity in the space in which they (the components) exist by specifying the topological domain of its relation as such a network.[41]

The self-reflexivity of these autopoietic machines is isomorphic with the self-reflexivity of Kant's self-organizing being as well as with Hegel's self-developing whole. Indeed, Maturana and Varela seem to be offering a gloss on Hegel's account of the dialectical relation between parts and whole when they argue that "an autopoietic machine continuously generates and specifies its own organization through its operation as a system of production of its own components, and does this in an endless turnover of components under conditions of continuous perturbations and compensation of perturbations." Machines, which produce their components, which, in turn, produce the machines producing the components, display a circularity that sometimes seems vicious: "This circular organization constitutes a homeostatic system whose function is to produce and maintain this very same circular organization by determining that the components that specify it be those whose synthesis or maintenance it secures."[42]

But this circle that seems to be closed is not necessarily complete; upon closer inspection, apparently self-contained autopoietic systems prove to be necessarily implicated with other entities, systems, and networks. While Maturana and Varela concentrate on biological systems, Niklas Luhmann extends their insights to social systems and communication networks. In an important essay,

"The Autopoiesis of Social Systems," Luhmann goes so far as to declare that "the theory of autopoietic social systems requires a conceptual revolution within sociology: the replacement of action theory by communication theory as characterization of the elementary operative level of the system." When understood as networks of communication, Luhmann argues, social systems display the characteristics Maturana and Varela identify in autopoietic systems: "Social systems use communication as their particular mode of autopoietic reproduction. Their elements are communications that are recursively reproduced by a network of communications that cannot exist outside of such a network."[43] While the subtleties of Luhmann's rich analysis need not be considered in this context, it is important to note that he insists that the circularity of autopoietic systems does not necessarily preclude their openness. Such systems are both closed and open, or, in his own terms, "the concept of autopoietic closure has to be understood as the recursively closed organization of an open system."[44] The interplay of openness and closure in autopoietic systems is a result of the necessary interaction between system and environment. The structural closure of the system presupposes its *difference* from its environment. Luhmann's monumental study, *Social Systems,* turns on this critical point:

There is agreement within the discipline [of sociology] today that the point of departure for all systems-theoretical analysis must be the difference *between system and* environment. *Systems are oriented by their environment not just occasionally and adaptively, but structurally, and they cannot exist without an environment. They constitute and maintain themselves by creating and maintaining a difference from their environment, and they use their boundaries to regulate this difference. Without difference from an environment, there would not even be self-reference, because difference is the functional premise of self-referential operations. In this sense,* boundary maintenance is system maintenance.*[45]

It is precisely the necessary relation to the environment that keeps the recursivity of the autopoietic system from completely closing in on itself. This is an important point, because Luhmann believes that it prevents autopoietic systems from becoming repressive totalizing structures. "The complexity of the system and of the environment," he argues, "excludes any totalizing form of dependence in either direction."[46]

There are, however, problems with Luhmann's effort to keep autopoietic systems open. These difficulties are the result of his tendency to confuse oppositional and dialectical notions of difference. Luhmann's central claim is that the identity of an autopoietic system necessarily entails its *difference* from its environment. Difference, he assumes, *is* opposition: the system *is not* the environment. The purported openness of the system results from its relation to difference, which appears to be an opposition. Such openness, however, remains external to the autopoietic system as such. Turning in on itself in its difference from the environment, this self-reflexive structure is not *in itself* open but remains closed even when differentiating itself from its opposite. The second problem with Luhmann's argument is that his analysis of the relation between identity and difference in autopoietic systems is self-contradictory. What he takes to be oppositional difference is actually dialectical difference. Insofar as identity constitutes itself in and through difference and vice versa, identity and difference are joined in a dialectical relation that is mutually constitutive and self-reflexive. If self-reference presupposes reference-to-other, then difference is not oppositional but is a necessary condition of identity. This is, of course, precisely Hegel's point when he argues that "the reflection-into-self . . . is equally reflection-into-an-other, and vice versa."[47] When Luhmann's argument is read through Hegel's dialectical analysis of identity and difference, it becomes clear that the relation between system and environment involves the very self-reflexivity characteristic of autopoietic systems. As we shall see in greater detail in a later chapter, the environment is not an indifferent difference but is itself a differential network comprised of a group of systems. If the differences that establish the identities of these systems display the self-reflexive structure of the systems themselves, then the relations between and among systems form something like an autopoietic system. Within this system of systems, parts and whole are not only reciprocally related but are virtually fractal—the structure of the part mirrors the structure of the whole and vice versa.

Hegel further complicates the self-reflexive loops of his reflection by arguing that scientific inquiry, as well as the objects it investigates, is a self-organizing system, which, like autopoietic machines, is circular and thus appears to be closed. In concluding his *Logic,* he summarizes his approach: "Science exhibits itself as a *circle* returning upon itself, the end being wound back into the beginning, the simple ground, but the mediation; this circle is moreover a *circle of circles,* for each individual member ensouled by the method is reflected into itself, so that in returning to the beginning it is at the same time the beginning of a new member."[48] As a result of this circularity and closure, neither Maturana and Varela's autopoietic system nor Hegel's dialectical one represents a complete break with the intrinsically stable systems characteristic of the mechanistic universe. It is clear that in both autopoietic and dialectical systems parts are mutually constitutive and internally related. Moreover, the whole is certainly more than the sum of its parts in these systems. The reciprocity of the relation among the parts as well as the relation between the parts and the whole creates recursive feedback loops that complicate causal relations. Such recursivity does not, however, transform causality in a way that makes the operation of the systems irreversible. While for Hegel, dialectical development is both directional and progressive, for Maturana and Varela, autopoiesis follows a course from creation through transformation to destruction, which moves from less to more complexity both within and among organisms. In spite of such directionality, autopoietic and dialectical systems remain closed. As a result of this closure, these systems share two other important characteristics with intrinsically stable systems. First, autopoietic and dialectical systems seek to overcome disequilibrium and thus tend toward homeostasis or equilibrium. Maturana and Varela go so far as to assert that autopoietic "machines are homeostatic machines."[49] For Hegel, dialectical relations involve a play of forces in which contending opposites attempt to balance each other. Since the operation of the system presupposes the *difference* between and among forces, the system works only as long as the ideal balance it seeks is deferred. Perfect harmony or complete equilibrium would be the death of the system. It is important to note that Hegel does not draw this conclusion because he is convinced that "absolute knowledge" issues in an integration in which differences are balanced in a way that maintains both their distinction and circulation. With the discovery of the principle of entropy and the second law of thermodynamics, however, equilibrium appears to be death-dealing rather than life-giving. The second remaining similarity with intrinsically stable systems is the determinism that characterizes homeostatic and equilibrium systems. Since everything in such systems is internally related, there seems to be no possibility of chance, accident, or contingency. Hegel attempts to include contingency in his system, but chance always appears illusory when its truth is comprehended through philosophical reflection. If chance is actually illusory, temporality and history are in some sense penultimate or even unreal. What appears to be temporal emergence or historical development is actually the unfolding of a necessary end or prescribed program. Even if the program can only be decoded retrospectively, the oak is always in the acorn from the beginning. For chance to have a chance, for time to be more than an illusion, and for history to be the site of aleatory emergence, systems must be open.

UNDECIDABLE OPENINGS

In an effort to open systems that seem closed, let us return to the Gödelian knot Hofstadter detects in Magritte's paintings. What if self-reflexive structures are open even more radically than Luhmann imagines? Such an opening would not be external to the system but would have to be folded into the system itself. Derrida explores precisely such a possibility in his reading of Mallarmé in *Dissemination*. Hegel haunts *Dissemination*—as well as everything else Derrida writes; indeed, Derrida is *always* reading both with and against Hegel even when he seems to be analyzing others. Enacting the problem of reflexivity in the very structure of the text, *Dissemination* begins with something like a preface ("Hors Livre Prefacing") devoted to the problem of prefaces and introductions in the works of Hegel. Derrida concentrates primarily on the *Phenomenology of Spirit*, which Hegel intended to be the preface to his system as a whole. Never satisfied with preliminaries, Hegel continued to nest prefaces within prefaces in all of his works; the prefatory *Phenomenology*, as well as most of his other books, includes a preface as well as an introduction. Derrida realizes that these textual supplements create problems for the purported circularity and completion of the Hegelian system.[50] Though his argument is complex, Derrida's point is simple: if Hegel's system needs a preface, it is not, on its own terms, truly a system. If the preface is inside the system, it is not a preface, and if it is outside, the system is both open and incomplete and, therefore, is not a system. The preface functions like Luhmann's boundary, which is necessary for "system-maintenance." But this boundary is not part of the system; nor is it simply outside the system. As the condition of the possibility of the system, it is also the condition of the impossibility of its closure and completion. Irreducibly liminal, it is neither inside nor outside the system. Derrida insists that such liminality eludes both binary opposition and dialectical differentiation. What does not conform either to the binary principle of noncontradiction or to the rule of dialectical logic remains "undecidable." In his discussion of Mallarmé, which is actually an extended interrogation of Hegel, Derrida credits Gödel with the discovery of undecidability. Indirectly restaging the shift from reference to relation that characterizes Baudrillard's account of simulacra, Derrida writes:

The Mime mimes reference. He is not an imitator; he mimes imitation. The hymen interposes itself between mimicry and mimesis *or rather between* mimesis *and* mimesis. *A copy of a copy, a simulacrum that simulates the Platonic simulacrum— the Platonic copy of a copy as well as the Hegelian curtain have lost here the lure of the present referent and thus find themselves lost for dialectics and ontology, lost for absolute knowledge. . . . In this perpetual allusion being performed in the background of the* entre *[between] that has no ground, one can never know what the allusion alludes to, unless it is to itself in the process of alluding, weaving its hymen and manufacturing its text. Wherein allusion becomes a game conforming only to its own formal rules. As its name indicates, allusion* plays. *But that this play should in the last instance be independent of truth does not mean that it is false, an error, appearance, or illusion. Mallarmé writes "allusion," not "illusion." Allusion, or "suggestion" as Mallarmé says elsewhere, is indeed that operation we are here* by analogy *calling undecidable. An undecidable proposition, as Gödel demonstrated in 1931, is a proposition which, given a system of axioms governing a multiplicity, is neither an analytic nor deductive consequence of these axioms, nor in contradiction with them, neither true nor false with respect to those axioms.* Tertium datur, *without synthesis.*[51]

Gödel's 1931 paper is entitled "On Formally Undecidable Propositions in *Principia Mathematica* and Related Systems I." Though he eventually became a professor of mathematics, Gödel always remained interested in philosophical problems. Born in Moravia in 1906, he studied at the University of Vienna, where he participated in the discussion group that became the famous Vienna Circle. Among those Gödel met in this elite circle were the philosopher Rudolf Carnap and the mathematician Kurt Menger. Their discussions introduced Gödel to a broad range of problems in philosophy and mathematical logic and immersed him in the writings of Wittgenstein. Gödel did not share the philosophical outlook of the Vienna Circle but quickly realized the similarities between the self-referentiality involved in the investigation of language through language and the mathematical formalization of mathematical systems.

While Gödel's theorem is directed specifically against arguments advanced in *Principia Mathematica,* its implications extend far beyond any particular mathematical dispute. Russell and Whitehead announced that they had derived all mathematics from logic without any contradictions. Intrigued by this claim but not convinced of its truth, the well-known mathematician David Hilbert issued a challenge to prove its veracity in a lecture delivered on September 9, 1930, in Kant's hometown of Königsberg. Is it possible, he asked, "to demonstrate rigorously—perhaps following the very methods outlined by Russell and Whitehead—that the system defined in *Principia Mathematica* was both *consistent* (contradiction-free), and *complete* (i.e., that every true statement of number theory could be derived within the framework drawn up in *Principia Mathematica*)?"[52] If the propositions advanced by Russell and Whitehead could be shown to be both consistent and complete, the age-old dream of logical and mathematical certainty would become a reality. But Gödel proved once and for all the impossibility of this dream by demonstrating that "all consistent axiomatic formulations of number theory include undecidable propositions."[53] This theorem involves exactly the kind of strange loop we have already discovered in both Magritte's paintings and the Liar's Paradox. Gödel extends the paradoxes of self-reflection, which he encountered in the debates about language in the Vienna Circle, to number theory. "The proof of Gödel's Incompleteness Theorem," Hofstadter explains, "hinges upon the writing of a self-referential mathematic statement, in the same way as the Epimenides paradox is a self-referential statement of language. But whereas it is very simple to talk about language in language, it is not at all easy to see how a statement about numbers can talk about itself. In fact, it took genius merely to connect the idea of self-referential statements with number theory."[54] Gödel uses a statement that approximates the Liar's Paradox to prove his claim that "all consistent axiomatic formulations of number theory include undecidable propositions." Loops become strange when, in turning back on themselves, they generate irresolvable paradoxes. The consequences of Gödel's proof are not limited to the arguments of *Principia Mathematica* but apply to any system that conforms to Whitehead and Russell's description. When interpreted in this way, Gödel's theorem suggests the unavoidable limits of human certainty and the inevitable incompleteness of all apparently complete and consistent systems.

It is important to understand precisely what is at stake in the notion of undecidability. For Gödel, axiomatic systems of sufficient complexity necessarily include propositions that are consistent inside the system but whose truth or falsity is not ascertainable in terms of the system itself. While seeming to appropriate Gödel's theorem, Derrida subtly shifts its meaning in order to expand its significance. For Derrida, every system necessarily entails propositions or factors that are demonstrably *neither* true *nor* false. Gödel's insistence that the propositions within a system can be true even if they cannot be proven within that system renders them too decidable for Derrida.[55] In the 1971 interview to which I have already referred, Derrida comments on the passage in which he invokes Gödel:

*Henceforth, in order better to mark this interval (*La dissemination, *the text that bears that title . . . is a systematic and playful exploration of the interval—"écarte," carré, carrure, cartee, charte, quarte, etc.) it has been necessary to analyze, to set to work, within* the text of the history of philosophy, *as well as within* the so-called literary text *(for example, Mallarmé), certain marks, shall we say . . . that by analogy . . . I have called undecidables, that is, unities of simulacrum, "false" verbal properties (nominal or semantic) that can no longer be included within philosophical (bi-*

nary) opposition, resisting and disorganizing it, without ever *constituting a third term, without ever leaving room for a solution in the form of speculative dialectics.*[56]

Several points in this telling remark deserve emphasis. Since undecidables cannot "be included within philosophical (binary) opposition" and leave no "room for a solution in the form of speculative dialectics," neither structuralism nor Hegelianism can comprehend them. This does not mean that undecidables are simply outside binary structures or dialectical systems. If they were clearly *outside* the structure or system, undecidables would be incorporated *within* an oppositional structure and thus would be part of the system they are supposed to elude. Such structuring would make the undecidable determinate and thus decidable. Like the preface, which the system needs even at the price of its own completion, or the boundary, which structures need even if it interrupts their reflexive self-relation, undecidables haunt all systems and structures as an exteriority, which is in some sense within even though it cannot be incorporated. In Derrida's reworking of Gödel's theorem, every system or structure includes as a condition of its own possibility something it cannot assimilate. This "outside," which is "inside," exposes the openness of every system that seems to be closed. Unlike the exteriority Luhmann attributes to autopoietic systems, this openness is not extraneous but is *"within"* the system itself.

The "interval" creating this opening involves a spacing that insinuates time into every structure and system constructed to exclude it. Attempting to clarify his best-known and most notorious undecidable, Derrida writes that *différance* is "the becoming-time of space and the becoming-space of time."[57] Since there can be no structural organization without the timely interval of this spacing, systems and structures are neither static nor eternal but emerge historically and are always changing. Such change is not prefigured or programmed but is subject to the uncertainty and unpredictability of chance. In the time-space of the interval, the aleatory emerges and what emerges is aleatory. The ceaseless play of chance in the gaps of systems eventually upsets every equilibrium and makes homeostasis finally impossible.

With the exposure of the opening *in the midst* of binary structures as well as autopoietic and dialectical systems, it becomes possible to imagine different structures and systems that are neither closed nor stable. Always open and forever subject to chance, such systems emerge at the edge of order and operate far from equilibrium. Such complex structures and systems cannot be understood in terms of either binary or dialectical logic but can be grasped by what might be called paralogic. Paralogic, which captures the logic of parasitism, is the logic of undecidability. As J. Hillis Miller explains,

Para is a double antithetical prefix signifying at once proximity and distance, simi-
larity and difference, interiority and exteriority, something inside a domestic economy
and at the same time outside it, something simultaneously this side of a boundary
line, threshold, or margin, and also beyond it. . . . A thing in "para," moreover, is not
only simultaneously on both sides of the boundary line between inside and out. It is
also the boundary itself, the screen, which is a permeable membrane connecting in-
side and outside. It confuses them with one another, allowing the outside in, making
the inside out, dividing them and joining them. It also forms an ambiguous transi-
tion between one and the other.[58]

Neither this nor that, the parasite remains undecidable. The strange logic of the
parasite makes it impossible to be sure who or what is parasite and who or what
is host. Caught in circuits that are recursive and reflexive yet not closed, each
lives in and through the other. In these strange loops, nothing is ever clear and
precise; everything is always ambiguous and obscure. For comprehensive systems
that are supposed to illuminate everybody and everything, the paradoxes inherent in parasiticism create a darkness that must be dispelled. But this darkness
cannot be easily dismissed, for it is more obscure than the darkness of the night
that precedes and follows the day. For those who are willing to linger patiently,
this darkness finally becomes illuminating. Grand Tortue implies but does not
quite explain that far from resolving the paradoxes discovered in strange loops,
enlightenment renders them more profound. In all of its obscurity, the moment
of enlightenment is a moment of complexity when "it's nighttime, black":

What happens would be the obscure opposite of conscious and clear organization,
happening behind everyone's back, the dark side of the system. But what do we call
these nocturnal processes? Are they destructive or constructive? . . . There is no system
without parasites. But how so?[59]

There is a hint of an answer to this puzzling question in the duplicity of the
word "parasite." If reflection is not simple but loops back around itself to become complex, "parasite" can also be read as *parasite,* which means, among other
things, *static* or *interference.* For systems that claim to communicate clearly, such
interference is noise.

Here is page 31 again, page 32 . . . and then what comes next? Page 17 all over again, a third time! What kind of book did they sell you anyway? They bound together all these copies of the same signature, not another page in the whole book is any good.

You fling the book on the floor, you would hurl it out of the window, even out of the closed window, through the slats of the Venetian blinds; let them shred its incongruous quires, let sentences, words, morphemes, phonemes gush forth, beyond recomposition into discourse; through the panes, and if they are of unbreakable glass so much the better, hurl the book and reduce it to photons, undulatory vibrations, polarized spectra; through the wall, let the book crumble into molecules and atoms passing between atom and atom of reinforced concrete, breaking up into electrons, neutrons, neutrinos, elementary particles more and more minute; through the telephone wires, let it be reduced to electronic impulses, into flow of information, shaken by redundancies and noises, and let it be degraded into a swirling entropy.

italo calvino

noise in formation

BUZZ OF INFORMATION

In the spring of 1989, the Santa Fe Institute sponsored a conference on "Complexity, Entropy, and the Physics of Information." The ambitious goal of the meeting was to explore "not only the connections between quantum and classical physics, information and its transfer, computation, and their significance for the formulation of physical theories" but also to consider "the origins and evolution of the information-processing entities, their complexity, and the manner in which they analyze their perceptions to form models of the Universe." The announcement of the conference included "A Manifesto," written by Wojciech Zurek, which begins by invoking the famous opening lines of *The Communist Manifesto:* "The specter of information is haunting the sciences." Having begun on such an ominous note, Zurek proceeds to underscore the growing importance of information in a variety of scientific fields: "Thermodynamics, much of the foundation of statistical mechanics, the quantum theory of measurement, the physics of computation, and many of the issues of the theory of dynamical systems, molecular biology, genetics, and computer science share information as a common theme."[1] This calculated displacement of the communist revolution by the information revolution suggests one of the distinctive features of emerging network culture. The social and economic problems Marx and Engles detected and the cures they prescribed reflect an industrial society and its corresponding form of capitalism, which are passing away in the moment of complexity. One of the most revealing symptoms of the changes currently occurring is the virtual disappearance of Marx and Marxism as relevant voices in cultural analysis. Other than in certain corners of the university where the news of 1989 does not seem to have arrived, Marx has become irrelevant. As mechanical processes of production give way to electronic processes of reproduction, new modes of interpretation are required. Though Zurek and his colleagues focus on the natural sciences, the analyses they develop suggest that we cannot appreciate the significance of emerging network culture without understanding the relationship of information and information theory to complexity and complexity theory. In the concluding sentence of his manifesto, Zurek reports: "*Complexity,* its meaning, its measures, its relation to entropy and information, and its role in physical, biological, computational, and other contexts have become an object of active research in the past few years."[2]

Though we are in the midst of what many describe as an "information revolution," we do not really understand what information is and surely do not know what makes it revolutionary. What many people do know about information is that there is *too much* of it. Constantly bombarded by proliferating information transmitted on ever more sophisticated yet less and less obvious networks, what purports to be information is often experienced as noise. The buzz of information is the white noise of today's culture. For many people, the much-hyped information glut creates both confusion and a debilitating sense of vertigo. Adrift in the noisy sea of information, more and more people suffer a feeling of nausea that brings them close to what seems to be the edge of chaos. *Noise,* it is instructive to note, derives from the Latin *nausea,* which originally meant seasickness.[3] When information becomes the noise that engenders nausea, distinctions, differences, and oppositions that once seemed to fix the world and make it secure become unstable. Lines of separation become permeable membranes where transgression is not only possible but unavoidable:

Information	Noise
Difference	Indifference
Order	Disorder
Organization	Disorganization
Form	Chaos
Improbability	Probability
Negentropy	Entropy
Heterogeneity	Homogeneity 4

As these polarities slip and slide, they eventually reverse themselves to disclose the specter of dynamics that appear to be fluid.

This is not, of course, the first time we have encountered the specter of specters. In our consideration of Baudrillard's "spectral and cyclical world of by-gone signs," which echoes in Derrida's *Specters of Marx,* we have seen that the specter complicates the relationship of opposites. Neither present nor absent, inside nor outside, living nor dead, the specter can be understood—if at all—only through a logic that is parasitical. This should not be surprising, for a specter, after all, is itself something of a parasite. Indeed, another name for a specter is "ghost." A ghost is never present as such but only draws near in a haunting absence. As a present absence and absent presence, the identity of the ghost is undecidable. This undecidability is the source of the ghost's attraction as well as repulsion. Incorporated yet not assimilated, the ghost is consumed and also consumes. "If the host is both eater and eaten, he also contains in himself the double antithetical relation of host and guest, guest in the bifold sense of friendly presence and alien invader. The words 'host' and 'guest' go back in fact to the same etymological root: *ghos-ti,* stranger, guest, host."5 Guest/host: either/or, both/and, neither/nor? It is precisely such irreducible undecidability that makes ghostly specters so unsettling.

Since the undecidability created by ghostly specters and strange loops cannot be overcome, its implications must be understood more adequately by further consideration of the dynamics generated by paralogic. If information sometimes can be noisy and noise informative, the logic of *le parasite* (i.e., interference) might help to clarify the fluid dynamics of information. Michel Serres begins his book *The Parasite* with an interruption: "*Interrupted Meals—Logics.*" The first chapter opens with "*Rats' Meals—Cascades*" in which Serres de-

velops an unlikely interpretation of one of La Fontaine's fables by using information theory to read the apparently simple fable and by using the fable to explain the complexities of information theory:

The city rat invites the country rat onto the Persian rug. They gnaw and chew left-
over bits of ortolan. Scraps, bits and pieces, leftovers: their royal feast is only a meal
after a meal among the dirty dishes on a table that has not been cleared. The city rat
has produced nothing and the dinner invitation costs him almost nothing. Boursault
says this in his Fables d'Esope, *where the city rat lives in the house of a big tax*
farmer. Oil, butter, ham, bacon, cheese—everything is available. It is easy to invite
the country cousin and to regale oneself at the expense of another.[6]

At this dinner party, the city rat is apparently the host and the country rat the
guest. And yet, both host and guest are parasites who live off the "scraps, bits,
and pieces, leftovers" of others. The situation becomes even more complicated
when we realize that the unwitting host of these parasites is a tax farmer, who
lives "off the fat of the land," and thus is himself a parasite. As the fable becomes
more convoluted, the host turns into a parasite and the parasites turn out to be
hosts. No sooner has the meal begun than the rats are interrupted by "a noise at
the door":

The tax farmer produced neither oil nor ham nor cheese; in fact, he produced noth-
ing. But using power or the law, he can profit from these products. Likewise for the
city rat who takes the farmer's leftovers. And the last to profit is the country rat. But
we know that the feast is cut short. The two companions scurry off when they hear a
noise at the door. It was only a noise, but it was also a message, a bit of information
producing panic: an interruption, a corruption, a rupture of information. Was the
noise really a message? Wasn't it, rather, static, a parasite? A parasite who has the last
word, who produces disorder and who generates a different order.[7]

If the interruption of the parasite "produces disorder," which "generates a different order," the last word is also the first word and vice versa. Within the strange
loops of this chain where parasites are hosts and hosts are parasites, noise, Serres
insists, is the "ultimate parasite." Serres might well have been a guest at the conference Zurek hosted when, elaborating this point, he writes:

Hosts and parasites. We live, in the city or in the country, in the space of the two rats.
Their fabulous feast is this book. A book that is oral and aural, about famine and

murderers, about knowledge and bondage. Both in the fable and in this book, it is a
question of physics, of certain exact sciences, of certain techniques of telecommunica-
tions, a question of biophysics and of certain life sciences, of parasitology, a question
of culture and of anthropology, of religions and literatures, a question of politics, of
economics.[8]

Serres's retelling of the fable of the city rat and the country rat is a parable
of the so-called Information Age. Though not immediately obvious, this parable
enacts the paralogic it is designed to investigate. *Parable* derives from the Greek
parabola (juxtaposition, comparison), which, in turn, can be traced to *para-*
ballein (*para*, beside + *ballein*, to throw). Accordingly, a parable is "a comparison, a similitude; any saying or narration in which something is expressed in
terms of something else." Parables throw together things usually held apart to
create something like a verbal montage in which disparate elements interact in
novel ways to create new meanings that provide unexpected insights. The interactions at work in parables extend beyond the words on the page to engage the
reader in creative interactivity. The juxtapositions of parables are undeniably
noisy. For those who can bear neither ambiguity nor uncertainty, such noise
must be eliminated; for those with more open minds, however, noise is a wel-
come guest whose interruptions and disruptions are as creative as they are destructive. Whether resisted or welcomed, noise is, as Serres insists, "a sign of the
increase in complexity."[9] No sign of the current moment is more telling than the
increasing complexity created by the noise of information.

INFORMATION AGES

It is difficult to know where to draw the boundaries of the Information Age. If
the information revolution is associated with the computer, its origins can be
traced to Charles Babbage's nineteenth-century invention called the Difference
Engine. This machine, Babbage claimed, could "generate mathematical tables of
many kinds by the 'method of differences.'"[10] The Difference Engine is not only
the precursor of modern computers but is an early information processing machine. Before Babbage had the opportunity to build the Difference Engine, he
became obsessed with designing and constructing a more complex machine
modeled after the card-operated Jacquard loom. This remarkable device used a
prototype of the punch cards that controlled early computers to create extremely
complex woven designs. Comprised of a "store" (memory) and a "mill" (CPU),
the Analytic Engine is, in effect, the first digital computer. This machine was to

be built from thousands of cylinders with interlocking gears, which were to be controlled by a *program* recorded on punched cards. Babbage's friend and collaborator, Lady Ada Lovelace, daughter of Lord Byron and arguably the world's first computer programmer, described the invention in characteristically poetic terms: "the Analytic Engine *weaves algebraic patterns* just as the Jacquard-loom weaves flowers and leaves."[11]

While remarkably advanced for its time, the Analytic Engine, had it been built, obviously would have been a mechanical device and, as such, remains a product of the industrial revolution. For some analysts, the association of the information revolution with mechanical calculating devices represents far too broad a perspective, while for others it is much too narrow. In *Information Ages: Literacy, Numeracy, and the Computer Revolution,* Michael Hobart and Zachary Schiffman argue that there is not merely one Information Age but many. The first and most significant information revolution, they contend, was the discovery of writing:

For the invention of writing actually gave birth to information itself, engendering the first information revolution. Writing created new entities, mental objects that exist apart from the flow of speech, along with the earliest, systematic attempts to organize this abstract mental world. Here we find the roots of the activity that would ultimately lead the Greeks to correlate the order of the mental world with nature. Thus, when we tear ourselves away from the engrossments of electronic culture, we discover that our information age is but the latest of several. From a historical perspective, perhaps the only "information age" truly deserving the title is the original, primeval one some five thousand years ago.[12]

From this point of view, writing functions like a code or program whose quasi-algorithmic operations produce information. Though Hobart and Schiffman do not realize it, their argument rests upon two principles of binary opposition, which lie at the foundation of Western metaphysics. First, they draw a sharp distinction between speech and writing, or oral and literate cultures. The assumption that information is necessarily related to writing leads to a remarkable conclusion: "The information revolution born of literacy is all the more stunning and revolutionary when seen in stark relief against an oral world where information did not exist."[13] Information, in other words, is impossible apart from writing. If, however, speech and writing are not as different as Hobart and Schiffman assume, their argument collapses. Two of the primary aims of Derrida's deconstructive analysis are, of course, to expose the far-reaching implications of the traditional distinction between speech and writing, and to demonstrate the untenability of their opposition. Derrida argues persuasively that speech, like writing, is an endless play of differences through which all signs are articulated.[14] If this understanding of writing is combined with Gregory Bateson's definition of information as "a difference which makes a difference," information appears to be a function of differential inscription.[15] Since there can be no culture apart from processes of differentiation, there is no culture without either "writing" or information. Accordingly, if there is such a thing as an information revolution, it must involve something more than the mere existence of information. The second metaphysical presupposition underlying Hobart and Schiffman's analysis is that form and matter are opposites and thus can be separated. This duality, which is as old as philosophy itself, is reborn in every age. Drawing on etymology, they interpret information in terms of form, which can be abstracted from its "material substrate":

The term itself traces back to the Latin verb informare, *which for Romans generally meant "to shape," "to form an idea of," or "to describe." The verb, in turn, supplied action to the substantive,* forma, *which took varied, cognate meaning that depended mostly on context. The historian Livy used* forma *as a general term for "characteristic," "form," "nature," "kind," and "manner." Horace applied it to a shoelast, Ovid to a mold or stamp for making coins, while the wily Cicero, among other uses, extended it to logic as "form" or "species," his rendering of the Greek* eidos kai morphe, *a philosophical expression denoting the essence or form of a thing as distinguished from its matter or content.*[16]

This is a very important point because it inevitably leads to a variety of influential misunderstandings of information. Many contemporary cultural theorists interpret the classical distinction between form and matter to imply that information is the *opposite* of matter or, in some cases, energy, and, therefore, can be separated from its "material substrate." Theodore Roszak goes so far as to assert that the "Platonic dream survives, and no place more vividly than in the cult of information."[17] As we will see in chapter 7, from popular surveys like Mark Dery's *Escape Velocity* and Erik Davis's *Techgnosis* to sophisticated analyses like Hans Moravec's *Robot* and Katherine Hayles's *How We Became Posthuman*, information is interpreted as formal, abstract, and, finally, immaterial.[18] The Information Age, then, is characterized by a loss or repression of the material conditions supporting life as we know it. While for theorists like Moravec, the prospect of dematerialization promises liberation, for critics like Hayles, the vision of bodiless minds is a nightmare. Hayles concludes that the repression of the body and materiality in the era of cybernetics, informatics, and virtuality creates a "posthuman" condition. "The posthuman view," she explains, "privileges informational pattern over material instantiation, so that embodiment in a biological substrate is seen as an accident of history rather than an inevitability of life."[19] According to Hayles, when humanity and machines become indistinguishable, we lose our distinctive humanity. While resisting any simple nostalgia that would dream of removing the machine from the garden, she sounds a warning:

Information, like humanity, cannot exist apart from the embodiment that brings it into being as a material entity in the world; and embodiment is always instantiated, local, and specific. Embodiment can be destroyed but it cannot be replicated. Once the specific form constituting it is gone, no amount of massaging data will bring it back. This observation is as true of the planet as it is of an individual life-form. As we rush to explore the new vistas that cyberspace has made available for colonization, let us also remember the fragility of a material world that cannot be replaced.[20]

The opposition between form and content, or information and matter, upon which such arguments rest is no more defensible than the opposition between speech and writing. It is becoming increasingly obvious that information is, in important ways, material, and matter is informational. From this expanded point of view, neither information nor materiality is what it seems to be when it is interpreted in simple oppositional terms. Thus, the movement into the Information Age should not be conceived in terms of growing abstraction and increasing dematerialization, but as the complication of the relation between information and the so-called material conditions of life. As the line between the material and the informational becomes permeable, information processes become considerably more extensive.

The parameters of the current Information Age become clear when we understand the information revolution not only as a major sociocultural change but also as something like an orbital movement in which information revolves in such a way that it begins to act on itself. The information revolution occurs when information turns on itself and becomes self-reflexive. This turn has been made possible by new electronic and telematic technologies, through which information acts on information to form feedback loops that generate increasing complexity. This is why the information revolution issues in the moment of complexity.

The information technologies currently transforming society and culture trace their origins to research and development conducted during and after World War II. In 1935, Alan Turing attended lectures at Cambridge University in which M. H. A. Newman, summarizing the implications of Gödel's theorem, asked if it is possible to devise a *mechanical* process that could determine whether or not a mathematical problem could be solved. In attempting to answer this question, Turing developed the logical prototype for today's computers. With the outbreak of the war, the development of a Turing Machine that actually worked became an urgent necessity. A decisive turning point in the research that eventually led to the computer was reached when Turing, along with John von Neumann, encountered the inherent limitations of mechanical devices and recognized the necessity of using electronic switches. When logical operations were linked to electronic circuitry, the computer as we now know it was born.

Another vital area of research and development that was closely tied to the war effort was cybernetics. The word *cybernetics,* coined by Norbert Weiner, derives from the Greek *kubernetes,* which means pilot or governor. The field of cybernetics involves the investigation of regulatory processes in mechanical, electronic, and biological systems. From at least the middle of the nineteenth century, mechanical systems had been stabilized by different kinds of feedback devices.[21] In steam engines, for example, thermostats and governors create negative feedback loops that regulate the temperature and thus allow the machines to run smoothly. During the Second World War, the growing complexity of weapons systems required the development of much more sophisticated means of control. The emergence of cybernetics in the 1930s and 1940s led to the invention of information devices that could regulate both mechanical and electronic machines. In some cases, machines were linked to servo-mechanisms, and in other cases machines and humans were joined in feedback loops. After the war, a group of scientists met under the auspices of the Josiah Macy Foundation for over a decade to develop a theory of information and communication that could be used to develop technologies capable of regulating both human and nonhuman systems.[22]

One of the major contributors to the Macy Conferences was Claude Shannon. In 1948, Shannon published an extraordinarily influential paper entitled "The Mathematical Theory of Communication" in the relatively unknown *Bell System Technical Journal.* This essay, along with an article by Warren Weaver intended to be an "expository introduction" to Shannon's complex argument, ap-

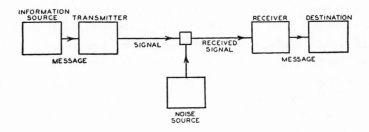

FIGURE 15 Claude Shannon, Schematic diagram of a general communication system. (Reprinted from *The Mathematical Theory of Communication*. Copyright 1949, 1998 by the Board of Trustees of the University of Illinois. Used with the permission of the University of Illinois Press.)

peared as a book a year later.[23] Though Shannon formulated his theory to determine the requirements for the transmission of messages on telephone or telex lines, his analysis, along with Weaver's explanation, has played a decisive role in defining information and determining the conditions necessary for communication. The specific technical problems Shannon attempted to solve led him to interpret communication as the transmission of a message from an information source (sender) along a channel to a destination (receiver). Within this scheme, noise designates anything that interrupts the transfer of the signal. Shannon provides a diagram to illustrate this scheme (fig. 15). On the first page of his introduction, he summarizes his approach to the problem:

The fundamental problem of communication is that of reproducing at one point either exactly or approximately a message selected at another point. Frequently the messages have meaning; *that is they refer to or are correlated according to some system with certain physical or conceptual entities. These semantic aspects of communication are irrelevant to the engineering problem. The significant aspect is that the actual message is one* selected *from a set* of possible messages. *The system must be designed to operate for each possible selection, not just the one which will actually be chosen since this is unknown at the time of design.*[24]

In this passage, Shannon makes two points that proved decisive for all later considerations of communication and information theory. First, he is not concerned with the questions of meaning and thus brackets all semantic issues in order to concentrate on what are in effect syntactic problems. This strategy has led to widespread misunderstandings of his theory and to countless confusions about its implications for information theory. As Weaver explains: "The word *information,* in this theory, is used in a special sense that must not be confused with its ordinary usage. In particular, *information* must not be confused with meaning."[25] Second, Shannon defines communication in terms of the *selection* from a set of *possible* messages. The most rudimentary form of information is the bit, which represents a choice between only two alternatives (0 or 1). As possibilities proliferate, the situation becomes more complex and the amount of information that can be communicated increases. Weaver's commentary once again clarifies Shannon's point:

> *This word information in communication theory relates not so much to what you* do *say, as to what you* could *say. That is, information is a measure of one's freedom of choice when one selects a message. If one is confronted with a very elementary situation where he has to choose between two alternative messages, then it is arbitrarily said that the information associated with this situation is unity. Note that it is misleading (although often convenient) to say that one or the other message conveys unit information. The concept of information applies not to the individual messages (as the concept of meaning would), but rather to the situation as a whole, the unit of information indicating that in this situation one has an amount of freedom of choice, in selecting a message, which it is convenient to regard as a standard or unit amount.*[26]

Insofar as information is a measure of the freedom of choice within a set of possible messages, it is a function of probabilities. More precisely, in Shannon's theory, information is a quantitative measure of improbability. If confusion is to be avoided, it is very important to understand the multiple implications of this claim.

Though at first glance the association of information with improbability seems puzzling, further reflection suggests that Shannon's theory is actually a mathematical formulation of a simple insight. As everyone who reads the morning paper or watches the nightly news knows, information is news. When we read or hear what we expect or already know, little or no news is conveyed. Information involves what is unexpected and, thus, is necessarily related to improbability. *Information, in other words, is inversely proportional to probability: the more probable, the less information, the less probable, the more information.* The definition of information in terms of improbability establishes its difference from redundancy. Redundancy functions as a constraint that increases the likelihood of the message's arrival at its destination. While such constraints can take different forms, they tend to function as rules, codes, or syntactic structures through which messages to be transmitted can be selected. These structures also serve as more or less determinate parameters of expectation for the reception of transmitted information. The more rigid the rule, the greater the redundancy, and thus the less information is conveyed; the less rigid the rule, the less the redundancy, and thus the more information is conveyed. Insofar as it is inversely related to probability and directly related to improbability, information is inseparable from *uncertainty* or, more precisely, the resolution of uncertainty. By reducing the number of ways in which various parts of a system can be arranged and thereby reducing the number of possible messages, redundancy increases certainty while at the same time decreasing information. For information to be conveyed, there must be neither too much nor too little redundancy. If everything is predictable, no information is conveyed; if there is no redundancy, which can determine the parameters of possibility and probability, uncertainty cannot be resolved and once again no information is conveyed.

The implications of this point can be clarified by recasting the question of information in terms of difference. If information is "a difference which makes a difference," then the domain of information lies *between* too little and too much difference. On the one hand, information *is* a difference and, therefore, in the absence of difference, there is no information. On the other hand, information is a difference that *makes a difference*. Not all differences make a difference because some differences are indifferent and hence inconsequential. Both too little and too much difference creates chaos. Always articulated *between* a condition of undifferentiation and indifferent differentiation, information emerges at the two-sided edge of chaos.

When considering information in terms of difference, it is important to recall that difference is necessarily relational. If there were such a thing as an absolute difference, it could not be recognized as such; it would be something like a pure noise that can never be heard. But noise is parasitical and thus is never pure. There can no more be noise apart from the signal with which it interferes than there can be signal apart from the noise it excludes.[27] This implies that the parasite, which is, among other things, interference or static, is as much a host as the host is a parasite. Noise and information, in other words, are thoroughly relative; what is noise at one level or in one location is information in another moment or in another location:

At the feast everyone is talking. At the door of the room there is a ringing noise, the telephone. Communication cuts the conversation, the noise interrupting the messages.

As soon as I start to talk with this new interlocutor, the sounds of the banquet become noise for the new "us." The system has shifted. If I approach the table, the noise slowly becomes conversation. In the system, noise and message exchange roles according to the position of the observer and the action of the actor, but they are transformed into one another as well as a function of time and of the system. They make order or disorder.[28]

Noise becomes information, which becomes noise, which becomes information, which becomes noise . . . As observer and actor are transformed into each other to generate noise that is information and information that is noise, we return to the problem of self-reflexivity.

TIME AND UNCERTAINTY

As Calvino's remarkable *If on a winter's night a traveler* opens, actor and observer as well as author and reader are caught in a reflexive play of mirrors: "You are about to begin reading Italo Calvino's new novel, *If on a winter's night a traveler.* Relax. Concentrate. Dispel every other thought. Let your world fade." Pages, which initially seem to unfold, turn back on themselves creating redundancies that seem random and thus obscure rather than disclose patterns of meaning. "Here is page 31 again, page 32 . . . and then what comes next? Page 17 all over again, a third time!" As frames enframe frames, the narrative seems to unravel into "sentences, words, morphemes, phonemes" and the "flow of information" is "shaken by redundancies and noises" until it is finally "degraded into a swirling entropy":

The order of the words in the text of Calizto Bandera, preserved in the electronic memory to be brought to light at any moment, has been erased in an instant of de-magnetization of the circuits. The multicolored wires now grind out the dust of dis-solved words: the the the, of of of of, from from from from, that that that that, in columns according to their respective frequency. The book has been crumbled, dis-solved, can no longer be recomposed, like a sand dune blown away by the wind.[29]

When the order of words dissolves into blowing sand, swirling entropy increases.

In Shannon's influential theory, entropy and information are the measure of each other. The decision to use the notion of entropy in communication theory was influenced by John von Neumann who, in jest, advised Shannon to use this term since "no one knows what entropy is, so in a debate you will always have the advantage."[30] Explicating Shannon's argument, Weaver explains that the link joining information and entropy is probability:

The quality which uniquely meets the natural requirements that one sets up for "information" turns out to be exactly that which is known in thermodynamics as entropy. It is expressed in terms of the various probabilities involved—those of getting to certain stages in the process of forming messages and the probabilities that, when in those stages, certain symbols be chosen next. . . . That information be the measure of entropy is, after all, natural when we remember that information, in communication theory, is associated with the amount of freedom of choice we have in constructing messages. Thus for a communication source one can say, just as he would also say it of a thermodynamic ensemble, "This situation is highly organized, it is not characterized by a large degree of randomness or of choice—that is to say, the information (or the entropy) is low."[31]

When understood in relation to entropy, the boundaries of information expand considerably. Physical systems governed by the laws of thermodynamics appear to be rudimentary information systems. If information is in some sense entropic, information systems differ from the systems characteristic of Newton's world even more decisively than do dialectical and autopoietic systems.

Though Newton's universe is mechanical, there are, of course, many different kinds of machines. Intrinsically stable systems, which are reversible, represent an idealized world of frictionless devices that seem to remain perpetually in motion. As one moves from the world of mechanical clocks and watches to the world of steam engines and other thermal machines, different laws begin to appear. In contrast to machines in which energy seems to be conserved and processes appear reversible, thermal devices convert usable energy into less usable forms, thereby diminishing and eventually exhausting available resources. The operation of a thermodynamic system depends on a temperature differential between the machine and its environment as well as among its different components. Just as a waterwheel operates only when water flows from a higher to a lower level, so a thermal engine functions only when one part of the system, which is at a lower temperature, absorbs heat from an energy source that is at a higher temperature. In closed systems, the direction of the energy flow is always from higher to lower and inevitably tends toward equilibrium. In 1811, Baron Jean-Joseph Fourier summarized phenomena of heat conduction in a surprisingly simple formula: "heat flow is proportional to the gradient of the temperature." Ilya Prigogine confidently claims that the "science of complexity" was born with this discovery.[32]

Over four decades later, William Thompson's recognition of the implications of Fourier's principle led to the discovery of the second law of thermodynamics, according to which entropy increases over time in any closed system. This process is not only inevitable but also *irreversible*. When energy is converted into motion, some energy is unavoidably lost as waste heat. There is, in other words, an expenditure without any possibility of a return of or on the energy spent.[33] Prigogine clarifies the importance of this insight:

The world of Laplace [and Newton] was eternal, an ideal perpetual-motion machine. Since Thompson's cosmology is not merely a reflection of the new ideal heat engine but also incorporates the consequences of the irreversible propagation of heat in a world in which energy is conserved. This world is described as an engine in which heat is converted into motion at the price of some irreversible waste and useless dissipation. Effect-producing differences in nature progressively diminish. The world uses up its differences as it goes from one conversion to another and tends toward a final state of thermal equilibrium, "heat death." In accordance with Fourier's law, in the end there will no longer be any differences of temperature to produce mechanical effect.[34]

In 1865, Rudolf Clausius translated the second law of thermodynamics into the cosmological principle of entropy, which he summarized in two well-known axioms:

The energy of the universe is a constant.
The entropy of the universe tends to a maximum.

In order to affirm both of these principles, Clausius draws a distinction between conservation and reversibility. As we have discovered, in equilibrium and homeostatic systems, the principle of conservation seems to imply reversibility, which occurs when forces and energies are caught in a play of action and reaction. "Unlike mechanical transformations, where reversibility and conservation coincide, a physiochemical transformation may conserve energy even though it cannot be reversed. This is true, for instance, in the case of friction, in which motion is converted into heat, or in the case of heat conduction as it was described by Fourier."[35] In actual as opposed to ideal systems, friction and heat loss are unavoidable and therefore the processes by which they operate are inevitably irreversible. Entropy, according to Clausius, is the measure of the availability or unavailability of energy that can be used in a system. The less the entropy, the more usable energy is available, and vice versa.

The irreversibility of thermodynamic processes introduces into systems what Arthur Eddington dubbed "the arrow of time." All isolated systems evolve in the direction of increasing entropy. The telos of this process is the state of thermal equilibrium constituting "heat death." If the operation of a system depends upon differences in heat and energy distribution, the state of equilibrium would make energy flow impossible and thereby would mark the destruction or death of the system. Unlike a clock, which can always be rewound, closed systems inevitably run down and cannot be revived. The trajectory of this development is from more to less difference. If information is a difference that makes a difference and noise is a condition in which differences become indifferent, then the arrow of time moves from information to noise. This process culminates when, in Calvino's terms, "the book [or any other object or system] has been crumbled, dissolved, can no longer be recomposed, like a sand dune blown away by the wind." But what are these grains of sand that have been "degraded into a swirling entropy?"

One of the many curiosities in the history of science is that the laws of thermodynamics were formulated *before* anyone really understood what heat is. The nature of heat was not determined until the Austrian physicist Ludwig Boltzmann began investigating James Clerk Maxwell's revolutionary theories about the relationship between electrical and magnetic phenomena. Maxwell discovered that "light is electromagnetic radiation—shifting electrical and magnetic fields that travel out into the universe, alternating forever at right angles to the direction of their dispersal."[36] In developing his ideas, Maxwell made use of atomic theory to argue that matter is made up of countless tiny particles that are in constant motion. In what is perhaps his most radical departure from the principles of classical physics, he applied statistical methods and probability, as developed in the analysis of average behavior by sociologist Adolphe Quetelet, to physical phenomena. "The innovation," according to Prigogine, "was to introduce probability in physics not as a means of approximation but rather as an explanatory principle, to use it to show that a system could display a new type of behavior to which the laws of probability could be applied."[37] This was the first time that a law of nature was cast as a probability rather than absolute. Boltzmann extended Maxwell's analysis by arguing that heat is the function of the movement of the atoms and molecules comprising matter. Temperature, Boltzmann concluded, is a function of molecules' *average* rate of movement. At any particular temperature individual molecules move at different speeds. If more molecules move at high speeds, the temperature is correspondingly high, and if more move slowly, the temperature is lower. What must be stressed in this context is that it is impossible to determine temperature (i.e., the macrostate) from the activity of any individual molecule (i.e., the microstate). Temperature is a statistical phenomenon and as such is subject to probability. In his investigation of heat and temperature, Boltzmann's primary concern was to determine the mechanisms by which molecules move toward a state of equilibrium and systems thus become entropic. Using Maxwell's statistical methods, he argued that, while the precise activity of individual particles cannot be ascertained, the overall trajectory of a population of molecules can be determined.

The formulation of the laws of thermodynamics and the recognition of the role of probability in natural processes marks the beginning of the end of the dream of certainty that fueled the modern imagination for over three centuries. In classical physics, the principles of repeatability, predictability, and determinism lead to the conviction that it is possible to be certain about the objective truth of natural phenomena. Since effects are proportionate to causes, if the initial conditions of a system can be determined, its future course of development can be projected with certainty. This search for certainty is not, of course, limited to those who believe in realism and objectivity. We have already seen that modern philosophy begins with Descartes's fateful turn to the subject. For the self-conscious subject, truth is *self*-certainty. While objective and subjective quests for certainty have frequently been attacked for a variety of reasons, no criticism is as devastating for claims of certainty than the scientific developments we have been considering. If some natural processes at the microlevel are probabilistic rather than deterministic, uncertainty is woven into the very fabric of things themselves. Uncertainty, in other words, is not a result of ignorance or the partiality of human knowledge but is a characteristic of the world itself. In the years since Boltzmann's work, it has become undeniable that "instability destroys the equivalence between the individual and statistical levels of description." As classical physics is reconsidered in light of "the physics of populations," the dream of certainty becomes a distant memory.[38]

It is important to distinguish the uncertainty involved in statistical phenomena from the uncertainty described by Heisenberg and at work in quantum mechanics. Quantum physics calls into question scientific and philosophical realism and renders claims of objective certainty suspect. According to Heisenberg's famous uncertainty principle, it is impossible to determine accurately both the position and the momentum of subatomic particles. The laws of physics, therefore, entail relative rather than absolute certainties. More important, Heisenberg's insight can be understood in terms of the problem of self-reflexivity. Since observer and observed are mutually implicated in the same system, knowledge of the object is conditioned by the subject. In less philosophical terms, observation or description of any object requires the selection of instruments of observation, which inevitably influence how the object is understood. Knowledge, therefore, is relative to the multiple perspectives from which it is constituted. This multiplicity of perspectives cannot be reduced to a single angle of vision that is true for all observers. While Heisenberg deepens the perplexities of our relationship to the world for those who remain committed to the pursuit of objective certainty, the uncertainty he identifies is not the same as the uncertainty involved in the probabilistic processes. Prigogine writes:

The role of the observer was a necessary concept in the introduction of irreversibility, or the flow of time, into quantum theory. But once it is shown that instability breaks time symmetry, the observer is no longer essential. In solving the time paradox, we also solve the quantum paradox, and obtain a new realistic formulation of quantum theory. This does not mean a return to classical deterministic orthodoxy; to the contrary, we go beyond the certitudes associated with the traditional laws of quantum theory and emphasize the fundamental role of probabilities. In both classical and quantum physics, the basic laws now express possibilities. We need not only laws, *but also* events *that bring an element of radical novelty to the description of nature.*[39]

With the emergence of events that are not completely determinable by laws, we are brought to the edge of the aleatory, which has no place in intrinsically stable systems.

One of the most significant breaks with the principles of classical physics is the rejection of the notion of reversibility and the insistence that at least some natural processes are irreversible. Our consideration of the phenomenon of entropy has shown that irreversibility introduces time into systems and structures that otherwise seem unchanging. It is important to note that, while quantum mechanics claims to establish probability in microscopic physics, it stops short of affirming irreversibility. Indeed, Einstein not only famously asserted that "God doesn't play dice with the universe," but also often stated his belief that "time is an illusion."[40] It can be argued, however, that the recognition of stochastic processes secures the trajectory of the arrow of time and makes reversibility so unlikely as to be effectively impossible.

The far-reaching implications of irreversibility have provoked resistance on the part of many distinguished scientists. Nobel laureate Max Born, one of the leading figures in quantum mechanics, went so far as to declare: "Irreversibility is the effect of the introduction of ignorance into the basic laws of physics."[41] For many others, however, a growing appreciation of the importance of stochastic processes makes it impossible to deny the reality of irreversibility. If certain systems display "intrinsic randomness," there must be "intrinsic irreversibility." Time, in other words, is a measure of increasing randomness. Accordingly, the accumulation of random events introduces a distinction between past and future that lends time its decisive direction. As randomness increases, entropy grows and vice versa.

The inescapability of irreversibility, which is the correlate of randomness, has only gradually gained acceptance. From the time Maxwell posed the puzzle of his infamous demon in 1867, some scientists have tried to find ways to avoid the conclusion that certain processes are irreversible. By postulating a form of intelligence that could sort high-speed and low-speed molecules, Maxwell attempted to overcome entropy by reestablishing the temperature differential necessary to make useful energy available. While everything seems to be falling away from order, Maxwell insisted that confusion and disorder are not absolute but are relative to the capacity to apprehend and manage the elements comprising matter. Though human beings might not be able to determine the differences among molecules equally distributed within a closed volume, it is in principle possible to imagine a higher form of intelligence with greater apperceptive powers who could distinguish and redistribute the molecules. Maxwell illustrates his point with a deceptively simple example:

A memorandum-book does not, provided it is neatly written, appear confused to an illiterate person, or to the owner who understands it thoroughly, but to any other person able to read it appears to be inextricably confused. Similarly the notion of dissipated energy could not occur to a being who could not turn any of the energies of nature to his own account or to one who could trace the motion of every molecule and seize it at the right moment. It is only to a being in the intermediate stage, who can lay hold of some forms of energy while others elude his grasp, that energy appears to be passing inevitably from the available to the dissipated state.[42]

Were it real, Maxwell's demon would seem to contradict the second law of thermodynamics and thus to create the possibility of escaping time.

But matters are considerably more complex. While physicists had attempted since 1871 to resolve the conundrum Maxwell posed, it was not until recent developments in research on the energy requirements of computers that a persuasive refutation of the implication of his argument was formulated. In the early 1960s, Rolf Landauer of IBM analyzed the thermodynamics of data processing and concluded that some data operations are "thermodynamically costly but others, including, under certain conditions, copying data from one device to another, are free of any fundamental thermodynamic limit."[43] Landauer's discovery that "clearing memory is a thermodynamically irreversible operation" proves decisive for Maxwell's argument. Commenting on the implications of this insight for Maxwell's demon, Charles Bennett explains:

In order to observe a molecule, [the demon] must forget results of previous observations. Forgetting results, or discarding information, is thermodynamically costly. If the demon had a very large memory, of course, it could simply remember the results of all its measurements. There would then be no logically irreversible step and the engine would convert one bit's worth of heat into work in each cycle. The trouble is that the cycle would not then be a true cycle: every time around, the engine's memory, initially blank, would require another random bit. The correct thermodynamic interpretation of this situation would be to say that the engine increases the entropy of its memory in order to decrease the entropy of its environment.[44]

Bennett's conclusion not only reinforces the second law of thermodynamics but also reaffirms the complex interrelationship between information and entropy. Summarizing his argument, he underscores its unexpected implications:

Another source of confusion is that we do not generally think of information as a liability. We pay to have newspapers delivered, not taken away. Intuitively, the demon's record of past actions seems to be a valuable (or at worst a useless) commodity. But for the demon "yesterday's newspaper" (the result of a previous measurement) takes up valuable space, and the cost of clearing that space neutralizes the benefit the demon derived from the newspaper when it was fresh. Perhaps the increasing awareness of environmental pollution and the information explosion brought on by computers have made the idea that information can have a negative value seem more natural now than it would have seemed earlier in this century.[45]

As we will see in chapter 7, discarding information involves a process of screening through which knowledge emerges. Knowledge, it seems, always has a price and this price lends time its direction.

SHIFTY STATIC

This understanding of the intricate interrelation of entropy, probability, randomness, disorder, and temporality brings us back to the problems of communication and information. Information, we have discovered, is inversely related to probability and directly related to improbability. The more improbable a phenomenon or event, the less it is anticipated, and thus the more information it communicates when it occurs. The most probable state is one in which differences become indifferent and thus in which a signal dissolves in noise. If information is understood as a difference that makes a difference, noise can be interpreted as either the lack of differentiation or the profusion of indifferent differences. Since the arrow of time follows a trajectory from information to noise, the movement of history is from heterogeneity to homogeneity. Wiener points out the cosmological significance of this point.

As entropy increases, the universe, and all closed systems in the universe, tend naturally to deteriorate and lose their distinctiveness, to move from the least to the most probable state, from a state of organization and differentiation in which distinctions and forms exist, to a state of chaos and sameness. In [Willard] Gibbs' universe order is least probable, chaos most probable.[46]

Wiener's formulation underscores an important qualification: entropy increases only in *closed* systems. If there are extrinsic sources of energy or information, the progressive realization of entropy is not inevitable. Since the universe as a whole appears to be a closed system, there seems to be no way to avoid the ultimate triumph of entropy. Within this all-encompassing system, however, there are countless open systems, which are counterentropic. Wiener continues:

But while the universe as a whole, if indeed there is a whole universe, tends to run down, there are local enclaves whose direction seems opposed to that of the universe at large and in which there is a limited and temporary tendency for organization to increase. Life finds its home in some of these enclaves.[47]

Serres describes these "local enclaves" as "quasi-stable eddies," which are "the erratic blinking of aleatory mutations" that form a "local flow upstream toward negentropic islands—refuse, recycling, memory, increase in complexities."[48]

Negentropy is, as the term implies, negative entropy. Whereas entropy represents the loss of differences constitutive of organizational structure, negentropy designates the temporary reversal of this process, which occurs when differentiated structures and systems emerge in the midst of disorder. The counterthrust of negentropy cannot, of course, reverse the overall increase in entropy in the universe as a whole. The entropic condition that marks the end of time can at best be delayed or deferred. Prigogine labels the islands of negentropy that form in the river of time "dissipative structures." This term can be confusing and must be understood very precisely. Since entropy refers to the degradation of heterogeneity into homogeneity, increasing entropy would seem to dissipate the differences upon which organization as well as order depends. Prigogine's dissipative structures dissipate this dissipation in a process that approximates a dialectical negation of negation. Dissipative structures, which always emerge spontaneously in conditions far from equilibrium, complicate the relation between information and noise.[49]

When Shannon's mathematical theory of communication is read through thermodynamics, information and entropy appear to be directly related. Accordingly, information increases with entropy. A correlate of this claim would, of course, seem to be that information *decreases* as negentropy increases. The argument once again turns on the interplay between probability and improbability. By temporarily slowing the slide into chaos, negentropic processes generate organized structures, which create redundancies. This increase in order decreases randomness and increases probability. It now becomes apparent that there is a further paradox in this already complicated situation. Within the overall scheme of things, order is less probable than disorder. Consequently, the movement from order to disorder, which is the mark of expanding entropy, is characterized by an increase in a certain aspect of probability. This is a subtle point that must be formulated precisely. When entropy increases, the *range* of possible microscopic states expands, and, thus, the *probability of the range* increases. Since any given state occurs somewhere within this range, the *probability of any particular state* is very small. As the probability of the range of possibilities *increases,* the probability of a particular state *decreases.* In other words, probability can be seen as increasing or decreasing, depending on which order of probability one is considering.

As if this situation were not confusing enough, Wiener further complicates things by offering a definition of information that appears to be the precise inverse of Shannon's position. "Just as entropy is a measure of disorganization," Wiener argues, "the information carried by a set of messages is a measure of organization. In fact, it is possible to interpret the information carried by a message as essentially the negative of entropy, and the negative logarithm of its probability."[50] When information is understood as "a measure of organization," noise appears to be the dissolution or breakdown of organization. From this point of view, the movement from noise to information is the emergence of organization from disorganization, or order from chaos. Insofar as Wiener maintains that the message is "essentially the negative of its entropy," information is *negentropy*. This conclusion appears to be the precise opposite of Shannon's contention that information and entropy are directly related. But the difference between Shannon and Wiener is not as deep as it initially appears. Whereas Shannon focuses more on the *information one lacks,* Wiener focuses on the *information one gains* upon receiving a message. One's state of knowledge is characterized by having higher entropy before receiving the message, and lower entropy after receiving it. From this point of view, information *reduces* entropy and, thus, appears to be negentropic. If the information associated with negentropy is, as Wiener maintains, the information gained upon receiving a message, Shannon would agree that it reduces entropy.

If Shannon and Wiener's analyses of the complex interplay of information, noise, entropy, and negentropy are read through Prigogine's notion of dissipative structures, the relationship between noise and information can be reconfigured in interesting ways. By momentarily resisting entropy, dissipative structures dissipate the dissolution of the differences necessary for organization. This dissipation of dissipation, or negation of negation, corresponds to negentropy, which is, in Nørretranders's apt phrase "dis-disorder, i.e., order."[51] However, if order is dis-disorder, then order and disorder are necessarily related. Disorder does not simply destroy order, structure, and organization but is also a condition of their formation and reformation. Describing the emergence of dissipative structures, Prigogine writes: "The interaction of a system with the outside world, its embedding in nonequilibrium conditions, may become in this way the starting point for the formation of new dynamic states of matter—dissipative structures."[52] These new dynamic states emerging in conditions far from equilibrium tend to be characterized by greater complexity than the states from which they come.

It should be evident that the relation between information and noise can be understood in terms of the interplay between order and chaos at work in dissipative structures. If understood as what Serres labels "the ultimate parasite," noise is the interference that is simultaneously disruptive and creative. There can be no information without noise and vice versa. Noise can no more be silenced in the world than parasites can be exterminated; life depends on parasites as much as information depends on noise. "Mistakes, wavy lines, confusion, obscurity are part of knowledge; noise is part of communication, part of the house."[53] Though undeniably a part of every domestic economy, noise is always unexpected—it comes from the outside, like a thief or a knock on the door in the middle of the night. Such interruptions can, of course, be destructive. Rats scatter, dishes shatter, and, as the feast comes to an end, the books and bodies crumble, dissolving into swirling grains of sand. But this is not the only story to be told—at least for a while. For those who are willing to host unexpected guests, life is often enriched. Such enrichment does not make life simpler; to the contrary, unanticipated disruptions complicate life and usually make things considerably more difficult. The voice of the stranger remains noisy until it is translated. Forever unsolicited, noise is given as if from without and remains merely data until it is processed.[54] Processing, however, leaves neither processor nor processes unchanged. When programs are flexible and codes adaptable, noise can be processed in ways that allow novelty to emerge. In the open-ended revolution of information and noise, noise transforms the systems and structures that transform noise:

The parasite invents something new. Since he does not eat like everyone else, he builds a new logic. He crosses the exchange, makes it into a diagonal. He does not barter; he exchanges money. He wants to give force for matter, (hot) air for solid, superstructure for infrastructure. People laugh, the parasite is expelled, he is made fun of, he is beaten, he cheats us; but he invents anew. This novelty must be analyzed.[55]

If novelty is analyzed, it immediately becomes clear that invention cannot, by definition, be redundant; the emergence of the new reconfigures the old in unexpected ways that allow differences not previously articulated to emerge. When differences do not remain indifferent but are organized, albeit temporarily, noise becomes "responsible for the growth of the system's complexity":

Who is the parasite here, who is the interrupter? Is it the noise, the creaking of the floorboards or of the door? Of course. It upsets the game, and the system collapses. If it stops, everything comes back, is reformed and the meal continued. Think of another noise: the chain is broken again and everything vanishes in the bewildered flight. The noise temporarily stops the system, makes it oscillate indefinitely. To eliminate the noise, a nonstop signal would be necessary; then the signal would no longer be a signal and everything would start again, more briskly than usual. Theorem: noise gives rise to a new system, an order that is more complex than the simple chain. This parasite interrupts at first glance, consolidates when you look again.[56]

The oscillation of interruption and consolidation captures the rhythms of noise and information.

Noise is always in formation; there can be neither form nor formation without noise. When information is understood as a process rather than a product, the line separating it from noise is difficult to determine. Noise is not absolute but is relative to the systems it disrupts and reconfigures, and, conversely, information is not fixed and stable but is always forming and reforming in relation to noise. Forever parasitic, noise is the static that prevents the systems it haunts from becoming static. Static makes systems shifty. If, on the one hand, structures become too rigid to adapt to changing circumstances, the systems they support collapse; if, on the other hand, there are no systems to process data, noise becomes fatal. Life is lived on the shifting margin, boundary, edge between order and chaos, difference and indifference, negentropy and entropy, information and noise. The interplay of noise, which is informative, and information, which is noisy, creates the conditions for emerging complexity, which is the pulse of life.

five

And what of our Author? I see now that there is no Author—or else, there are many Authors, so that our story was not one but many, and we ourselves are fictions whose apparent complexity and subtlety of meaning is something which has emerged from simple matter, from the humblest origins. We are a thought which passed through a mind in an instant; a gesture hardly worthy of elucidation.

CARTOGRAPHER: Must I abandon this vision?

LEOPOLD: Only if you wish to go back and live amongst men.

CHOROUS LOGICO-PHILOSOPHICUS:

Whereof we were silent
Now becomes spoken
No longer must words
Stand in place of pure thought.
Language and meaning
Are hollow inventions;
Emergent complexity
Bear us aloft!

andrew crumey

emerging complexity

REAL CLOSE

The prefatory image accompanying Robert Storr's introductory catalog essay for the Museum of Modern Art's 1998 exhibition of Chuck Close's paintings is in effect a visual representation of Andrew Crumey's philosophical reflections in his novel *Pfitz* (fig. 16). The photograph pictures Close, seated in his wheelchair beside tubes of paint and brushes and in front of an easel on which one of his characteristically large canvases rests. The artist is gazing at a photograph of himself, which he is in the midst of painting. The work in progress eventually bears the name *Self-Portrait.* As images fold into images, self-reflexive loops begin to become strange, thereby making Close's seemingly transparent work more

125

FIGURE 16 Chuck Close in his studio working on *Self-Portrait*, 1993. Oil on canvas, 72" × 60". (© 1993 Chuck Close. *Portrait of the Artist with Work in Progress*, photograph by Ellen Page Wilson, courtesy of Pace Wildenstein.)

and more complex. What appears to be a self-portrait is a painting of a photograph—a representation of a representation, a sign of a sign. As both the subject and the object of the work of art, Close seems to be caught in a play of images that resembles the uncanny world charted in *Pfitz*. Crumey, who studied theoretical physics and mathematics at St. Andrews University and Imperial College and did postdoctoral research on nonlinear dynamics at Leeds University, realizes that the complications that emerge when images fold into each other bring one to the edge of madness. Close might well be the character in the novel named Spontini, who

began the Aphorisms shortly before the onset of illness which deprived him of his reason. . . . The nature of Spontini's insanity was such that he became overwhelmed by the delusion, first of all, that he was in fact a character in his own book, rather than its author. He saw himself variously as the Prince and as the accused servant, and then imagined a series of authors struggling to gain control of his soul. Finally, he was led to believe that he did not exist at all, except in the minds of others.[1]

If, in fact, Close's *Self-Portrait* is a self-portrait, the self it portrays is, if not mad, then surely very strange.

Self-Portrait, completed in 1993, was not the first time Close painted a self-portrait, nor was it the last. Close has been doing self-portraits throughout his career: 1967–68, 1968, 1975, 1976–77, 1979, 1986, 1991, 1995, 1996, 1997. . . . Surveying the course of Close's career, Kirk Varnedoe observes: "Like other artists of his generation—Sol Le Witt, Richard Serra, Philip Glass—Chuck Close does the same damned thing over and over. The constrictive boxes he puts around his options, and his nonstop recycling of motifs and methods, might seem to guarantee monotony."[2] But, as Varnedoe realizes, the differences among the works are more important than their similarities. By framing an intricate interplay between sameness and difference, Close creates paintings that are anything but monotonous. Like so many of the trajectories we have been tracing, Close began his series of self-portraits in 1968, or, more precisely, 1967–68. On the face of it, his earliest and most recent self-portraits could not be more different. Measuring 107½ × 83½ inches, *Big Self-Portrait* lives up to its name (fig. 17). Following a procedure he would use throughout his career, Close depicts only the head from the neck up. It is difficult to know how to classify these works: realism/abstraction, minimalism/realism, formulaic/inventive, academic/avant-garde, programmed/spontaneous, standardized/ singular, impersonal/expressive, grid/network? If none of the categories fits precisely, it is because Close calls into question all such oppositions by working at their edge.

The scale of the works combines with their extraordinary resolution and breathtaking detail to create a sense of overwhelming realism. These paintings are not just realistic; they are hyperreal. Their hyperreality has led many critics to classify Close's work as photo-realistic. There are, of course, obvious similarities between Close and such artists as Philip Pearlstein, Richard Estes, Robert Cot-

FIGURE 17 Chuck Close, *Big Self-Portrait,* 1967–68. Acrylic on canvas, 108" × 84" (273.1 × 212.1 cm). (© 1968 Chuck Close. Collection of the Walker Art Center Center, Minneapolis. Art Center Acquisition Fund, 1969.)

tingham, and Michael Morely. Like the photo-realists, Close works from photographs, and his paintings have an undeniable photographic quality. Indeed, often it is virtually impossible to detect even the trace of a brushstroke. But Close has always resisted being associated with any form of realism. At the precise moment he was completing *Big Self-Portrait,* realism was enjoying something of a renaissance in the art world. The renewed interest in representation and figurative art was a reaction against what many believed to be the growing vapidness of excessive abstraction. In 1968, Linda Nochlin signaled the importance of these developments by mounting an influential exhibition—*Realism Now*—which included a painting by Close. Five years later she explained the stakes of the debate between realism and abstraction in her essay "The Realist Criminal and the Abstract Law":

Yet on the whole, realism implies a system of values involving close investigation of particulars, a taste for ordinary experience in a specific time, place and social context; and an art language that vividly transmits a sense of concreteness. . . . To ask why realist art continues to be considered inferior to nonrealist art is really to raise questions of a far more general nature: Is the universal more valuable than the particular? Is the permanent better than the transient? Is the generalized superior to the detailed? Or more recently: Why is flat better than three-dimensional? Why is truth to the nature of the material more important than truth to nature or experience? Why are the demands of the medium more pressing than the demands of visual accuracy? Why is purity better than impurity?[3]

Once again, however, Close's work subverts traditional binary oppositions: nonrealist/realist, universal/particular, general/detailed, flat/three-dimensional, purity/impurity. The reason Close's paintings are so elusive begins to appear as one approaches the canvases and examines them carefully. In the blink of an eye, analog images switch to digital as tiny bits begin to appear. This method is not, of course, new; from ancient mosaics to nineteenth-century pointillism, artists have used bits to generate forms and figures. With the development of televisual and digital technologies, bits form pixels to create images whose complexity increases exponentially. Close's mature art is more closely related to twenty-first-century digital technology than it is to the work of a nineteenth-century painter such as Georges Seurat. Close's works are distinguished not only by the use of pixels but also by his deployment of the *grid*. What is explicit in the later work is implicit in the early work: images appearing to be hyperreal are actually generated by an abstract grid. A study for *Big Self-Portrait* exposes Close's strategy (fig. 18). The complex image emerges from the formal structure of a grid. Once again, Close's tactics are not original but borrowed; indeed, it is as if his paintings were composed of lines drawn from Ad Reinhardt's "25 LINES OF WORDS ON ART":

11. *PAINTING AS CENTRAL, FRONTAL, REGULAR, REPETITIVE.*
18. *BRUSHWORK THAT BRUSHES OUT BRUSHWORK.*

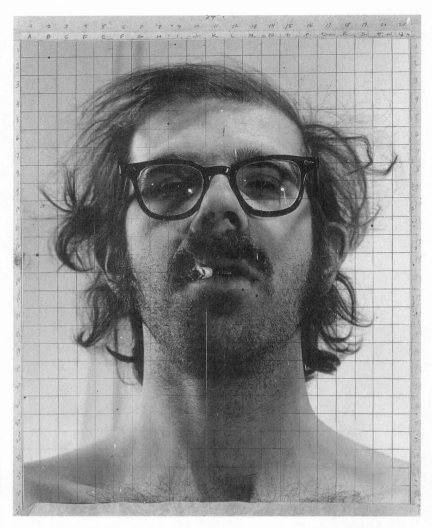

FIGURE 18 Chuck Close, Study for *Self-Portrait*, 1968; photographed 1967. Photograph, pen and ink, pencil, masking tape, synthetic polymer paint, wash, and blue plastic strips on cardboard, 18⅝ × 13⅜ (47.2 × 33.9 cm.). (© 1968 Chuck Close. The Museum of Modern Art, New York. Gift of Norman Dubrow. Photograph © 2000 The Museum of Modern Art, New York.)

20. *THE STRICTEST FORMULA FOR THE FREEST ARTISTIC FREEDOM.*

21. *THE EASIEST ROUTINE TO THE DIFFICULTY.*

22. *THE MOST COMMON MEAN TO THE MOST UNCOMMON END.*

23. *THE EXTREMELY IMPERSONAL WAY FOR THE TRULY PERSONAL.*

24. *THE COMPLETEST CONTROL FOR THE PUREST SPONTANEITY.*
25. *THE MOST UNIVERSAL PATH TO THE MOST UNIQUE. AND VICE VERSA.*[4]

To create *Big Self-Portrait,* Close first superimposes a grid on the photograph as well as the canvas, and then proceeds methodically to transfer the visual information contained in each square of the photograph to the painted surface. What emerges is something that can be neither apprehended in the individual squares nor predicted from their interaction.

From Warhol's serial silk screens to Le Witt's geometric squares, the grid was used throughout the 1960s to develop different responses to the overblown egos and overwrought gestures of abstract expressionism. For many artists, the abstraction of the grid provided a way to move from the personal to the impersonal by transforming the creative process into a mechanical activity. These artists seem to take Walter Benjamin's account of the work of art in the age of mechanical reproduction as prescriptive rather than descriptive. Le Witt expresses a machinic fantasy shared by many artists at the time: "When an artist uses a conceptual form of art, it means that all of the planning and decisions are made beforehand and the execution is a perfunctory affair. The idea becomes a machine that makes the art."[5] Rejecting the romantic myth of the artist as a creative genius, which culminates in heroic abstract expressionism, artists attempt to enact what Foucault and Barthes eventually label "the death of the author." Storr underscores this point when commenting on the relation between Close and Gerhard Richter:

Richter quite literally began by reducing painting to a tabula rasa, *while Close banished gesture for an uninflected pictorial factualness equivalent to what the French semiotician Roland Barthes described as writing at "degree zero." Under these conditions, Close effectively removed himself as the author of the picture in order to remain as the agency of a picture in the process of structuring itself.*[6]

When the picture becomes "the process structuring itself," it begins to approximate the principle of self-organization, which Kant identified in his analysis of the beautiful work of art and scientists have recently explored in research on emerging self-organizing systems.

Following a 1980–81 retrospective exhibition, Close stopped using the grid for four years. When it reappeared in 1985, the grid was completely transformed. While Close had always appreciated the force of Reinhardt's insistence on "the

FIGURE 19 Chuck Close, *Big Self-Portrait* (detail), 1967–68. Acrylic on canvas, 108" × 84" (273.1 × 212.1 cm). (© 1968 Chuck Close. Collection of the Walker Art Center Center, Minneapolis. Art Center Acquisition Fund, 1969.)

FIGURE 20 Chuck Close, *Self-Portrait* (detail), 1997. Oil on canvas, 8' 6" × 7' (259.1 × 213.4 cm.). (© 1997 Chuck Close. The Museum of Modern Art, New York. Gift of Agnes Gund, Jo Carole, and Ronald S. Lauder, Donald L. Bryant, Jr., Leon Black, Michael and Judy Ovitz, Anna Marie and Robert F. Shapiro, Leila and Melville Strauss, Doris and Donald Fisher, and purchase. Photograph by Ellen Page Wilson, courtesy of Pace Wildenstein.)

FIGURE 21 Chuck Close, *April* (detail), 1990–91. Oil on canvas, 100" × 84". (© 1991 Chuck Close. The Eli and Edythe L. Broad Collection. Photograph by Ellen Page Wilson, courtesy of Pace Wildenstein.)

strictest formula for the freest artistic freedom," the return to the grid made it possible for him to create works of considerably greater complexity. The new grid, which is greatly enlarged, is twisted and turned in ways that suggest Gehry's architectural diagrams. In some works, the grid is angled to create dia-

FIGURE 22 Chuck Close, *Self-Portrait*, 1997. Oil on canvas, 102" × 84" (259.1 × 213.4 cm.). (© 1997 Chuck Close. The Museum of Modern Art, New York. Gift of Agnes Gund, Jo Carole, and Ronald S. Lauder, Donald L. Bryant, Jr., Leon Black, Michael and Judy Ovitz, Anna Marie and Robert F. Shapiro, Leila and Melville Strauss, Doris and Donald Fisher, and purchase. Photograph by Ellen Page Wilson, courtesy of Pace Wildenstein.)

monds, and in others rotated to generate a circular matrix. Rather than subtly concealed, the grid is foregrounded in a way that makes its compositional function obvious. The effect is no longer photo-realistic; what is important now is what happens within each individual square on the grid as well as how they interact. Gone are homogeneous points and in their place are tiny paintings within paintings, which are as different from each other as the faces they figure (figs. 19, 20, 21). By changing the grid, Close transforms the work of art. No longer static images, his paintings become dynamic interactive processes. When examining these works at close range, it is impossible to discern any overall pattern; they appear to be a cacophony of colors and shapes creating the visual equivalent of noise. But as one slowly backs away from the painted surface, noise is retextured and a significant change occurs. Eventually, one reaches what can best be described as "the tipping point," where everything shifts through a process of coarse graining, and a pattern suddenly emerges (fig. 22). Varnedoe captures the far-reaching implications of the changes in Close's strategy:

As the paintings' grids have become steadily larger, each box has come to contain a more complex subworld of colors and shapes, vividly teeming with a variegated organicism largely removed from any specific correlation with local descriptive features. As a result, these recent pictures can seem to have less to do with the processes of photography than with those of thought itself: by treading more boldly along the edge of incoherence, they visibly play up the way in which our neural networks form a complex, uniquely specific whole—in this case, an identity, a likeness—from the synchronous firings of dissimilar, fractiously disconnected units.[7]

This tipping point, where "the synchronous firings of dissimilar, fractiously disconnected units" generate new forms, is the visual equivalent to the process Serres describes: "noise gives rise to a new system, an order that is more complex than the simple chain. The parasite interrupts at first glance, consolidates when you look again." "Variegated organicism," marked by emerging complexity, is, we have come to suspect, the result of the interplay of noise and information occurring "along the edge of incoherence," or, in terms of complexity theory, "at the edge of chaos." When understood in this way, Close's paintings provide graphic illustrations of emerging complexity.

CRYSTALS AND SMOKE

Serres begins his prescient essay "The Origin of Language: Biology, Information Theory, and Thermodynamics" with a seemingly simple assertion: "An organism is a system." There are, of course, many different kinds of systems. If an organism is a system, the way in which it is understood will depend on how systems are interpreted. A system, for example, can be logico-mathematical, as it is for Descartes, Spinoza, and Leibniz, or it can be mechanical, as it is for Newton and Laplace. While the former type of system is deductive, the latter is reversible.

With the advent of the industrial revolution and the discovery of the laws of thermodynamics, the classical ideal of knowledge collapses and new understandings of systems must be developed. During the last half of the twentieth century, recasting the relation between information theory and the laws of thermodynamics has created a communications revolution that once again transforms the way systems are conceived. Extending the model drawn from communications and information theory to biology, Serres concludes that the organism is "a hypercomplex system":

It is an information and thermodynamic system. Indeed, it receives, stores, exchanges, and gives off both energy and information—in all forms, from the light of the sun to the flow of matter which passes through it (food, oxygen, heat, signals). This system is not in equilibrium, since thermodynamic stability spells death for it, purely and simply. It is in a temporary state of imbalance, and it tends as much as possible to maintain this imbalance. It is hence subject to the irreversible time of the second law, since it is dying. But it struggles against time.[8]

Life, in other words, continues only as long as things are *out of balance.* But how is imbalance maintained and how is equilibrium deferred? Serres's answer is deceptively simple: *noise.* Noise simultaneously disrupts order and creates the condition of the possibility of the emergence of a new and more complex order:

Consider any level of an interlocking system. Locally . . . it operates like a series of chemical reactions at a certain temperature. . . . Let us consider only the energy conditions at this one level. It mobilizes information and produces background noise. The next level in the interlocking series receives, manipulates, and generally integrates the information-background noise couple that was given off at the preceding level. . . . In a certain sense, the next level functions as a rectifier, in particular, as a rectifier of noise. What was once an obstacle to all messages is reversed and added to the information. This discovery is all the more important since it is valid for all levels. It is a law of the series, which runs through the system of integration.[9]

The rectification of noise is not, of course, its elimination. What is information in one context is noise in another context. Noise, as we have discovered, is always in formation.

Though not mentioned in the text proper, Serres indicates in an inconspicuous footnote that he borrows his central insights from the work of Henri Atlan, who is professor of biophysics at the University of Paris and scholar in residence for studies in philosophy and ethics of biology at the Hadassah Hebrew University Medical Center in Jerusalem. Atlan's two most important studies are *L'organisation biologique et la théorie de l'information* (1972), and *Entre le cristal et la fumée: Essai sur l'organisation du vivant* (1979). They unfortunately remain untranslated and thus are little known in the English-speaking world.[10] Atlan's wide-ranging theoretical speculations and extensive mathematical calculations rest on his guiding principle of "order from noise." In formulating and elaborating this principle, he draws on a variety of sources, ranging from Shannon's theory about the generation of information from noise and Prigogine's account of order through fluctuation in dissipative structures, to von Neumann's research on cellular automata and von Foerster's work on self-organizing systems. Atlan begins his contribution to the International Symposium on Order and Disorder, held at Stanford University in 1981, by asserting: "What I have to say may be summarized in two sentences: One is that *randomness is a kind of order, if it can be made meaningful;* the second is that *the task of making meaning out of randomness is what self-organization is all about.*"[11] By recasting the relation between order and noise in terms of self-organization, Atlan creates new openings for theoretical reflection about the structure and function of systems.

Insofar as an organism is understood as a communications system, noise plays a critical role in both its functioning and disfunctioning. For systems to work, Atlan maintains, there must be neither too little nor too much noise. Poised between disorder and order, self-organizing systems "imply a transmission between substructures but with ambiguity or equivocation. We arrive, then, at the apparently paradoxical idea that organization is proportionally greater as ambiguity increases, up to a certain limit where there is no more transmission at all and where organization disappears."[12] Noise, according to Atlan, is an "aleatory aggression" on the part of the environment, which, on the one hand, increases the disorder of the system, and, on the other hand, provides the occasion for the emergence of a more complex order. Atlan draws on the work of von Neumann to summarize his argument:

Generalization: theory of organization by the diminution of redundancy under the effect of factors of noise.

Following von Neumann, we have arrived at the idea according to which in a system of "extremely high complexity," the property of self-organization should consist in that the factors of noise in the environment produce two opposite effects: on the one hand, they increase the quantity of information of the total system by augmenting the autonomy among the parts; on the other, they diminish this quantity of in-

formation by the accumulation of errors in the structure of these parts. In order for these effects to be possible, that is to say for them to be able to coexist without the system ceasing to function, it is necessary for the system to be of "extremely high complexity," that is, composed of a great number of parts interconnected in multiple ways.[13]

Complexity, for Atlan, is nothing other than the "property of being able to react to noise in two opposed ways without ceasing to function."[14] Since order emerges from noise and noise is an "aleatory aggression" directed toward any system or organizational structure within a system, self-organization necessarily entails "organizational chance." If the aleatory cannot be integrated or assimilated, the system is in danger of dissolving. In this case, noise is, obviously, destructive for the system. If, however, the aleatory is effectively appropriated, noise can be creative. Accordingly, the system displays "the capacity to utilize aleatory phenomena, to integrate them into the system, and to make them function as positive factors, creators of order, structures, functions."[15] The process of appropriation results in an internal proliferation of differences, which, when integrated, form a more complex system.

Atlan's use of communication and information theory to understand biological organization leads to three insights that illuminate the dynamics of emerging complexity. First, complexity is "composed of a great number of parts interconnected in multiple ways"; second, complexity is an *emergent* phenomenon whose occurrence cannot be accurately predicted; and third, the negentropic processes at work in living organisms "produce an evolution apparently *oriented* toward more complexity."[16] In his later book *Entre le cristal et la fumée,* Atlan identifies the state between rigid structure (the crystal) and vanishing structure (smoke) as the domain of living organisms. This interstitial condition marks the moment of complexity.

IMPLICATIONS OF COMPLEXITY

It should not be surprising that there is no simple definition of complexity. One of humankind's most ancient dreams is to reduce complexity to simplicity. If the foundation of reality is simple, complexity is, in the final analysis, penultimate. The desire for simplicity has long inspired efforts to explain the world in terms of simple systems that function smoothly and simple laws that can be reduced to simple equations. While advertised as so-called hard science, this vision is actually metaphysical or even theological. It is notable that Newton, for example, wrote more theological treatises than scientific works. In science as well as theology, Newton adheres to the fundamental tenet of all major Western religions—monotheism. God, he believes, is one, or, alternatively, oneness is God. If ultimate reality is one, simplicity is epistemologically truer and ontologically more real than complexity. From this point of view, to remain entangled in the complexities of life is to be trapped in a world of error and illusion. Obviously, the religious belief in simplicity does not die easily; indeed, for many people, the more complex the world becomes, the more they long for simplicity.

But things are never simple or are never merely simple; complexity, we are discovering, is inescapable. The word *complex* means "consisting of interconnected or interwoven parts; composite; compound; involved or intricate, as in structure." Etymologically considered, *complexity* derives from the past participle of the Latin *complectere, complexus,* which means to entwine together (*com-*, together + *plectere,* to twine braid). The stem *plek* (to plait) forms the Latin suffix *-plex,* which means to fold. Complexity, then, is formed by interweaving, interconnecting, and folding together different parts, elements, or components. Complexity not only harbors multiple implications but is actually an intricate process of implication; complication implicates and implications complicate.[17]

While simplicity has been privileged throughout the Western tradition, in recent years there has been a growing interest in the issue of complexity. One of the reasons for the preoccupation with the problem of complexity is surely the rapid development of information technology during the past five decades. High-speed computers and recent advances in parallel distributed processing now make it possible to perform calculations that disclose previously unimaginable orders of complexity.[18] The impact of technology, however, extends far beyond the computer lab. The rapid spread of information and telecommunication technologies is making increasing complexity an unavoidable part of everyday life for more and more people throughout the world. As complexity grows, the need to understand the implications of its dynamics becomes urgent.

Information technology has a significant impact on the definition as well as the investigation of complexity. In the 1960s, three people working independently—one Russian, Andrei N. Kolmogorov, and two Americans, Ray Solomonoff and Gregory Chaitin—developed a computational definition of complexity. Complexity, they argued, can be understood in terms of "algorithmic compression." An algorithm is, of course, a rule used in calculation, which can serve as a program for computing. According to Chaitin, "The complexity of something is the size of the smallest program which computes it or a complete description of it. Simpler things require smaller programs."[19] To illustrate this point, it is instructive to consider two sequences of letters:

1.   A B C D A B C D A B C D
2.   Z K T D Q R F Z E A R X

Intuitively, the first sequence appears to be less complex than the second. "This elementary observation," John Casti points out, "forms the basis for most characteristics of the complexity of an object: The complexity is directly proportional to the length of the shortest possible description of that object. As a corollary, we can give a rather clear-cut condition for something to be random (i.e., maximally complex): A string of letters is random if there is no rule for generating it whose statement is appreciably shorter—that is, requires fewer letters to write down—than the string itself. So an object or pattern is random if its shortest possible description is the object itself. Another way of expressing this is to say that something is random if it is *incompressible.*"[20] When understood in this way, complexity and compressibility are indirectly proportional: the less compressible, the more complex, and vice versa. The first sequence of letters is compressible because it displays redundancy, while the second sequence is incompressible because it is random. The title of Chaitin's book, *Information, Randomness, and Incompleteness,* suggests that the investigation of the relationship between redundancy and information can help to clarify the notion of complexity. If the insights of information theory are combined with the notion of algorithmic complexity, it is possible to conclude: *the less redundancy, the less compressibility; the less compressibility, the more complexity; the less redundancy, the more information; the more information, the less compressibility, and thus the more complexity.* This is why algorithmic complexity is sometimes also defined as "algorithmic information content."[21]

The relation of algorithmic complexity and algorithmic information content to the widely accepted definition of complexity as "consisting of interconnected or interwoven parts," or to Atlan's variation of this insight when he describes complexity as "composed of a great number of parts interconnected in multiple ways," is not immediately obvious. In both of these formulations, complexity is directly related to connectivity. As connections or interconnections proliferate, complexity expands and, correlatively, information increases. The logic of the relation between algorithmic information content and complexity, understood in terms of connections among multiple parts, begins to emerge when we recall that information can be defined as a difference that makes a difference. Inasmuch as information is differential, it increases with an increase in differences. Differences, however, can multiply only as interconnections grow. The greater the connectivity, the more the differences, and the more the differences, the more the information and the greater its complexity. Differences that are not indifferent, we have discovered, are constituted and sustained by interconnections, which presuppose reciprocal relations. These relations entail both the reciprocity between and among parts and between parts and the whole. While parts are constituted through their mutual relations as well as their relation to the whole, the whole emerges from the interplay of the parts. This mutual implication of parts with parts and parts with whole generates the condition of complexity.

The relational structure that forms the differences constituting complexity transforms part and whole into an open system. Any theory of complexity must, therefore, also be a theory of systems. In 1950, Ludwig von Bertalanffy published an article entitled "The Theory of Open Systems in Physics and Biology," which, during the ensuing decade, gave rise to a field known as systems theory. Von Bertalanffy summarizes the aim of this new area of inquiry in his 1968 book, *General System Theory: Foundations, Development, Applications*:

These considerations lead to the postulate of a new scientific discipline, which we call general systems theory. Its subject matter is the formulation of principles that are valid for "systems" in general, whatever the nature of their component elements and the relations or "forces" between them.

General system theory, therefore, is a general science of "wholeness" which up until now was considered a vague, hazy, and semi-metaphysical concept. In elaborate form it would be a logico-mathematical discipline, in itself purely formal but applicable to the various empirical sciences.[22]

Von Bertalanffy traces the origin of this new discipline to what he describes as "the organismic revolution," which results from a change of focus from the physical to the biological sciences. The "core" of this revolution is "the notion of *system*." In a manner reminiscent of the shift from mechanistic to organic concepts and metaphors at the end of the eighteenth century, von Bertalanffy argues that classical atomistic and mechanistic models are not adequate for understanding living organisms. As a result of technological advances in the twentieth century, the line between machines and organisms must be redrawn.

The 19th and first half of the 20th century conceived of the world as chaos. *Chaos was the oft-quoted blind play of atoms which, in mechanistic and positivistic philos-*

ophy, appeared to represent ultimate reality, with life as an accidental product of physical processes and mind as an epiphenomenon. . . . Now we are looking for another basic outlook—the world as organization. . . . *This trend is marked by the emergence of a bundle of new disciplines such as cybernetics, information theory, general system theory, theories of games, of decisions, of queuing and others; in practical application, systems analysis, systems engineering, operations research, etc. They are different in basic assumptions, mathematical techniques and aims, and they are often unsatisfactory and sometimes contradictory. They agree, however, in being concerned in one way or the other with "systems," "wholes" or "organization"; and in their totality they herald a new approach.*[23]

While claiming to reformulate metaphysical concepts in scientific terms, general system theory actually perpetuates the dream of oneness and unity, which is inseparable from Western metaphysics. According to von Bertalanffy, the longing for oneness expresses itself in his effort to establish the unity of individual sciences and to integrate the natural and social sciences. He believes such unification and integration are possible because "systems display structural similarities or isomorphisms in different fields." For von Bertalanffy, as for Hegel, systems are not merely epistemological constructs but are also ontological realities. As systems theory becomes reflexive, it bends back on itself to systematize theoretical reflections in different fields. "Developing unifying principles running 'vertically' through the universe of the individual sciences," von Bertalanffy concludes, "this theory beings us nearer the goal of the unity of science."[24]

Although complexity theory differs from general system theory in ways that will become evident, it nonetheless shares some of its guiding ambitions. Like general system theory, complexity theory attempts to identify common characteristics of diverse complex systems and to determine the principles and laws by which they operate. Moreover, students of complexity share the conviction that the systems they investigate are not limited to natural phenomena but can also be discerned in social, economic, political, and cultural life. In a prescient series of lectures delivered at the Massachusetts Institute of Technology during the turbulent spring of 1968, Herbert A. Simon explored what he described as "The Architecture of Complexity." His definition of complex systems is still quite useful. A complex system, he argues, is "one made up of a large number of parts that interact in a nonsimple way. In such systems, the whole is more than the sum of the parts, not in an ultimate, metaphysical sense, but in the important pragmatic sense that, given the properties of the parts and the laws of their interaction, it is not a trivial matter to infer the properties of the whole."[25] It is obvious that Simon's cautious formulation is designed to hedge his bets on several critical issues. Though the whole is more than the sum of its parts, it remains undecidable whether or not the principles and operations of a complex whole can be reduced to the parts from which they emerge. It is, nevertheless, clear that the interactions of the parts that comprise the whole are "nonsimple." In the years since Simon posed his argument, extensive research in a variety of fields has resulted in a considerably more sophisticated understanding of complex systems. While the multiple branches of these investigations cannot be reduced to simple insights, common threads running through different arguments can be identified. In an extensive survey of the history of evolution, David Depew and Bruce Weber provide a definition of complex systems that is unusually concise yet very comprehensive:

Complex systems are not just complicated systems. A snowflake is complicated, but the rules for generating it are simple. The structure of a snowflake, moreover, persists unchanged, and crystalline, from the first moment of its existence until it melts, while complex systems change over time. It is true that a turbulent river rushing through the narrow channel of rapids changes over time too, but it changes chaotically. The kind of change characteristic of complex systems lies somewhere between the pure order of crystalline snowflakes and the disorder of chaotic or turbulent flow. So identified, complex systems are systems that have a large number of components that can interact simultaneously in a sufficiently rich number of parallel ways so that the system shows spontaneous self-organization and produces global, emergent structures.[26]

With these insights in mind, it is possible to identify the following characteristics of complex systems:

1. Complex systems are comprised of many different parts, which are connected in multiple ways.
2. Diverse components can interact both serially and in parallel to generate sequential as well as simultaneous effects and events.
3. Complex systems display spontaneous self-organization, which complicates interiority and exteriority in such a way that the line that is supposed to separate them becomes undecidable.
4. The structures resulting from spontaneous self-organization emerge from but are not necessarily reducible to the interactivity of the components or elements in the system.

5. Though generated by local interactions, emergent properties tend to be global.
6. Inasmuch as self-organizing structures emerge spontaneously, complex systems are neither fixed nor static but develop or evolve. Such evolution presupposes that complex systems are both open and adaptive.
7. Emergence occurs in a narrow possibility space lying between conditions that are too ordered and too disordered. This boundary or margin is "the edge of chaos," which is always far from equilibrium.

E - M E R G E N C E

Systems operating far from equilibrium are typically nonlinear. Microscopic and macroscopic operations and events are implicated in loops that involve both negative and positive feedback. Left to itself, negative feedback tends toward equilibrium by counterbalancing processes, which, if unchecked, can destroy the system. As we have seen, negative feedback is most obvious in cybernetic systems regulated by various governing devices. Positive feedback, by contrast, tends to disrupt equilibrium by increasing both the operational speed and the heterogeneity of the components of a system. When positive feedback increases the speed of interaction among more and more diverse components, linear causality gives way to recursive relations in which effects are disproportionate to the causes from which they emerge. In order to understand the dynamics of emerging complexity, it is instructive to return to the paintings of Chuck Close and to reexamine them though the operations of cellular automata.

Cellular automata were first proposed by von Neumann in a 1948 lecture, "General and Logical Theory of Automata." Drawing on Warren McCulloch and Walter Pitts's work on neural networks, von Neumann argued that computers and biological systems process data in analogous ways. To support this claim, he and his colleague Stanislaw Ulam developed the concept of cellular automata, which operate on themselves with iteratively recursive rules. A cellular automaton is "a computer program or piece of hardware consisting of a regular lattice or array of cells. Each cell is assigned a set of instructions by means of an algorithm that tells it how to respond to the behavior of adjacent cells as the automaton advances from one discrete step to the next. Cellular automata are inherently parallel computing devices."[27] The lattice structuring the cells takes the form of a grid. In the absence of any programmatic direction or overall design, each cell evolves according to simple rules that respond to altering circumstances

FIGURE 23

Cellular

Automata

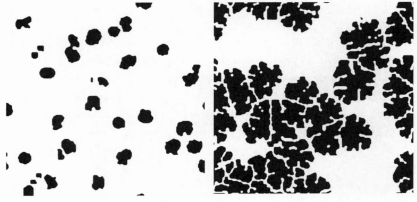

created by changes in surrounding cells. As the cells interact, complex forms begin to emerge. In some cases, these patterns appear to be remarkably lifelike (fig. 23).

While von Neumann was the first to propose cellular automata and to prove they can, in principle, reproduce, he never actually made a self-replicating program or machine. As computers became more powerful, many theorists and mathematicians who were fascinated by von Neumann's speculations attempted to produce cellular automata. It was not, however, a sophisticated computer program but a remarkably simple game that disclosed the far-reaching implications of von Neumann's discovery. In 1968, John Conway, a mathematician who was then at Cambridge University, became fascinated by a game involving cellular automata.[28] After repeatedly playing the game, he was convinced that von Neumann's proof, which postulated 200,000 cells that could be in one of twenty-nine states, could be greatly simplified. Conway put his theory into practice by creating the Game of Life, whose rules are as simple as its results are complex. The space of the game is a grid with each square forming a cell that is either occupied or empty. Every cell is governed by rules, which determine the parameters for responding to the state of surrounding cells. In contrast to the crystalline structure of a snowflake, local interactions of relatively simple rules, Conway discovered, produce complex dynamic global patterns that emerge, evolve, and disappear unpredictably. Context functions as the constraint necessary for the articulation of complex forms. When the rules of the game are followed, the first shapes to emerge settle into patterns rather quickly. Conway and his colleagues named these initial forms: block, ship, longboat, beehive, loaf, canoe, and pond. Depending on the initial configuration, other patterns, labeled toads, blinkers, and traffic lights, might eventually emerge and oscillate at periodic intervals be-

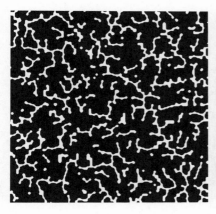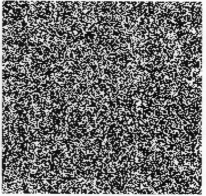

tween different shapes.[29] The most intriguing forms are gliders, which display intricate patterns that change in seemingly unpredictable ways. As gliders intersect and collide, they sometimes devour each other and sometimes produce new gliders, which, in turn, morph unexpectedly. When the Game of Life is played, complex networks of mobile forms bearing an uncanny resemblance to living organisms gradually emerge. If played on computers, these shapes move like strange creatures floating in some primeval biotic soup.

The grids of Close's portraits, formed by paintings within paintings, resemble the squares in the Game of Life. Each miniature painting functions like a cellular automaton with its own rules programmed to respond to surrounding squares. These local interactions generate global events. As the viewer moves closer or farther away, the painted surface is set in motion until it is teeming with mutating shapes. When the critical tipping point is reached, the resolved organic form of the face suddenly emerges.

The interplay between information and form displayed in cellular automata can be understood in ways that suggest certain similarities to living organisms. On the one hand, life appears to be informational, and, on the other hand, information seems to be lifelike.[30] In the early 1970s, Chris Langton, a college dropout who was working as a conscientious objector at the Massachusetts General Hospital in Boston, became fascinated with the Game of Life. He saw in cellular automata rudimentary components he thought could be used to create a computer, which would include data storage, information processing, and computational capabilities. Langton was not particularly interested in building an all-purpose computer but was intrigued by the possibility of making a computational machine with the properties of life. By the late 1980s, his work had given rise to a field bearing the paradoxical name "artificial life." Recognizing that cellular automata are remarkably similar to nonlinear dynamic systems, Langton set out to evolve life in a von Neumann universe. In 1984, working at the University of Michigan under John Holland, who had been a student of von Neumann, and in collaboration with physicist Stephen Wolfram, Langton made an important discovery. Wolfram had found that cellular automata could be classified according to four typical patterns of behavior: (1) rigid structures that do not change; (2) oscillating patterns that change periodically; (3) chaotic activity that exhibits no stability; and (4) patterns that are neither too structured nor too disordered, which emerge, develop, divide, and recombine in endlessly complex ways. Langton found the fourth category of automata particularly interesting. Though the principles governing behavior in this intermediate domain remained obscure, he suspected activity in this region would be characterized by "a phase transition between highly ordered and highly disordered dynamics, analogous to the phase transition between the solid and fluid states of matter."[31] As Langton studied the dynamics of phase transition in more detail, he discovered that the four patterns of behavior fall into a regular sequence. The surprising activity occurring between order and disorder Langton described as "complex."

. . . Order → Complexity → Chaos → Complexity → Order . . .

The domain of complexity, in other words, lies *between* order and disorder or, in Langton's preferred description, "at the edge of chaos."

This account of complexity in terms of "an ordered regime, a chaotic regime, and a transition regime" has proven remarkably suggestive for understanding the operation of information systems as well as for interpreting the dynamics of living organisms. Summarizing Langton's conclusions, Stuart Kauffman writes: "Just between, just near this phase transition, just at the edge of chaos, the most complex behavior can occur—orderly enough to ensure stability, yet full of flexibility and surprise. Indeed, this is what we mean by complexity."[32] Having become persuaded that life takes shape between too much and too little order, Kauffman was drawn into an examination of precise dynamics of emergence. During the late 1980s and early 1990s, Langton and Kauffman became deeply involved in the debates about complexity that were taking place at the recently founded Santa Fe Institute. One of the most insightful and provocative participants in these discussions was Langton's erstwhile teacher, John Holland, whose seminal work on genetic algorithms led to a sustained investigation of what he describes as the phenomenon of emergence. The problem of emergence raises the ancient question of the interrelation of parts and whole, which we have already encountered. In his recent book, *Emergence: From Chaos to Order,* Holland explains: "Emergence . . . occurs only when the activities of the parts do *not* simply sum to give the activity of the whole."[33] So understood, emergence refers to phenomena as different as Gehry's architecture, Close's portraits, or the Game of Life, in which "the interactions among system components generate unexpected global system properties not present in any of the subsystems taken individually."[34] Not content with a phenomenological description of emergence, Holland attempts to define the rules by which it occurs. Components or agents interacting according to recursive rules generate metarules whose operation can form natural, social, and cultural systems.[35]

Kauffman borrowed Holland's understanding of emergence to develop an account of biological organisms. To explore how emergence takes place, he imagined an experiment using a network of light bulbs connected to each other, with each bulb turned on or off by inputs from the bulbs connected to it, following a set of Boolean rules. Kauffman describes his procedure:

We might, for example, study the pool of networks with 1,000 bulbs (we'll call this variable N) and 20 inputs per bulb (the variable K). Given N = 1000 and K = 20, a vast ensemble of networks can be built. We sample this ensemble by randomly assigning to each of the 1,000 bulbs 20 inputs and, again at random, one of the possible Boolean functions. Then we can study the network's behavior, counting the number of attractors, the lengths of attractors, the stability of attractors to perturbations and mutations, and so forth. Throwing the dice again, we can randomly wire another network with the same general characteristics and study its behavior.[36]

After running many simulated trials of such networks, in which he varied *N* and *K,* Kauffman discovered the emergence of the same patterns Wolfram and Langton had found in their work with cellular automata. When *K* = 1, the networks fell into a short "state cycle," and often into a frozen pattern of illumination. With values of *K* as low as 4 or 5, the networks displayed chaotic behavior: their attractors had long state cycles and were unstable. In other words, as the lightbulbs blinked on and off, one pattern after another would follow its predecessor, without repetition; and interference in the network, by flipping a bulb on or off, or by changing the Boolean rule for one bulb, would set the network on a completely different evolutionary path. When *K* = 2, however, chaos gives way to the emergence of order. This critical point of transition occurs when the network or networks within the network reach a condition that Kauffman labels "combinatorial optimization." If connections are too few, networks are frozen and no change occurs, and if connections are too many, there is no stability and networks remain chaotic. Along the margin between too little and too much connection, "the spontaneous emergence of self-sustaining webs" occurs.[37] Since these webs are not preprogrammed but are "randomly assembled," Kauffman describes the result of emergence as "order for free." Though unplanned and unpredictable, he is convinced such order is not incomprehensible.[38]

When a network or system reaches the condition of combinatorial optimization, there is a "combinatorial explosion."[39] This critical transition takes place at the tipping point where quantitative change suddenly leads to qualitative change.[40] Kauffman suspects that in biological organisms, "communities of cells might evolve to the subcritical-supracritical boundary where major changes can occur."[41] The easiest way to understand the dynamics of the tipping point is to imagine a sand pile to which individual grains of sand are gradually added. As the grains accumulate one by one, the pile approaches a critical state in which the addition of one more grain of sand can unleash an avalanche in the pile as a whole. Physicist Per Bak has analyzed such events in great detail and has developed a theoretical explanation of what he describes as "self-organized criticality." Bak goes so far as to propose a new "science of self-organized criticality," which will disclose previously unsuspected laws that illuminate the nonlinear dynamics of complex systems. "Complex behavior in nature," Bak argues,

reflects the tendency of large systems with many components to evolve into a poised, "critical" state, way out of balance, where minor disturbances may lead to events, called avalanches, of all sizes. Most of the changes take place through catastrophic events rather than by following a smooth gradual path. The evolution of this very delicate state occurs without design from any outside agent. The state is established solely because of the dynamical interactions among individual elements of the system: the critical state is self-organized. Self-organized criticality is so far the only known general mechanism of complexity.[42]

Four points in this description of self-organized criticality must be emphasized at this juncture. First, as the term implies, self-organized criticality displays the typical features of other self-organizing systems. The state results from interactions among components rather than from the intervention of any external forces or agents.[43] Second, in the state of self-organized criticality, nonlinear events can have effects disproportionate to their causes. Third, the dynamic interactions among individual elements of the system generate global events that require a holistic description, which cannot be reduced to an account of the individual elements. Finally, at the tipping point, the effect of individual events is unpredictable. While it is possible to know that at some point an avalanche will occur, it is never possible to be sure which particular grain of sand will tip the balance and thereby upset the equilibrium.

There is considerable disagreement about how this unpredictability is to be understood. While Casti, for example, equates unpredictability with "chance as a cause,"[44] Kauffman, by contrast, insists that unpredictability notwithstanding, it is possible to describe phase transitions accurately with mathematical formulas. Bak, perhaps predictability, takes a middle position by insisting that in a sense both of these alternatives are correct. "In hindsight one can trace the history of a specific large avalanche that occurred" in such a way that the event seems, if not necessary, at least not surprising. However, citing the criticism his fellow Dane, Kierkegaard, leveled at Hegel's system, Bak observes: "Life is understood backward, but must be lived forwards."[45] When looking ahead rather than behind, unpredictability can never be overcome. Bak does not mean to suggest that lawfulness and chance are merely two different perspectives on the same events; his point is considerably more subtle and complex. Law and chance are not exclusive opposites but are folded into each other in a way that makes them inseparable. "Randomness," Bak avers, "does not preclude the system's evolving to the delicate critical state, with well-defined statistical properties."[46] Derrida might well be commenting on Bak's principle of self-organized criticality rather than Mallarmé's *Coup de dés* when he writes: "The game here is the unity of chance and rule, of program and its leftovers or extras."[47] This game of chance, which does not rule out necessity, is, of course, the game of life.

Bak's passing reference to Kierkegaard harbors implications that are more suggestive than he realizes. The phase shift that Bak interprets in terms of self-organized criticality is structurally similar to Kierkegaard's moment of decision. Kierkegaard distinguishes three different forms of life, which he describes as the "stages on life's way." The stages, which are characterized by different principles governing thought and rules guiding conduct, are separated by gaps where change occurs unpredictably. Since there is no necessary relation between or among the aesthetic, ethical, and religious forms of life, one must move from one to the other by free decision or, in Kierkegaard's famous phrase, by a "leap of faith." Decision, as the word suggests, is a cutting off (*de,* off + *caedere,* to cut). In the moment of decision, as in the moment of complexity, some possibilities are realized and others are cut off. As possibilities are actualized, new patterns, which both impose new constraints and open new possibilities, emerge. Though free decisions are always unpredictable, they are never independent of a certain determinism. Decisions, therefore, are the outcome of the interplay of fate and freedom; circumstances beyond one's control bring one to a crossroads where a decision *must* be made. As the moment of decision draws near, certainty becomes a vague memory and equilibrium remains a distant dream. This decisive crossroads is what theorists of chaos and complexity call a "bifurcation point" where two distinct alternatives or "choices" are open to a system. In his analysis of nonequilibrium thermodynamics, Prigogine demonstrates that there is a direct relationship between the bifurcation point and symmetry-breaking in systems that had seemed to be in balance. Systems far-from-equilibrium approach the point of transition between two alternative patterns or orders, where change occurs when the organizational symmetry of the structure dissolves. Erich Jantsch explains the dynamics of this phase shift:

At each transition, two new structures become spontaneously available from which the system selects one. Each transition is marked by a new break of spatial symmetry. The path which the evolution of the system will take with increasing distance from thermodynamic equilibrium and which choices will be made in the branchings can-not be predicted. The farther the system moves away from thermodynamic equilibrium, the more numerous become the possible structures. The possible paths of evolution resemble a decision tree with branchings at each instability threshold. The obvious analogy to the macroscopic process of an autopoietic structure . . . poses the question whether a holistic description will become available here, a description not only of a single structure, but of the total system evolution through many structures. [48]

While the uncertainty associated with the "choice" or "decision" occurring at the bifurcation point remains irreducible, the realization of one rather than another alternative reduces uncertainty and thereby increases information. Once symmetry is shattered and equilibrium lost, bifurcations can cascade until they threaten to spin out of control. The more bifurcations proliferate, the more decisions become possible, and the more decisions are made, the greater the changes in the system. If a new order is not constituted from cascading bifurcations and broken symmetries, the system dissolves into chaos. If, however, equilibrium is reestablished, a more complex order emerges. Poised at the bifurcation point, it is impossible to predict the consequences of the choice of one rather than another alternative. Though we know decisions have consequences, we never know the consequences of a particular decision before the leap occurs. As Kierkegaard repeatedly stressed, no amount of retrospective reflection can erase the uncertainty of the moment of decision.

The moment of decision is always a moment of complexity in which old patterns give way to new. "Order parameters," J. A. Scott Kelso explains, "are found near nonequilibrium phase transitions, where loss of stability gives rise to new or different patterns and/or switching between patterns."[49] The development from one to another form, pattern, structure, or system to another is marked by "punctuated equilibrium."[50] This punctuation is the tipping point creating the gap between orders that are always less stable than they appear. Far from a simple void, such gaps are highly complex. Inasmuch as the moment of complexity lies between different orders, the border is a border within a border. Like images mirroring each other in strange loops, borders, margins, frames, and edges fold into each other to create endless complexity. The dynamics of emergence are complex and what emerges in networks that are edgy inevitably bears the trace of complexity.

Complexity can emerge in literary works as well as biological systems. The tale Andrew Crumey spins in *Pfitz* emerges from the subtle interweaving of multiple strands, which results in a narrative that is coherent but not precisely unified. Near the end of the book, Crumey, or one of the authors he authors, confesses: "I see now that there is no Author—or else, there are many Authors, so that our story was not one but many." A few lines later, he concludes the novel with a "Chorous Logico-Philosophicus," which parodies Wittgenstein while suggesting the point of the tale the authors have spun:

Emergent complexity
Bear us aloft!

Pfitz is not just *about* emergent complexity but is a brilliant enactment of it. One of the strategies Crumey and his coauthors use to generate complexity is to create multiple self-reflexive loops by folding authors and readers into each other until the line separating them becomes obscure. At one point, a dialogue between Pfitz and his master is interrupted by what seems to be a direct address to the reader reminiscent of the opening lines of Calvino's *If upon a winter's night a traveler*. But what initially appears to be an exchange between Author and Reader turns out to be the Author's account of a conversation about the novel between a woman and a man, who are characters in the novel. When the unnamed man insists that "the purpose of a book is not simply to go from one place to another, from a beginning to an end," the woman responds: "The Author says that if his story is to resemble the world in any way at all, then it must be formless and without logic, proceeding randomly from one moment to the next. Then gradually, patterns will emerge which may or may not indicate events, ideas or actions."[51]

When we return to the story after this interlude, Pfitz is explaining the circumstances of his conception to his master. His mother, he reports, came from a country village where nothing ever happens. One day when she was sixteen, a traveling fair came to town. In addition to the usual magicians, freaks, jugglers, and fire-eaters, the fair's main attraction was Fernando, "the incredible Bee-King." Deeply impressed by the bees that had been trained to perform all kinds of remarkable feats, his mother lingered near the tent to question Fernando. In response to her persistent inquiry, the Bee-King took her back into the tent and showed her a special trick he rarely performed in public. Removing the cover from the glass box full of bees, he stepped back, raised his arms, and waited. The bees

were beginning to fly out; only a few at first, but soon there was a great swarm of them buzzing in the air, some with their little paper angel wings still attached, a great swirling cloud of stirring bees. And they were all flying towards the head of Fernando, circling and landing, crawling, and searching. A mass of them was forming on his face, his hair; a great beard of them hung from his chin, and the place towards which they were moving, as if led by some unseen force, was his open mouth. They were jostling on his tongue; some were pushed aside and would fly back to try again. In they were all going—hundreds, thousands of them into the mouth of Fernando and down his throat into their human hive, so that you could hear the humming inside.[52]

Though neither Crumey nor his author ever says so directly, this "human hive" illustrates the dynamics of emergent complexity that structures the novel *as a whole.*

Swarms of bees, flocks of birds, schools of fish, and colonies of ants fascinate researchers who attempt to understand complexity.[53] The reasons for this fascination are clear: the same rules and principles operating in cellular automata seem to be at work in these natural phenomena. If one were to animate the Game of Life or to make a Chuck Close painting into a video, it would behave very much like a swarm of bees or flock of birds. At the third Workshop on Artificial Life, held at the Santa Fe Institute in June 1992, Mark M. Millonas, from the Complex Systems Group at the Center for Nonlinear Studies in the Los Alamos National Laboratory, presented a paper entitled "Swarms, Phase Transitions, and Collective Intelligence." "The swarming behavior of social insects," Millonas explains,

provides fertile ground for the exploration of many of the most important issues encountered in artificial life. Not only do swarms provide the inspiration for many recent studies of the evolution of cooperative behavior, but the action of the swarm on a scale of days, hours, or even minutes manifests a nearly constant flow of emergent phenomena of many different types. Models of such behavior range from abstract cellular automata to more physically realistic computational simulations. The notion that complex behavior, from the molecular to the ecological, can be the result of parallel local interactions of many simpler elements is one of the fundamental themes of artificial life. The swarm, which is a collection of simple locally interacting organisms with global adaptive behavior, is a quite appealing subject for the investigation of this theme.[54]

Millonas attempts to develop a mathematical model of "the collective behavior of a large number of locally acting organisms," which "move probabilistically between local cells in space."[55] Drawing inspiration from E. O. Wilson's work on ant colonies, Millonas develops an analysis of swarms of bees as well as other collective phenomena. Swarms provide a graphic illustration of emergent self-organized behavior; indeed, it is impossible to understand swarms without understanding the principles of emerging complexity. Like Crumey's authors who work individually yet create a coherent story, bees form swarms with dynamic patterns even though no antecedent plan or program directs individual behavior. The swarm as a whole operates according to a logic that cannot be discerned in any of the activities of individual bees. Complex behavior, Millonas argues, "is the result of the *interactions between* organisms as distinct from behavior that is a direct result of the actions of *individual* organisms."[56] Belgian chemist Jean-Louis Deneubourg formulated the principle governing these interactions and named it "allelomimesis." In an effort to understand the behavior of wasps recorded by E. O. Wilson, Deneubourg extended Prigogine's theory of nonlinear self-organization to the activity of insects.[57] In allelomimetic behavior, the conduct of each individual is influenced by the activity of its neighbors. When Deneubourg modeled the behavior of cellular automata according to the principle of allelomimesis, he discovered that even relatively simple interactions can generate surprisingly complex behavior. While no member or component of the group has any knowledge of access to the principles of operation for the system as a whole, each individual is responsive to the actions of surrounding individuals. In a swarm, for example, there is no pilot yet bees are able to fly in a stable formation. Local interactions establish lines of communication that issue in the emergence of complex global behavior. According to Millonas, such communication throughout the system creates "collective intelligence." The buzz of the swarm is, in effect, noise flying in formation.

Inasmuch as swarms form communication networks, the logic of the swarm illuminates the logic of networking. Millonas uses connectionist models derived from the study of neural networks to understand swarm networks. He summarizes three basic characteristics of connectionist networks:

[1] *Their **structure** consists of a discrete set of nodes, and a specified set of connections between the nodes. For example, neural networks, the archetypal connectionist systems, are composed of neurons (nodes), and the neurons are usually linked by synapses (connections).*

[2] *There are **dynamics** of the nodal variables. . . . The dynamics are controlled by the connection strengths, and the input-output rule of the individual neurons. The dynamics of the whole system is the result of the interaction of all the neurons.*

[3] *There is **learning**. In its most general sense, learning describes how the connection strengths, and hence the dynamics, evolve. In general, there is a separation of time scales between dynamics and learning, where the dynamical processes are much faster than the learning processes.*[58]

According to this model, networks consist of interconnected nodes, which are able to communicate with each other. Each node is constituted by its interrelations with other nodes and its place in the overall network. A node, as the word implies (*nodus,* knot; from *ned,* twist, tie, knot), is a knot in a web of relations. Knots function like switches and routers that send, receive, and transmit information throughout the network. Separation and connection, like identity and difference, are mutually constitutive. The ways in which connections intersect create the distinctive traits and functions that differentiate nodes. While the connections of each node ramify throughout the network, the relations that are most decisive are relatively localized. Communication-at-a-distance is always possible but often not necessary. As we have already discovered, if there are too few connections, the network freezes, and if there are too many, it becomes chaotic. Since the interrelations of nodes are both reciprocal and many-to-many, feedback loops can be both negative and positive. In the former case, interactions check and balance activity, and in the latter case, they tend to accelerate the operation of the system and further diversify its components. Insofar as local interactions generate global behavior, the network as a whole is a network of networks. The web of nodes forms a *distributed* network, which is radically *decentered*. Operations do not have to be ordered sequentially but can run in parallel. In the interactions of this distributed network, complexity is always emerging and emergence is inevitably complex.

With this understanding of the way in which complexity emerges through networking, we have solved one of the problems creating our current critical impasse. At the end of chapter 2, I claimed that any adequate interpretation of emerging network culture must be able to describe the nonlinear dynamics of systems that act as a whole but do not totalize. One of the primary reasons for the critical emergency we are facing is the insistence of deconstructive critics that systems and structures inevitably totalize and thus necessarily exclude otherness and repress difference. Having reached this conclusion, critics endlessly repeat the same point. The paradoxical result is that deconstruction's solicitation of difference ends up repeating the same gesture of totalization it condemns in others. This strategy ends in an interminable mourning that leaves differences fragmented without any hope of significant change. After considering the logic of networking, it should be clear that systems and structures—be they biological, social, or cultural—are more diverse and complex than deconstructive critics realize. Emergent self-organizing systems *do* act as a whole, yet *do not* totalize. Furthermore, emergence involves an irreducible unpredictability that creates the opportunity for aleatory events. Phase transitions occur at thresholds or margins, which both make networks possible and leave them open and thus permeable. Far from repressing differences, global activity increases the diversity upon which creative and productive life depends.

Though this understanding of complexity and the dynamics of complex systems marks a significant advance in the interpretation of emerging network culture, it is not yet sufficiently complex. The folds of networks and webs in which complexity emerges are even more convoluted than this analysis suggests. As John Holland observes: "Persistent emergent phenomena can serve as components of more complex emergent phenomena."[59] Always emerging in networks of local area networks, complexity develops and evolves over time. This insight leads Millonas to identify two additional principles of "swarm intelligence." The first is "the **principle of stability**":

The group should not shift its behavior from one mode to another at every fluctuation of the environment, since such changes take energy, and may not produce a *worthwhile return for the investment. The other side of the coin is the **principle of adaptability**. When rewards for changing a behavioral mode are likely to be worth the investment in energy, the group should be able to switch. The best response is likely to be a balance between complete order and total chaos, and, therefore, the level of randomness in the group is an important factor. Enough noise will allow a diverse response, while too much will destroy any cooperative behavior.[60]*

Emerging self-organizing systems are *complex adaptive systems*. For complex systems to maintain themselves, they must remain open to their environment and change when conditions require it. Complex adaptive systems, therefore, inevitably evolve, or, more accurately, coevolve. As the dynamics of evolving complexity are clarified, it not only becomes apparent that complex adaptive systems evolve, but it also appears that the process of evolution is actually a complex adaptive system.

six

Thus every Part was full of Vice,
Yet the whole Mass a Paradise;
Flatter'd in Peace, and fear'd in Wars,
They were th'Esteem of Foreigners,
And lavish of their Wealth and Lives,
The Balance of all other Hives.
Such were the Blessings of that State;
Their Crimes conspir'd to make them Great:
And Virtu, who from Politicks
Had learn'd a Thousand Cunning Tricks,
Was, by their happy Influence,
Made Friends with Vice: And ever since,

The worst of all the Multitude
Did something for the Common Good.
      This was the State's Craft that maintain'd
The Whole of which each Part complain'd:
This, as in Musick Harmony,
Made Jarrings in the main agree;
Parties directly opposite,
Assist each other, as 'twere for Spight;
And Temp'rance with Sobriety,
Serve Drunkenness and Gluttony.

bernard mandeville, *the fable of the bees*

evolving complexity

ANT SMARTS

The remarkable success of *Gödel, Escher, Bach* is largely the result of Douglas Hofstadter's effective combination of simplicity and complexity in his discussion of a vast range of topics. While the precision of analysis and clarity of expression are surprisingly simple, the ideas explored and the overall structure of his argument are unusually complex. The book begins with an introductory "Musico-Logical Offering" and ends with a "Six-Part Ricercar." Though the argument never quite comes full circle, the beginning explains the end, and the end stages a performance of the beginning. Hofstadter opens his work with a report of a 1747 meeting between Frederick the Great, King of Prussia, and Bach, who was in Potsdam visiting his son, Carl Philip Emmanuel Bach, the choirmaster at the royal court. While testing the new Silbermann pianos in the palace, the elder Bach asked the king to give him a subject for a fugue. Bach then responded to the king's request to improvise, on different instruments, various forms of imitative counterpoint with an astonishing extemporaneous performance, but when asked for a fugue with six obligato parts, Bach demurred, saying that not all themes were suited for such a treatment. Upon his return to Leipzig, Bach feared he had disappointed Frederick and thus proceeded to compose a work, which he entitled *Musical Offering,* on the theme the king had proposed. A sheet accompanying the composition bore the following inscription:

Regis Iustu Cantio Et Reliqua Canonica Arte Resoluta.
(At the King's Command, the Song and the Remainder Resolved with Canonic Art.)[1]

The first letters of each word spell "RICERCAR," which, Hofstadter explains, is an Italian word meaning "to seek." "Ricercar" was the original name of the musical form known as the fugue. By the time Bach was composing, it designated "an erudite kind of fugue, perhaps too austerely intellectual for the common ear." Between the introductory "Musico-Logical Offering" and the concluding "Six-Part Ricercar," Hofstadter includes twenty chapters. Between each chapter, there is an unnumbered intervention, dialogue, offering, contrafactus, canon, or fugue. Just as Bach inscribes the structure of his composition in the prefatory inscription, so Hofstadter indirectly explains the structure of his book in the introductory remarks that frame the book. *Gödel, Escher, Bach* is, in effect, a musical composition made up of fugues, canons, and crab canons. To appreciate how the multiple registers of the book work together, it is important to understand what its different strands mean. A fugue is a polyphonic musical form in which themes are started sequentially, repeated, and developed contrapuntally. One of the most interesting properties of fugues, according to Hofstadter, is that each voice is "a piece of music in itself; and thus a fugue might be thought of as a collection of several distinct pieces of music, all based on a single theme, and all played simultaneously. And it is up to the listener (or his subconscious) to decide whether it should be perceived as a unit, or as a collection of independent parts, all of which harmonize." Given this interplay of distinct parts and whole, a fugue is something like a musical version of a Chuck Close portrait; or, conversely, a Close painting visually contrapuntal to a fugue. In a canon, the same melody is repeated by one or more voices overlapping in time. "In order for a theme to work as a canon theme, each of its notes must be able to serve in a dual (or triple, or quadruple) role: it must first be part of a melody, and secondly it must be part of a harmonization of the same melody. When there are three canonical voices, for instance, each note of the theme must act in two distinct harmonic ways, as well as melodically. Thus, each note in a canon has more than one musical meaning; the listener's ear and brain automatically figure out the appropriate meaning, by referring to context." The complexity of the canon increases as the pitch of different voices or "copies" is staggered, their speed varied, or the theme inverted by making the melody jump down wherever the original jumps up. The most complex canonical structure results from the inclusion of "retrograde copies" in which "the theme is played backward in time." Named after the creature that moves backward in space, this type of composition is known as a "crab canon." What Hofstadter finds so fascinating about Bach's *Musical Offering* is the way in which the different parts of the score work together as a whole. He alerts listeners: "Notice that every type of 'copy' preserves all information in the original theme, in the sense that the theme is fully recoverable from any of the copies. Such an information-preserving transformation is often called an *isomorphism*."

By weaving together rigorous analyses and imaginative interpretations, Hofstadter creates what is, in effect, a written version of Bach's polyphonic fugue. In the mathematics of Gödel, the art of Escher, and the music of Bach, he discerns isomorphisms which are not only mutually illuminating but also clarify natural and cultural phenomena ranging from genetic processes and brain functions to fairy tales and artificial intelligence. The lines of the text form what Hofstadter describes in the subtitle of his book as *An Eternal Golden Braid*. As the whimsical diagram of his "semantic network" suggests, the threads of this braid are not always as evident in the body of the chapters as they are in the interstices of the text (fig. 24). The margins of the book are where the insights that give the work its edge emerge in surprising ways.

The second part of the book, EGB, is a variation of the first, GEB. Like letters in an ancient alphabet or signs in a cryptic code, meaning shifts as signals are shuffled. Similar themes are explored in a different order to create new variations that resonate in different ways. Part two is prefaced by "Prelude," which is neither inside nor outside the text proper. True to its name, there is something ludic about the Prelude. The text consists of a dialogue among Tortoise, Achilles, Crab, and Anteater. Tortoise and Achilles are, of course, characters drawn from the Greek philosopher Zeno of Elea, Crab suggests Bach's crab canon, and Anteater anticipates the intervention between the two following chapters. Zeno

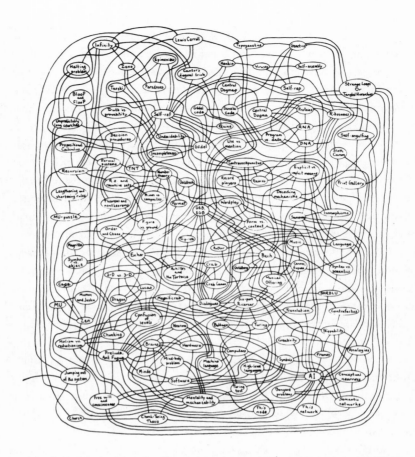

FIGURE 24   Douglas Hofstadter, *A tiny portion of the author's "semantic network."* (From *Gödel Escher, Bach: An Eternal Golden Braid*, by Douglas Hofstadter. Copyright © 1979 by Basic Books, Inc. Reprinted by permission of Basic Books, a member of Perseus Books, L.L.C.)

is best known for his famous paradox in which he attempts to demonstrate that motion is illusory. The argument against motion is an extension of Zeno's denial of the reality of plurality, which he appropriates from his mentor, Parmenides. For Zeno, as for Parmenides, the world of appearance or "The Way of Seeming," is characterized by dualities and oppositions, while "The Way of Truth" shows that all things are One.

The Prelude opens with Achilles and Tortoise visiting their friend Crab to meet his guest, Anteater. Achilles greets Crab with a gift of two records, which, he explains, derive from "Fermat's infamous 'Last Theorem.'" Tortoise claims not only to have solved Fermat's Last Theorem but also to have discovered a "COUNTEREXAMPLE, thus showing that the skeptics had good intuition."[2]

·Contrary to expectation, this theorem turns out to have a practical application. According to Achilles, the mathematics of the theorem makes possible "the retrieval of acoustic material from extremely complex sources." Using this novel technique, Tortoise and Achilles retrieved and recorded Bach actually playing his *Well-Tempered Clavier*. Each of the two records they gave to Crab contains "twenty-four preludes and fugues—one in each major and minor key." Achilles uses Escher's lithograph *Cube with Magic Ribbons* as a visual aid to explain the structure of Bach's work (fig. 25). Just as it is necessary to switch between seeing the bubbles on the magic ribbon as concave and convex, so one cannot hear Bach's fugue without alternating between hearing individual voices and all the voices together. As the Prelude draws to a close, or, more accurately, is interrupted by chapter 10, "Levels of Description, and Complex Systems," Achilles poses a question: "I wonder if, as I listen to this fugue, I will gain any more insight into the question, 'What is the right way to listen to a fugue: as a whole, or as the sum of its parts?'"

When the dialogue recommences between the end of chapter 10, "Mind *vs.* Brain," and the beginning of chapter 11, "Brains and Thoughts," it bears the title ". . . Ant Fugue." Achilles begins again where he left off by giving the answer to his own question about the whole and parts: "MU." With this seemingly simple word, the debate seems to shift from West to East, and from music to religion or philosophy. MU is the term in Zen Buddhism, which, according to Hofstadter,

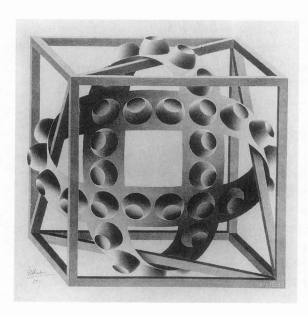

FIGURE 25 M. C. Escher, *Cube with Magic Ribbons*. Lithograph, 1957. (© 2000 Cordon Art B.V., Baarn, Holland. All rights reserved.)

provides a Parmenidean answer to the question of unity and plurality. In an earlier intervention, entitled "A MU Offering," Hofstadter plays with the words "Zen" and "Zeno" to suggest that the opposition between East and West is illusory:

Achilles: *[Okanisama] is my master and his teachings descend directly from those of the sixth patriarch. He has taught me that reality is one, immutable, and unchanging; all plurality, change, and motion are mere illusions of the senses.*

Hofstadter's interpretation of MU discloses the metaphysical vision that inspires his work. He believes that duality is merely apparent because all reality is, in the final analysis, one. Either intentionally or unintentionally, Hofstadter misreads this pivotal Zen notion. Rather than the oneness of all things, MU expresses the Buddhist doctrine of nonduality. In contrast to Hofstadter's belief that reality is *both* one *and* many, MU implies that the real is *neither* one *nor* many. The *difference* between both/and and neither/nor is as subtle as it is significant. In ways that Hofstadter does not fully appreciate, the neither/nor of difference is more important for his "Golden Braid" than the both/and with which he is so infatuated.

The issue around which the debate in the Ant Fugue revolves is the problem of reductionism and holism. We have, of course, already encountered this problem in our consideration of the interplay of parts and whole in organic and self-organizing systems. Crab argues in favor of holism, Anteater sides with reductionism, and Achilles tries to mediate the debate by drawing on the ancient wisdom of his Eastern and Western teachers, who, though worlds apart, seem to say the same thing. When faced with what he regards as specious oppositions, Zen(o) "unasks" the question they seem to pose:

Achilles: *You see, "MU" is an ancient Zen answer which, when given to a question, UNASKS the question. Here, the question seems to be, "Should the world be understood via holism, or via reductionism?" And the answer of "MU" here rejects the premises of the question, which are that one or the other must be chosen. By unasking the question, it reveals a wider truth: that there is a larger context into which both holistic and reductionistic explanations fit.*

Though Anteater claims to defend reductionism, his argument undergoes what amounts to a dialectical reversal and ends by unwittingly supporting holism. Far from a mere aggregate of individuals, he argues, ant colonies, like bee swarms and hives, are intelligent organisms "with their own qualities, at times including the mastery of language." The colony Anteater knows best is named Aunt Hillary. While the individual ants in her "are as dumb as can be," Aunt Hillary is surprisingly smart. When Achilles expresses puzzlement about how an intelligent whole can emerge from unintelligent parts, Tortoise comes to Anteater's aid by invoking the analogy of the way in which the mind emerges from the brain. Just as dumb brain cells connected in neural networks generate mental activity, so ants joined in distributed networks create intelligent behavior. And, though Tortoise fails to note it, vice versa. Since Aunt Hillary obviously cannot speak, Anteater must communicate with her by *writing*. The corresponding trails drawn in the soil by Anteater and the ants are lines of communication through which information and coded messages can be sent and received. In this text, ink is formed by chemicals secreted by Anteater and Aunt Hillary. Though claiming to be a friend of Aunt Hillary, Anteater's replies can sometimes be disastrous for individual ants. Convinced of his good relations with the colony, Anteater insists that he is a surgeon who corrects "nervous disorders of the colony by the technique of surgical removal." Over the course of time, the colony evolves a delicate distribution of responsibilities with the complexity necessary for the ability to converse internally and with the surrounding world. When this language capacity becomes impaired, surgery is usually the only cure. Anteater consumes the "defective" ants, thereby upsetting the faulty order and providing the occasion for a therapeutic reorganization. For Aunt Hillary to survive, the emergent order of the colony cannot be too rigid but must be flexible enough to adapt to changing circumstances. Anteater and Aunt Hillary are bound in a relationship in which each is both parasite and host. While Anteater's dependence on the colony is obvious, Aunt Hillary's dependence on Anteater is less evident but no less important. In a manner reminiscent of Serres's city and country rats, Hofstadter's Anteater is the condition of the possibility of Aunt Hillary's continuing evolution. Expressed in terms of information theory, which Hofstadter indirectly invokes, Anteater creates the interference or noise that disrupts the order of the colony. As we have discovered, however, order can be born of noise. The parasitical Anteater, therefore, not only interrupts but also provides information, which, when processed, issues in the increased organizational complexity necessary for the colony's ongoing survival. The relation of Aunt Hillary to Anteater is as complex as the relation between the ants and Aunt Hillary. Thus, it is not enough to insist that Aunt Hillary cannot be reduced to her ants; Aunt Hillary and Anteater adapt to each other to form a whole that is undoubtedly more than the sum of its parts. It is as if the intricate lines traced

by the ants, Aunt Hillary, and Anteater were braided together to form "An ant fugue!" (fig. 26).

COMPLEX ADAPTIVE SYSTEMS

What Hofstadter simultaneously analyzes and enacts in his Ant Fugue is what complexity theorists describe as a *complex adaptive system*. In *Hidden Order: How Adaptation Builds Complexity,* John Holland reports on his investigation of "the emergence of complex large-scale behaviors from the aggregate interactions of less complex parts":

An ant nest serves as a familiar example. The individual ant has a highly stereotyped behavior, and it almost always dies when circumstances do not fit the stereotype. On the other hand, the ant aggregate—the ant nest—is highly adaptive, surviving over long periods in the face of a wide range of hazards. It is much like an intelligent organism constructed of relatively unintelligent parts. Douglas Hofstadter's wonderful chapter "Ant Fugue" in his 1979 book makes this point better than anything else I have read. In it the ant nest provides a comprehensible version of more spectacular phenomena, such as the intelligence of large number of interconnected neurons, or the identity provided by a diverse array of antibodies, or the spectacular coordination of an organism made of myriad cell types, or even the coherence and persistence of a large city.3

As Holland's comment stresses, complex adaptive systems can be found in natural, social, and cultural phenomena. In an earlier essay entitled "The Global Economy as an Adaptive Process," which Holland acknowledges is indebted to members of the BACH group at the University of Michigan, he describes complex adaptive systems as "adaptive nonlinear networks."4 In addition to the economy, Holland cites the central nervous system, ecologies, immune systems, the development of multicellular organisms, and the processes of evolutionary genetics as other examples of complex nonlinear adaptive networks. Though the media vary, these networks appear to be isomorphic and seem to share a common logic.

To understand the distinctive features of this logic, it is important to recall the difference between complexity and chaos, and to differentiate recursively adaptive networks from the fluid dynamics of turbulence. The characteristic of chaos that seems most similar to complexity is its sensitivity to initial conditions. In a chaotic state, nonlinear dynamics can lead to positive feedback, which results in effects disproportionate to causes. This phenomenon appears to be similar to the self-organized criticality of complex systems in which quantitative changes reach a tipping point where qualitative change suddenly occurs. The dynamics of these changes seem similar, yet they differ in an important way. We have discovered that complex systems are not chaotic but are poised at the edge of chaos between too much and too little order. The degree of order in complex systems allows their feedback loops to be both nonlinear or recursive and adaptive. A chaotic condition, by contrast, can be reactive but not adaptive. For systems to be adaptive, they must be considerably more complex than the regime of chaos allows. At first glance, the nonlinear dynamics of turbulence seem to exhibit the kind of adaptiveness characteristic of complex systems. Consider, for example, a rushing stream in which eddies form, survive for a while, and then disintegrate. This appears to be a clear example of self-organization, as well as an adaptive interaction between an organized structure and its surroundings. But, as Murray Gell-Mann stresses, there is an important difference between the non-linear dynamics of turbulence and complex adaptive systems; they differ in "the way in which the information in the environment is recorded." Though best known for his work in physics, Gell-Mann has done extensive research on complex adaptive systems. Like his colleagues at the Santa Fe Institute, he is convinced that the dynamics of complexity help to explain phenomena ranging from prebiotic chemical and biological evolution to the evolution of both cultural and computer programs. Rather than preprogrammed and reactive, complex adaptive systems are organized to process information in ways that allow for surprise and even creativity. "In complex adaptive systems," Gell-Mann argues, "information about the environment . . . is not merely listed in what computer scientists would call a look-up table. Instead, the regularities of the experience are encapsulated in highly compressed form as a *model* or *theory* or *schema*. Such a schema is usually approximate, sometimes wrong, but it may be adaptive if it can make useful predictions including interpolation and extrapolation and sometimes generalization to situations very different from those previously encountered."[5] This is precisely the dynamic at work in the interaction between Anteater and Aunt Hillary. The action of Anteater is initially noise but becomes information as Aunt Hillary processes it and responds effectively. Survival depends on successful adaptation or, more precisely, coadaptation.

In his wide-ranging and sometimes speculative book, *The Quark and the Jaguar: Adventures in the Simple and Complex,* Gell-Mann provides a helpful diagram of the functional organization of complex adaptive systems (fig. 27). The formation and operation of complex adaptive systems involve five distinguishable yet closely related features or moments. First, the system must be able to identify regularities in its environment.[6] Every system is embedded in more or less extensive networks that provide streams of data that must be processed. For a system to respond effectively, it must be able to identify regularities, patterns, or redundancies in flows, which at first seem chaotic. Two kinds of errors must, as far as possible, be avoided: randomness must not be mistaken for order, and, conversely, order must not be mistaken for randomness. Second, once a regularity in the flux of data has been identified, the complex adaptive system must generate schemata that enable it to recognize the pattern if it occurs again. For a schema to be functional, relevant data have to be compressed as much as possible. In a manner consistent with algorithmic information theory, the degree of complexity is directly related to the amount of time or the number of steps required to unfold the compressed data. The greatest complexity occurs between

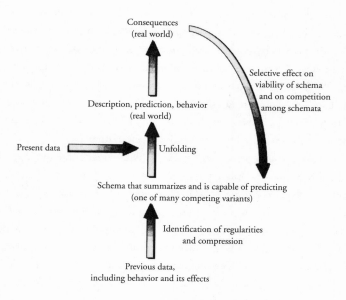

Consequences
(real world)

Selective effect on
viability of schema
and on competition
among schemata

Description, prediction, behavior
(real world)

Present data → Unfolding

Schema that summarizes and is capable of predicting
(one of many competing variants)

Identification of regularities
and compression

Previous data,
including behavior and its effects

FIGURE 27 Murray Gell-Mann, How a Complex Adaptive System Works. (Reprinted from *The Quark and the Jaguar: Adventures in the Simple and the Complex* [New York: W. H. Freeman, 1994], © 1994 by Murray Gell-Mann. Page 25.)

minimum and maximum algorithmic information content. The close relationship between complex adaptive systems and algorithmic information content underscores three important points. First, since schemata function like algorithms, complex adaptive systems operate by processing information. This requires the filtering of noise or, in Atlan's terms, the capacity to create order from noise. Second, rather than preprogrammed or fixed, the schemata or algorithms in complex adaptive systems are emergent and can change. Holland describes Gell-Mann's schemata as "internal models." In forming these models, "the agent must select patterns in the torrent of input it receives and then must convert those patterns into changes in its internal structure."[7] Such modification of the model, scheme, or algorithm is necessary for adaptation. Finally, the compression of data in the formation of schemata maximizes the storage capacity and thus extends the memory of the system. In this way, compression serves as an effective strategy for operating within unavoidable memory constraints.

The third distinguishing feature of complex adaptive systems is the ability of schemata to change in ways that produce variants, which compete with each other. Different schemata *within* a given system engage in something like an evolutionary process governed by the survival of the fittest. Fourth, schemata are not merely reactive but are deployed to describe and *anticipate* surrounding activities. The effectiveness of a schema is a function of the accuracy of its description and the reliability of its predictions of relevant events in its environment. As experiences accumulate, reliable schemata thrive and unreliable schemata disappear. Fifth, complex adaptive systems are bound to their environments through feedback loops that create the possibility of adaptive modifications and reorganizations. These changes can be both gradual and sudden. The success of any adaptation is measured by the survival of the system.

Inasmuch as complex systems are both emergent and adaptive, they are inescapably temporal. Unlike intrinsically stable systems in a Newtonian universe, complex adaptive systems develop over time and, therefore, are historical. Rather than a *nunc stans,* complexity presupposes a tensed moment poised between an accumulating past and a future that simultaneously unfolds and intrudes. As we have already discovered, the unpredictability and randomness associated with emergence create a decisive difference between past and future. The irreducibility of a complex whole to its constituent parts creates a gap between past and future that the present can frame but not bridge. Though there can be no oak without an acorn, the oak is not simply the result of the unfolding of the acorn. Nevertheless, the three modalities of time remain inextricably interrelated. Per Bak goes so far as to suggest that all systems characterized by self-organizing criticality have the functional equivalent of memory.[8] If qualitative change suddenly emerges from the gradual accumulation of quantitative changes, the present must in some way both recall and preserve the past. As feedback loops become more intricate, the relation between past and present becomes more complex. In complex adaptive systems, the past can be traced in the present through the modifications of the schemata. The temporality of these systems is not, however, limited to this interrelation of the past and present; for systems to be adaptive, they must "anticipate" the future as well as "learn" from the past. Here, as elsewhere, the use of terms like "experience," "memory," "anticipation," and "learning" does not, of course, imply either consciousness or self-consciousness. Consciousness is no more necessary for these operations than it is for the processing of information. It is, however, possible that the logic and operation of complex adaptive systems can disclose similarities between nonhuman and ostensibly human information processes.

The question of anticipation in complex adaptive systems raises the long-standing problems of design and teleology. Though I will consider this issue in more detail below, two important points must be stressed at this juncture. First, even though complex adaptive systems are neither preprogrammed nor designed, their development can be directed. Holland actually argues that complex adaptive systems are "the product of *progressive* adaptations."[9] Progress in this context is measured by the increase of diversity and, correlatively, the growth in complexity. In the strange loops of these systems, complexity breeds diversity, which increases complexity, which breeds diversity, . . . Second, it is possible to account for the directed increase in complexity in systems without recourse to teleology or finalism. To see why intentionality does not entail finality, it is helpful to draw a distinction that Henri Atlan borrows from Colin S. Pittendrigh. A biologist working on the problem of adaptation and natural selection, Pittendrigh distinguishes teleology from teleonomy. In contrast to teleological systems, which are purposeful, teleonomic structures are "non-purposeful, end-seeking systems."[10] Atlan elaborates this distinction in *Entre le cristal et la fumée:*

A teleonomic process does not, then, function by virtue of final causes even though it seems as if it were oriented toward the realization of forms, which will appear only at the end of the process. What in fact determines it [i.e., a teleonomic process] are not forms as final causes but the realization of a program, as in a programmed machine whose function seems oriented toward a future state, while it is in fact causally determined by the sequence of states through which the preestablished program makes it pass. The program itself, combined in the characteristic genome of the species, is the result of the long biological evolution during which under the simultaneous influence of mutations and natural selection, it transformed itself and adapted to its environment.[11]

As end-directed but not purposeful, teleonomic processes are neither linear nor circular. On the one hand, system and environment are joined in recursive circuits that create both unexpected and disproportionate changes, and, on the other hand, the openness of complex adaptive systems leads to aleatory changes in schemata that distinguish the point of departure from the point of arrival.

The reason complex adaptive systems must be coadaptive should now be clear. Like Anteater and Aunt Hillary, complex adaptive systems adapt to systems that are adapting to them. These systems comprised of systems are, as Gell-Mann explains, "collectivities of co-adapting adaptive agents, which construct schemata to describe and predict one another's behavior."[12] Such coadaptation compounds the complexity of the systems involved. Thus, Gell-Mann's diagram is not sufficiently complex and must be modified (fig. 28). But even this suggestive representation is too simple. Every complex system is folded into more or less extensive networks of other complex systems. Each system, therefore, must

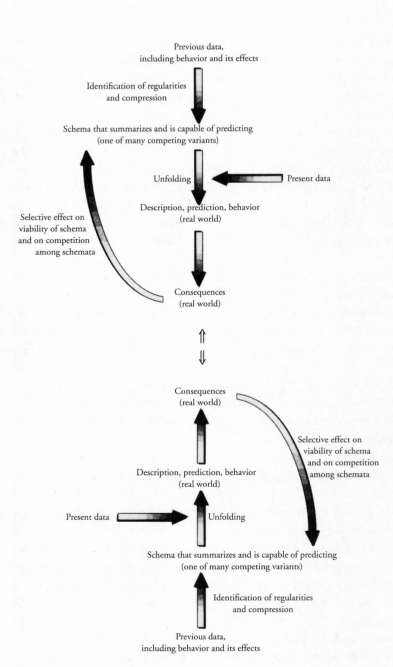

FIGURE 28 Coadaptive Complex Systems

adapt to multiple systems, which are adapting not only to it but also to the other systems to which it is related. Moreover, these networked systems are constantly evolving. This complex coadaptive network creates what E. O. Wilson describes as "the coevolutionary circuit, which is not fixed but constantly evolves. This capacity is itself the result of evolution."[13]

E·VOLUTION

For complex systems to be adaptive and thus coevolve, they must evolve the ability to evolve. In figure 29, Gell-Mann borrows terms from information theory, which we have already considered, to describe the evolution of evolvability in complex adaptive systems. To be able to adapt effectively, systems must move toward conditions far from equilibrium where they have enough flexibility to change yet enough stability to survive modifications and transformations. The analyses of complex adaptive systems that theorists like Gell-Mann and Holland develop are quite abstract, but the structures and processes they describe are remarkably concrete. While many specific examples could be cited, the work of their erstwhile colleague at the Santa Fe Institute, biologist Stuart Kauffman,

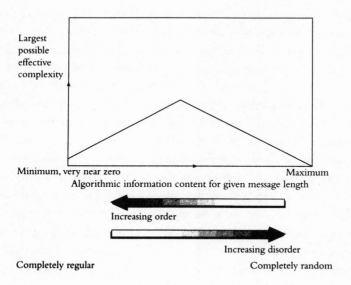

FIGURE 29 Murray Gell-Mann, A sketch showing roughly how the largest possible effective complexity varies with AIC (algorithmic information content). (Reprinted from *The Quark and the Jaguar: Adventures in the Simple and the Complex* [New York: W. H. Freeman, 1994], © 1994 by Murray Gell-Mann. Page 59.)

provides ample evidence of the ways in which complex adaptive systems operate in actual living organisms and processes. Kauffman begins his chapter on "Self-Organization and Adaptation in Complex Systems" in his *Origins of Order: Self-Organization and Selection in Evolution* by observing: "Eighteenth-century science, following the Newtonian revolution, has been characterized as developing the sciences of organized simplicity, nineteenth-century science, via statistical mechanics, as focusing on disorganized complexity, and twentieth- and twenty-first century sciences as confronting organized complexity. Nowhere is this confrontation so stark as in biology."[14] As the title of his book indicates, Kauffman interprets the evolutionary process in terms of self-organizing systems. The debates between biologists who use complexity theory and neo-Darwinians repeat important aspects of the arguments between Darwin and his critics in ways that reveal the continuing philosophical and theological significance of the controversy.

From the outset, evolution has been a *theological* issue. Throughout the seventeenth and eighteenth centuries, the success of science and increasing confidence in the power of reason led to sustained criticisms of religion. In response to growing doubt, many theologians attempted to develop rational defenses of religious belief. The centerpiece of this program was the revival and extension of the proofs of God's existence. There are three classic proofs: the ontological argument, which attempts to establish the identity of thought and being; the cosmological argument, which argues from the existence of the world to God as first cause; and the teleological argument, which argues from the order or design of the world to God as designer. These arguments represent different religious dispositions and philosophical orientations: continental rationalists tend to favor the ontological approach, and British empiricists are drawn to the cosmological and teleological arguments. While the latter two arguments are traditionally distinguished, they share the common strategy of arguing from effect to cause.[15] What appeals to many thinkers working in the British analytic tradition is the apparent empirical foundation of the cosmological and teleological arguments. Defenders of the faith argue that the only reasonable way to account for the existence and design of the world, both of which can be concretely experienced, is to postulate God as their necessary cause. In the England of Darwin's day, the argument from design enjoyed considerable support. Its chief proponent was William Paley, with whom Darwin studied at Cambridge. In formulating his argument, Paley presupposes a clockwork universe but gives this Newtonian world a distinctive twist. In *Natural Theology, or Evidences of the Existence and Attributes of the Deity Collected from the Appearances of Nature,* published in 1802, Paley develops what remains the classic version of the teleological argument. He imagines finding a carefully crafted watch while walking across a heath. Pondering how the existence of this watch can be explained, Paley concludes:

This mechanism being observed . . . the inference, we think, is inevitable that the watch must have a maker; that there must have existed, at some time, and at some place or other, an artificer or artificers, who formed it for the purpose of which we find it actually to answer; who comprehended its construction, and designed its use. [16]

Paley's argument betrays assumptions about intrinsically stable systems and structures that we have come to expect. Since the parts of a mechanism are externally related, order cannot arise from within but must be imposed from without. What distinguishes Paley's argument from countless other versions is his insistence that the watch is designed in such a way that its parts are perfectly *adapted* to each other:

To reckon up a few of the plainest of these parts, and of their offices, all tending to one result: —We see a cylindrical box containing a coiled elastic spring, which, by its endeavor to relax itself, turns round the box. We next observe a flexible chain (artificially wrought for the sake of flexure), communicating the action of the spring from the box to the fusee. We then find a series of wheels, the teeth of which catch in, and apply to each other, conducting the motion from the fusee to the balance, and from the balance to the pointer; and at the same time, by the size and shape of those wheels, so regulating that motion, as to terminate in causing an index, by an equable and measured progression, to pass over a given space in a given time. [17]

Twenty-five years after the appearance of Paley's argument, James Paxton published a short book entitled *Illustrations of Paley's Natural Theology.* The text consists of nothing more than a collection of remarkable images with descriptive notes. After an introductory representation of the parts of a watch, the rest of the images are of the human body and other living organisms (figs. 30, 31). Paxton selects the images to illustrate the way in which the parts in the watch as well as the human body are adapted to each other. The first example he selects is the human eye, which Darwin was never fully convinced could be explained by his theory of evolution. Though reflecting a very different philosophical and theological tradition, Paxton's illustrations of Paley's argument implicitly demonstrate the interrelation of parts and whole that is perfectly consistent with the reciprocity of means and ends that Kant argues defines the inner teleology of living organisms.

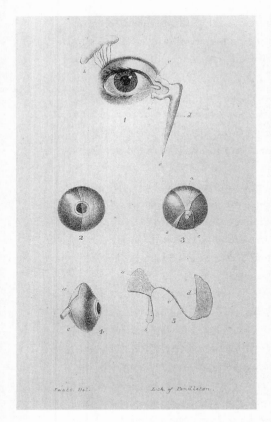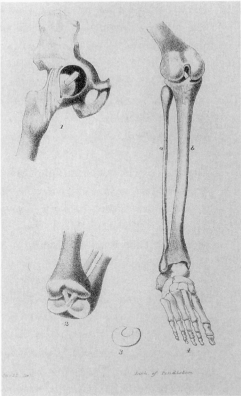

FIGURE 30 From James Paxton, *Illustrations of Paley's Natural Theology* (Boston: Hilliard, Gray, Little, and Wilkins, 1827).

FIGURE 31 From James Paxton, *Illustrations of Paley's Natural Theology* (Boston: Hilliard, Gray, Little, and Wilkins, 1827).

When Darwin set sail on the *Beagle* on December 27, 1831, he not only found Paley's arguments persuasive but actually assumed he would eventually become a country pastor. By the time he returned five years later, he had become convinced that the order of the world is the result of natural rather than divine causes. Nevertheless, traces of Paley's vision can be detected in Darwin's mature position. This is especially evident in his insistence on the importance of adaptation in natural selection.

It took Darwin more than two decades to process the data gathered on the voyage of the *Beagle*. The intellectual climate in which he began to formulate his theory was dominated not only by the arguments of creationists like Paley but also by rational morphologists such as Georges Cuvier, as well as arguments between uniformitarians and catastrophists about recent geological discoveries.

One of the most influential lines of analysis grew out of an alliance between creationists and morphologists. Cuvier, who was the permanent secretary of the French Academy of Sciences and a professor of comparative anatomy, attempted to rationalize Aristotle's analysis of natural organisms by identifying four distinct body types, which he called *embranchements*.[18] Different body plans are adapted to meet the demands of different environments. Cuvier insisted that it is impossible for *embranchements* to develop from each other or to cross from one to the other. When a species becomes extinct as the result of a natural catastrophe, God must intervene to create a new species. The theological appeal of this position is threefold. First, the creative role of God is preserved; second, the *embranchements* remain distinct; and third, one species cannot develop from another. The two latter points seem to make it possible to preserve the uniqueness of human beings by establishing their difference from animals. While Cuvier developed his position by refining Aristotelianism, his conclusions can be interpreted as a latter-day version of Platonism. Different *embranchements* resemble Platonic forms or archetypes, each of which is defined by certain characteristics that are gradually revealed over the course of time.

Cuvier's point of view gained a widespread public hearing during a series of debates he had with Étienne Geoffroy Saint-Hilaire in 1829–30. In contrast to Cuvier's insistence on the autonomy of different body types, Geoffroy contended that *embranchements* are not "impassable after all but form a unified and rational series of structural transformations from a single ground plan."[19] This argument called into question two central aspects of Cuvier's theory that many people found reassuring. If the transmutation of species is combined with progressive descent from a common ancestor, the line separating the human from the nonhuman is no longer clear. The far-reaching implications of Geoffroy's position become apparent in the work of Jean-Baptiste Lamarck. Though best known for his belief in acquired traits, Lamarck's viewpoint is considerably more sophisticated and subtle. Depew and Weber correctly point out that he developed "a biology in which adaptation is rational because organisms take their fate into their own hands. He blunted the dark vision of historical decay that occluded the shining picture of biological order by replacing a fixed world in which things come in preset kinds and then fall from grace when they are contaminated by mishaps, contingencies, and catastrophes of natural history, with a world in which organisms seem able, on the whole, to meet whatever challenges are thrown at them and, by meeting them, to transform themselves into new species. This process is based on an allegedly inherent tendency of living things to complexify and results in a directional thrust toward higher and higher organisms, culminating (so far) in humans."[20] Inasmuch as morphological differentiation, which is a function of environmental adaptation, naturally develops in the direction of increasing complexity, Lamarck's theory leaves no room for intervention by a divine creator. While he did not think that life had a common origin, Lamarck was convinced that the progressive development of living organisms emerges within the world itself and does not depend on an external cause. When viewed as a whole, the course of evolution seems to display self-developing complexity. Accidents and diversions occasionally occur, but they cannot thwart the overall evolutionary trajectory.

SPECIES OF ORIGIN

Without examining the social and political implications of Lamarck's theory in detail, it is important to note that the turbulent climate of nineteenth-century France created a tendency to associate evolution with atheism, anticlericalism, and materialism. Furthermore, many political reformers appropriated the notion of progressive evolution to promote the movement from a society of hierarchical classes to a democratic meritocracy. While sensitive to these issues, Darwin's concerns were more scientific and philosophical than political. Nevertheless, anxiety in England about developments across the Channel created a cautious, if not hostile, atmosphere for the evolutionary theory.

Darwin adamantly rejected the notion of the inheritance of acquired traits but accepted uniformitarianism and, correlatively, the transmutation of species. He also sided with Geoffroy on the question of descent from a common ancestor and opposed Lamarck's defense of multiple starting points for evolution. Darwin's most important disagreement with Lamarck, however, concerns his denial of the inherent propensity of the evolutionary process to move from simplicity to complexity. For Darwin, Lamarck's insistence on the "progressive complexification of organisms" was a vestige of the essentialism inherent in rational morphology. Though he never framed his arguments in philosophical terms, Darwin's suspicion of essentialism reflected the nominalistic proclivities of British thought dating back to the late Middle Ages. Medieval realists were actually Platonists for whom the universal form or idea *(eidos)* takes ontological and epistemological precedence over particular individuals and concrete sense experience. For nominalists, by contrast, only the individual is real and the universal is a mere name *(nom),* which serves a heuristic function but has no ontological reality. True knowledge derives from experience and is mediated by sensible intuition. The primordiality of individuality renders all relations within and among entities external and extrinsic rather than internal and intrinsic. From Darwin's point of view, Lamarck's claim that organisms have an *inherent* tendency to become more complex appeared to be an alternative version of Aristotle's doctrine of entelechy, in which Platonic forms are recast as immanent structures that unfold over time. Within this framework, inner drives rather than outer forces seem to determine development. Describing the significance of Darwin's contribution, Stephen Jay Gould invokes these philosophical and theological issues:

Darwin's revolution should be epitomized as the substitution of variation for essence as the central category of natural reality. (See, Mayr, our greatest living evolutionist, Animal Species and Evolution, *for a stirring defense of the notion that "population thinking," as a replacement for Platonic essentialism, forms the centerpiece of Darwin's revolution.) What can be more discombobulating than a full inversion, or "grand flip," in our concept of reality: in Plato's world, variation is accidental, while essences record a higher reality; in Darwin's reversal, we value variation as a defining (and concrete earthly) reality, while averages (our closest operational approach to "essences") become mere abstractions.*[21]

This change in orientation shifts attention from essence to existence, eternity to time, logic to history, and universality to individuality.

Darwin's emphasis on individuality is evident in the way he understood both the constitution of particular organisms and the relations among different organisms. What he found most interesting about specific organisms was their diverse traits and the changes they undergo. These variations are not inherent in the organism but issue from responses to factors in the environment. A species is comprised of separate organisms that have contingent rather than necessary relations to each other. It is as if organisms were atoms constantly acted upon by external forces in a universe that seems to be Newtonian. Indeed, in many ways, Darwin attempted to do for biology something like what Newton did for physics. The two foundational laws or principles he identified are chance variation and adaptive natural selection. Within a Darwinian framework, evolution is, in Jacques Monod's apt phrase, "chance caught on the wing." Darwin arrived at his understanding of selection while considering the effects of animal breeding practices of his time. Interactions between and among individuals as well as with the environment screen unpredictable changes in organisms. Changes that prove to be beneficial for survival are preserved and others die out. While Darwin formulated critical aspects of his theory relatively early in his career, the various strands of his investigation did not come together until he had two moments of illumination separated by almost two decades. When insight dawned, its source was unexpected. Writing in his *Autobiography,* he described one of the decisive moments in the development of his thinking:

Fifteen months after I had begun my systematic inquiry, I happened to read for amusement Malthus on Population, *and being well prepared to appreciate the struggle for existence, which everywhere goes on, from long continued observation of the habits of animals and plants, it at once struck me that under these circumstances favorable variations would tend to be preserved, and unfavorable ones would be destroyed. The result would be the formation of a new species. Here I had at last got a theory by which to work.*[22]

The second pivotal insight occurred eighteen years later, merely three years before the publication of *On the Origin of Species,* when Darwin suddenly realized the importance of Adam Smith's account of the division of labor. Darwin's mature theory of evolution results from an imaginative synthesis of his field observations and research with the population studies of Malthus and economic theory of Smith. This genealogy of ideas redraws generally accepted lines of descent in the history of ideas. While the recognition of the ways in which economists have long appropriated evolutionary theory to explain economic processes is widespread, few cultural critics realize the extent to which Darwin's theory is indebted to population studies and economic speculation. In the work of Malthus, Darwin saw a principle that helps to explain the logic of natural selection. Echoing Hobbes's *bellum omnium contra omnes,* he writes:

I do not doubt everyone till he thinks deeply has assumed that increase of animals exactly proportionate to the number that can live—population is increased at geometrical ratio in far shorter time than twenty five years—yet until the one sentence of Malthus no one clearly perceived the great check amongst men—there as a spring, like food used for other purposes as wheat for making brandy—Even a few years of plenty makes population in man increase and an ordinary crop cause a dearth. Take Europe: on an average every species must have same number killed year by year by hawks, by cold, etc.—even one species of hawk decreasing in number must affect instantaneously all the rest—the final cause of all this wedging must be to sort out proper structure and adapt it to changes—to do that for form, which Malthus shows is the final effect . . . of this populousness on the energy of man. One may say there is a force *like a hundred thousand wedges trying to* force *every kind of adapted structure into the gaps by* forcing *out weaker ones.*[23]

As Darwin understood it, Malthus's claim that population increases geometrically and food supply only arithmetically demonstrated that organisms are always subject to *external* pressures. The limitation of resources creates competition among individuals as well as different species.

What Darwin needed to complete his theory was Smith's account of the division of labor. Darwin, Depew and Weber point out, "rids the organic world of Aristotelian essences altogether by generalizing the individualist ontology of political economy."[24] Survival in a ruthlessly competitive world depends upon the ability of individuals and species to carve out niches where they can profit from their labors. As John Maynard Keynes stresses, the interpretation of the division of labor that Darwin borrowed from Smith was anticipated by Mandeville in his 1705 *Fable of the Bees*.[25] The subtitle of this infamous work summarizes Smith's guiding insight: *Private Vices, Publick Benefits*. A single couplet captures the substance of the argument:

Thus every Part was full of Vice,
Yet the whole Mass a Paradise

The invisible hand at work in beehives and ant colonies can also be detected in economic systems. Individuals pursuing their own self-interests unwittingly act together in ways that benefit the socioeconomic whole. Survival depends on a progressive division of labor, which creates competitive yet complementary niches:

Natural selection, also, leads to a divergence of character; for more living beings can be supported on the same area. The more they diverge in structure, habits, and constitution, . . . the more diversified these descendants become, the better will be their chance of succeeding in the battle for life. Thus the small differences distinguishing varieties of the same species, will steadily tend to increase till they come to equal the greater differences between species of the same genus, or even of distinct genera.[26]

While the division of labor leads to increased complexity, this is not, as for Lamarck, the result of an inherent tendency but is the consequence of environmental pressure. Populations, like industries and companies, must diversify and specialize to remain competitive. Fitness is not a matter of relative strength but is the ability to adapt and fit into an available niche in the competitive landscape. Marx was quick to realize the symbiotic relation between Darwin's theory of evolution and market capitalism, when, in an 1862 letter to Engels, he writes: "It is remarkable how Darwin recognizes among beasts and plants his English

179

society, with its division of labor, competition, opening up of new markets, 'inventions,' and the Malthusian struggle for existence. It is Hobbes' *bellum omnium contra omnes.*"[27]

Though Darwin's exposition of his theory is long and involved, its logic is straightforward. Depew and Weber offer a concise and accurate summary of Darwin's position:

Mathusian reproduction + Resource scarcity = Struggle for life
Struggle for life + Variation = Differential adaptedness of variants
Differential adaptedness + Strong inheritance = Adaptive natural selection
Adaptive natural selection + Niche diversification + Many generations = Branching taxa.[28]

When so understood, Darwin's natural selection is the biological version of Smith's invisible hand, which, in turn, is the doctrine of divine providence rewritten as economic theory. Theology, it seems, is often most influential where it is least obvious.

What is remarkable about Darwin's argument is that he was able to develop a descriptively persuasive theory without any knowledge of the actual mechanism responsible for inheritance. This situation changed dramatically in 1900, when biologists rediscovered Gregor Mendel's prescient 1865 paper entitled "Experiments in Plant Hybridization." The recognition of the role genes play in inheritance initially seemed to call into question Darwin's position, but evolutionary theory was quickly recast to reconcile natural selection with genetics. The resulting synthesis dominated research and speculation throughout the twentieth century. In one of its recent and highly controversial variations, Richard Dawkins extends the theory of evolution to the level of genes, by arguing that not only individual organisms but individual genes are selfish. In a variation of the age-old maxim that chickens are the way eggs multiply, Dawkins maintains that "A body is the genes' way of preserving genes unaltered."[29] Though individual bodies arise and pass away, genes endure. What makes genes more durable than the bodies they inhabit is their capacity to store and transmit coded information. Dawkins goes so far as to argue that in DNA "information passes through bodies and affects them, but it is not affected by them on its way through."[30] The struggle that really counts for survival in this rushing river of information takes place at the level of individual genes rather than individual organisms. Dawkins is the first to admit that "an uneasy tension disturbs the heart of the selfish gene theory":

It is the tension between gene and individual body as fundamental agent of life. On the one hand we have the beguiling image of independent DNA replicators, skipping like chamois, free and untrammeled down the generations, temporarily brought together in throwaway survival machines, immortal coils shuffling off an endless succession of mortal ones as they forge themselves as separate entities. On the other hand we look at the individual bodies themselves and each one is obviously a coherent, integrated, immensely complicated machine, with a conspicuous unity of purpose. A body doesn't look like *the product of a loose and temporary federation of warring genetic agents who hardly have time to get acquainted before embarking in sperm and egg for the next leg of the great genetic diaspora.*[31]

Though the body might be a "federation of warring genetic agents," it is, like Aunt Hillary, made up of networks of networks of agents acting in ways that are simultaneously competitive and cooperative. For Dawkins, what matters is, in the final analysis, the survival of the part rather than the whole.

Even though the "modern evolutionary synthesis" results in significant revisions of Darwin's position, it nonetheless represents an extension rather than rejection of the fundamental ideas informing his work. The most important principle running through different versions of Darwinism is radical individualism. Gould's penetrating observation once again is helpful:

This strict version [of Darwinism] went well beyond a simple version that natural selection is a predominant mechanism in evolution. . . . It emphasized a program for research that almost dissolved the organism into an amalgam of parts, each made as perfect as possible by the slow but relentless force of natural selection. This "adaptationist program" downplayed the ancient truth that organisms are integrated entities with pathways of development constrained by inheritance—not pieces of putty that selective forces of the environment can push in any adaptive direction. The strict version with its emphasis on copious, minute, random variation molded with excruciating but persistent slowness by natural selection, also implies that all events of large-scale evolution (macroevolution) were the gradual, accumulated product of innumerable steps, each a minute adaptation to changing conditions within a local population. This "extrapolationist" theory denied any independence to macroevolution and interpreted all large-scale evolutionary events (origin of basic designs, long-term trends, patterns of extinction and faunal turnover) as slowly accumulated microevolution. . . . Finally, proponents of the strict version sought the source of all change in adaptive struggles among individual organisms, thus denying direct causal status to other levels rich in hierarchy of nature.[32]

As this remark suggests, for Darwin and those who faithfully follow him, the whole *is* the sum of its parts. In the debate between Crab and Anteater, therefore, Darwin would side with Anteater's reductionism. If the whole is merely the sum of its parts, it can be reduced to its constitutive elements without significant loss. From this point of view, change within the whole likewise appears to be additive. Effects are proportionate to their causes and develop in ways that seem both linear and continuous. The continuity of change must not, of course, obscure its dependence on chance. In the absence of any inherent characteristics lending evolution direction, contingent changes do not seem to be limited by any developmental restrictions. Far from rationally designed, organisms are contingent constructions, which Kauffman suggestively describes as "Rube Goldberg machines."

MORPHING

It should not be surprising that neo-Darwinism has provoked strong critical reactions from a variety of perspectives. Some of the most thoughtful responses have grown out of recent work done in complexity studies. In the preface to his book, *How the Leopard Changed Its Spots: The Evolution of Complexity,* Brian Goodwin points out: "A striking paradox that has emerged from Darwin's way of approaching biological questions is that organisms, which he took to be primary examples of living nature, have faded away to the point where they no longer exist as fundamental and irreducible units of life. Organisms have been replaced by genes and their products as the basic elements of biological reality." The concentration on the constitutive parts of organisms, it seems, has led to the loss of an adequate understanding of the whole as such. Goodwin admits that "genocentric biology" is a logical extension of Darwinism but correctly insists that "organisms cannot be reduced to their genes or molecules."[33] In criticizing the radical nominalism implicit in Darwinism, Goodwin calls for a reconsideration of the virtues of rational morphology. A scientist with uncommon knowledge of the humanities, Goodwin realizes the philosophical heritage of the tradition he seeks to revive. He places particular emphasis on Kant's contribution to our understanding of living organisms: "Kant knew nothing about these dynamic processes, but he did correctly describe the emergence of parts in an organism as a result of internal interactions instead of as an assembly of preexisting parts, as in a mechanism or a machine. So organisms are not molecular machines. They are functional and structural unities resulting from a self-organizing, self-generating dynamic."[34] Goodwin not only situates his research program in relation to its historical precedents developed in the seventeenth and eighteenth centuries and elaborated in the twentieth century by theorists like D'Arcy Thompson and C. H. Waddington, but also orients it in terms of contemporary discussions in the humanities and social sciences.[35] In calling for a return to morphological analysis, Goodwin proposes "a structuralist programme in developmental biology." Extending Lévi-Strauss's insights from anthropology to biology, structuralism, he explains,

> is based upon the proposition that actual phenomena are particular realizations from a defined set of possibilities. The set is intelligible because its members show a common order arising from the operation of the same generative principles. Applied to the biological realm as a whole, whose basic phenomenological context is the morphological diversity of organisms that have arisen in the course of evolution, the structuralist enterprise is to discover the unity of dynamic processes that generates this ordered diversity of forms. This process is characterized by properties of wholeness, regulation and transformation that are characteristic of organismic life-cycles.[36]

Without in any way minimizing the profound similarities between Goodwin's morphology and Lévi-Strauss's structuralism, it is important to underscore an important difference between the two positions. While Lévi-Strauss sees the structures underlying biological, psychological, social, and cultural processes as universal and unchanging, Goodwin maintains that the forms of life are emergent as well as mutable.

Several important conclusions follow from the attempt to revive morphology. First, and most obviously, the organism cannot be understood as the sum of its parts but must be considered *as a whole*. The organic whole has characteristics that emerge from but are not reducible to its parts. Second, not everything is possible in the course of evolution. Morphological constraints define the parameters within which development takes place. "Metamorphosis," Goodwin argues, "is a process that has intrinsic properties of dynamic order so that particular forms are produced when the system is organized in particular ways."[37] Third, as a result of these intrinsic properties, evolution is not merely contingent but entails a circumscribed necessity. Accordingly, biology has a logic that is not simply reducible to historical circumstances and thus is not merely a historical science. And finally, as a result of the differentiation of morphological types, the course of evolution is discontinuous instead of continuous. There is "no clear evidence . . . for the gradual emergence of any evolutionary novelty," Goodwin argues. "New types of organisms simply appear upon the evolutionary scene, persist for various periods of time, and then become extinct. So Darwin's assumption that the tree of life is a consequence of the gradual accumulation of small hereditary differences appears to be without significant support. Some other process is responsible for the emergent properties of life."[38] In attempting to provide a corrective to what he regards as the reductive excesses of neo-Darwinism, Goodwin often stresses its irreconcilable differences with structural morphology. These differences, as Kauffman's work demonstrates, should not obscure the possibility of interpreting these two positions as complementary rather than antithetical.

Kauffman extends Goodwin's approach by arguing for a new synthesis that brings together neo-Darwinism with a carefully refined theory of complex self-organized adaptive systems. "Self-organization," he submits, "may be the *precondition* of evolvability itself. Only those systems that are able to organize themselves spontaneously may be able to evolve further."[39] The theory of complex adaptive systems provides an understanding of the interrelation of emergence and self-organization, which makes it possible to reinterpret the principles of rational morphology in a way that takes advantages of its insights while overcoming its limitations. Alluding to Gell-Mann's account of the role of algorithmic information content in complex adaptive systems, Kauffman argues that rather than "chance caught on the wing," evolution is a comprehensible process displaying orderly patterns and rational laws: "Even if it is true that evolution is such an incompressible process, it does not follow that we may not find deep and beautiful laws governing the unpredictable flow. . . . We can never hope to predict the exact branchings of the tree of life, but we can uncover powerful laws that predict and explain their general shape."[40] As we shall see, Kauffman's concerns are not only scientific; social, political, metaphysical, and even religious interests motivate his work. What must be understood in this context, however, is his conviction that "the revolution in complex systems dynamics is now making it possible to hope that complex, self-organized systems, including those investigated by evolutionary biology, can be more closely linked to physics and chemistry without reductionism or vitalism."[41]

The fundamental conviction informing all of Kauffman's work is that the emergence of order is *spontaneous but not accidental.* An inventive combination of computer modeling, biological research, and philosophical speculation leads to the conclusion that the principles of self-organization can explain the origin of life as well as the emergence and evolution of biological organisms. The early stages of his research were deeply influenced by Prigogine's work on dissipative structures in conditions far from equilibrium. These negentropic processes interest Kauffman because of their similarities to living systems. For Kauffman, the river of life is made up of "metabolic whirlpools" that differ significantly from Dawkins's "river out of Eden." Though streams of coded information rush through living organisms, life's turbulence is governed by nonlinear dynamics that operate, as we have discovered, at the edge of chaos. From his student days in medical school at the University of California, San Francisco, Kauffman has been persuaded that natural selection is not the only source of order. In the larger scheme of things, the emergence of life is not contingent but is, he believes, virtually inevitable. Though Kauffman's conclusions are grand, his approach is modest. In the early 1960s, Monod and Jacob had discovered that genes are regulated by feedback mechanisms that function like digital computers. For Kauffman, this insight pointed to a new line of inquiry. At the time, he was preoccupied with the mechanisms of embryonic cell differentiation. More specifically, he was trying to determine how the approximately 100,000 genes in the genome could possibly produce the 250 different cell types necessary for life. Since the potential activity states of the genes comprising the genome are about $10^{30,000}$, Kauffman thought the chances of natural selection developing the requisite types of cells was so infinitesimal as to be impossible. When he learned of Monod and Jacob's discovery of the feedback mechanisms by which genes switch on and off, he realized it might be possible to model genetic activity with Boolean networks. Named for the nineteenth-century British inventor of the mathematical logic underlying computer programming, Charles Boole, these networks are characterized by parallel distributed processes, which Kauffman believes approximate genetic activity. In research he was conducting at the time, he had found that Boolean networks, made up of different nodes interconnected in multiple ways, display the distinguishing features of emergent self-organizing systems. Each node in the network can be either on or off and can be affected by surrounding nodes. Kauffman discovered that when these networks receive random inputs, they tend to settle into orderly patterns known as "state cycles." "This repeated series of states," Roger Lewin explains, "is in effect an attractor in the system, like the whirlpool in the treacherous sea of complex systems dynamics. A network can be thought of as a complex dynamical system, and is likely to have many such attractors."[42] In other words, given the right parameters, Boolean networks generate spontaneously emerging webs, which are self-sustaining patterns. These webs within networks result in what Kauffman describes as "order for free." As surprising as the formation of such order is the suddenness with which it emerges. Kauffman recounts the impact and implications of his discovery:

The order arises, sudden and stunning, in K = 2 *networks. For these well-behaved networks, the length of state cycles is not the square root of the number of states, but, roughly, the square root of the number of binary variables. Let's pause to translate this as clearly as we can. Think of a randomly constructed Boolean network with* N = *100,000 lightbulbs, each receiving* K = 2 *inputs. The "wiring diagram" would be like a madhaterly scrambled jumble, an impenetrable jungle. Each lightbulb has also been assigned at random a Boolean function. The logic is, therefore, a similar mad scramble, haphazardly assembled, mere junk. The system has* $2^{100,000}$ *or* $10^{30,000}$ *states— megaparsecs or possibilities—and what happens? The massive network quickly and meekly settles down and cycles among the square root of 100,000 states, a mere 317.*

I hope this blows your socks off. Mine have never recovered since I discovered this almost three decades ago. Here is, forgive me, stunning order. At a millionth of a second per state transition, a network, randomly assembled, unguided by any intelligence, would cycle though its attractor in 317-millionths of a second. This is a lot less than billions of times the history of the universe. Three hundred seventeen states? To see what this means in another way, one can ask how tiny a fraction of the entire space the network squeezes itself into. A mere 317 states compared with the entire space is an extremely tiny fraction of that state space, about 1 divided by $10^{29,998}!$[43]

Kauffman is convinced that the principle of order for free makes it possible to explain the origin as well as the development of life. "I believe that this 'order for free,'" he confesses, "which has undergirded the origin of life itself, has also undergirded the order in organisms as they have evolved and has even undergirded the very capacity to evolve itself."[44] Kauffman eventually extends his analysis from biological organisms to society and culture. At the biological level of investigation, he identifies three regimes or levels of adaptive self-organization: the emergence of life, the development of organisms, and the evolution of different life forms.

Drawing on his experiments with Boolean networks, Kauffman argues that life emerges in "autocatalytic sets," which are chemical versions of Hofstadter's strange loops. These loops, as Hofstadter explains, occur "wherever, by moving upwards (or downwards) through the levels of some hierarchical system, we unexpectedly find ourselves right back where we started."[45] In his explanation of the strange loops of autocatalytic sets, Kauffman elaborates Kant's interpretation of the organism in terms of the notion of intrinsic purpose or inner teleology:

Immanuel Kant, writing more than two centuries ago, saw organisms as wholes. The whole existed by means of the parts; the part existed because of and in order to sustain the whole. This holism has been stripped of a natural role in biology, replaced by the image of the genome as the central directing agency that commands the molecular dance. Yet an autocatalytic set of molecules is perhaps the simplest image we can have of Kant's holism. Catalytic closure ensures that the whole exists by means of the parts, and they are present both because of and in order to sustain the whole. Autocatalytic sets exhibit the emergent property of holism.[46]

Though philosophically abstract, Kauffman intends this analysis to describe very specific chemical processes. From its simplest to its most complex forms, life emerges in networks comprised of webs of interconnected webs. For life to originate, the chemicals in the prebiotic "soup" must become sufficiently diverse and adequately but not overly interconnected. In describing this process, Kauffman provides a network diagram that is remarkably similar to Hofstadter's chart of his "semantic network" (fig. 32; cf. fig. 24). As the number of chemicals or nodes proliferates and interconnections grow, networks "evolve to a natural state between order and chaos, a grand compromise between structure and surprise."[47] Like other theorists we have considered, Kauffman maintains that self-organization emerges between too much and too little order. In terms of network structure, this means that the site of emergence falls between too little connectivity, where systems are frozen, and too much connectivity, where they are chaotic. At a critical juncture, *more becomes different.* This is the tipping point where order emerges from disorder and patterns develop from noise. Describing the way in which autocatalysis occurs in "the networks of life," Kauffman writes:

At its heart, a living organism is a system of chemicals that has a capacity to catalyze its own reproduction. Catalysts such as enzymes speed up chemical reactions that might otherwise occur, but only extremely slowly. What I call a collectively autocatalytic system is one in which the molecules speed up the very reactions by which they themselves are formed: A makes B; B makes C; C makes A again. Now imagine a whole network of these self-producing loops. Given a supply of food molecules, the network will be able constantly to re-create itself. Like the metabolic networks that inhabit every living cell, it will be alive. What I aim to show is that if a sufficiently diverse mix of molecules accumulates somewhere, the chances that an autocatalytic system—a self-maintaining and self-reproducing metabolism—will spring up becomes a near certainty. If so, then the emergence of life may have been much easier than we have supposed.[48]

When Kauffman charts this process graphically, it looks like a diagram of the strange loops in Escher's *Waterfall* (figs. 33, 34).

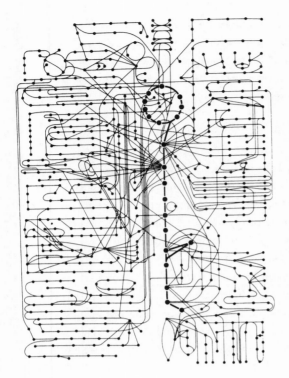

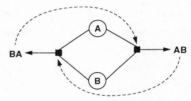

As I have noted, Kauffman argues that emerging self-organization operates from the simplest to the most complex levels of life. Moving from the strictly chemical to the genetic level of life, it becomes apparent that Kauffman rejects both Darwin's individualism and nominalism as well as the genocentric reductionism characteristic of much neo-Darwinism. Far from a simple aggregate of individual genes responsible for particular traits, the genome, Kauffman argues, is a complex adaptive system. When researchers triumphantly announce the discovery of the gene responsible for monogamy or even anxiety on the nightly news, the problems with such reductive genocentrism becomes obvious. Instead of viewing genes as separate entities programmed in different ways, Kauffman sees the genome as comprised of "networks of genes and their products interacting with one another in enormous webs of regulatory circuitry."[49] Accordingly, the genome is not a simple alphabet arranged according to fixed syntactic laws, but a complex web of webs whose interactions generate patterns that establish parameters of development for organisms.

Since complex systems are adaptive, their evolution tends to be coevolution. When systems as well as networks adapt to systems and networks that are

FIGURE 34 M. C. Escher, *Waterfall*. Lithograph, 1961. (© 2000 Cordon Art B.V., Baarn, Holland. All rights reserved.)

adapting to them, change is necessarily correlative. Gell-Mann underscores the importance of co-adaptation in his analysis of the interrelation of the genotype and the environment in the creation of the phenotype:

The genotype satisfies the criteria for a schema, encapsulating in highly compressed form the experience of the past and being subject to variation through mutation. The genotype itself does not usually get tested directly by experience. It controls, to a great extent, the chemistry of the organism, but the ultimate fate of each individual de-

189

pends also on environmental conditions that are not at all under the control of genes. The phenotype, in other words, is co-determined by the genotype and by all those external conditions, many of them random. Such an unfolding of schemata, with input from new data, to produce effects in the real world is characteristic of a complex adaptive system.[50]

Development at the level of the individual as well as the species takes place in webs within networks that are always morphing by adapting to each other. This coadaptation not only sets up a reciprocal interrelation between parts and whole within a system, but also establishes something like an ecological network with niches that form "fitness landscapes." Though particular changes and developments within these niches are contingent, the landscapes themselves are "nonrandom." This "nonrandomness" Kauffman argues, "is critical to the evolutionary assembly of complex organisms. We will find reasons to believe that it is not natural selection alone that shapes the biosphere. Evolution requires landscapes that are not random. The deepest source of such landscapes may be the kind of principles of self-organization that we seek. Here is one part of the marriage of self-organization and selection."[51]

Kauffman's argument that fitness landscapes are not random is an extension of his work on Boolean networks. The fitness landscape for a given individual or species does not develop in isolation but is always correlated with the other fitness landscapes. Each landscape, therefore, is like a node in a web of competing landscapes. Since different landscapes determine the constraints for each other, the network establishes the conditions of the possibility and impossibility of development. It is important to stress that constraints are not only negative; as new niches form through intersecting constraints, new opportunities unexpectedly emerge for individuals and species that are able to adapt. A change in one niche or landscape effectively transforms the ecology of the whole network. As Gell-Mann explains: "Ecological communities made up of many complex individuals, belonging to a large number of species, all evolving schemata for describing and predicting one another's behavior, are not systems likely to reach or even closely approach an ultimate steady state. Each species evolves in the presence of constantly changing congeries of other species."[52] Drawing on Per Bak's work, Kauffman argues that coadapting landscapes evolve to a point of self-organized criticality where minor changes in one landscape can trigger an avalanche of changes that ripple throughout the entire network. As a result of the complexity of the web and intricacy of the interrelations among landscapes, not all things are possible within a given network. Networks of fitness landscapes, like Boolean networks, tend to settle into rhythms with recurrent patterns. These are the patterns that function as something like attractors in dynamical systems. Pursuing the arguments of other complexity theorists, Kauffman identifies an additional attractor beyond the three described in classical dynamics. In addition to stable point attractors, which mark the end of change, as well as periodic and chaotic attractors, there appears be a fourth attractor poised at the edge of chaos. As always, this is where critical developments occur:

> The edge-of-chaos image arises in coevolution as well, for as we evolve, so do our competitors; to remain fit, we must adapt to their adaptations. In coevolving systems, each partner clambers up its fitness landscape toward fitness peaks, even as that landscape is constantly deformed by the adaptive moves of its coevolutionary partners. Strikingly, such coevolving systems also behave in an ordered regime, a chaotic regime, and a transition regime. It is almost spooky that such systems seem to coevolve to the regime at the edge of chaos. As if by an invisible hand, each adapting species acts according to its own selfish advantage, yet the entire system appears magically to evolve to a poised state where, on average, each does as best as can be expected.[53]

Lying along a margin of difference that is constantly changing, networks self-organize in patterns that remain relatively stable until another phase transition occurs.

PUNCTUATION

Shortly after discovering self-organization in Boolean networks in the early 1960s, Kauffman stumbled upon a book written by a student of C. H. Waddington. Like earlier morphologists, Waddington, who held that organisms must be considered as a whole, maintained that the task of biology is to study the genesis of form. In a 1963 book, *Temporal Organization in Cells,* Waddington's student, Brian Goodwin, extended his mentor's argument by demonstrating the way in which order spontaneously emerges in complex systems. At first, Kauffman feared Goodwin had anticipated his discovery; however, when he realized that their research was actually complementary, competition transformed into cooperation. In ensuing years, this cooperation has led to the emergence of many productive insights.

Kauffman never intended his analysis of complex adaptive systems ranging from the genome to ecological environments to replace the theory of natural selection. By bringing together the analysis of the dynamics of complex adaptive systems with natural selection, he attempts to develop an explanatory model that mediates apparent opposites such as whole and part, species and individual, necessity and chance, and reason and history. In any adequate theory, neither pole in these oppositions can be reduced to the other. Kauffman's account of self-organization and selection in evolution creates the possibility of rethinking Goodwin's reinterpretation of morphology and morphogenesis, and conversely Goodwin's morphological investigations shed new light on Kauffman's notion of self-organization. Echoing Kauffman, whom in certain instances he actually anticipates, Goodwin explores the implications of

the sciences of complexity *as they apply to our understanding of the emergence of biological forms in evolution, particularly the origin and nature of the morphological characteristics that distinguish different types of organism. These questions overlap those addressed by Darwin, but they focus on the large-scale, or global, aspects of biological form rather than on small-scale, local adaptations. As a result, there is no necessary conflict between the approaches, nor with the insights of modern biology into the genetic and molecular levels of organisms. These contribute to the construction of dynamical theories from which emerge higher-level properties of biological form and the integrated behavior of organisms. Conflict arises only when there is confusion about what constitutes biological reality. I take the position that organisms are as real, as fundamental, and as irreducible as the molecules out of which they are made. They are a distinct level of emergent biological order, and the one to which we most immediately relate.*[54]

While urging that the examination of macrosystems should be neither exclusive nor reductive, Goodwin freely admits he is more interested in the dynamics of "large-scale, or global, aspects of biological form" than in changes and adaptations at the microlevel. Kauffman, by contrast, is more attentive to the complex feedback and feed-forward loops between parts and whole. The process of emergent self-organization issues in morphological speciation and extinction, which constitute the parameters for the development of individual organisms. Mutations in individuals affect the structure of the species, and transformation at the species level redefine the evolutionary possibilities and impossibilities for individuals. When morphology is reinterpreted in terms of complex adaptive systems, and vice versa, natural selection appears to operate "in a fairly narrow range of possibility space, since it selects among entities that are already self-organized modules and that are in the process of spontaneously organizing into still higher levels of self-organization."[55] Since each level of organization is coadapting to other emerging self-organizing systems and networks, the structure of morphological networks is as complex as the architecture of a Frank Gehry building. When viewed as a whole, the overall dynamic of evolution is itself a complex adaptive system.

Two important changes in Darwin's theory follow from this conclusion. Darwin, we have seen, both affirms natural selection as purely a matter of chance and denies Larmarck's contention that the evolutionary process tends in the direction of greater complexity. If, however, everything from the genome and individual organisms to the species and the overall course of evolution conforms to the structure of complex adaptive systems, these claims become questionable. The emergence of a new morphological type is a bifurcation, which is unpredictable but not unintelligible. Mutations in individuals and changes in other species are, in effect, noise, which accumulates until a major structural reorganization occurs. Since gradual changes lead to sudden transformations, the path of evolution is not continuous but is best described by what Stephen Jay Gould labels "punctuated equilibrium." A new morphological type emerges, remains relatively stable for a period, and then unexpectedly transforms or disappears. The unpredictability of this process does not necessarily mean it has no direction. Rather than a teleological process, self-organizing systems can be understood as following a teleonomic trajectory tending toward increasing complexity. While each organism "is itself on the boundary between sub- and supracriticality, by trading their stuff, they collectively produce a supracritical biosphere, one that inexorably becomes more complex."[56] Since development is punctuated rather than continuous, the growth of complexity is episodic as well as unpredictable.

Kauffman's extensive research and speculation are inspired by deeply held philosophical, metaphysical, and religious beliefs, which often stand in tension with his scientific investigations. His obvious rejection of the existence of a Judeo-Christian God cannot disguise the profound longing for unity and reconciliation that lies at the heart of his work.[57] If Darwin and his followers are right when they claim that evolution is a matter of chance, human life would seem to be an accident. For Kauffman, such a vision renders life meaningless and makes it impossible to feel "at home in the universe." If, however, there is an emergent order to things that lends evolution a discernible order and probable direction, life has a logic that makes human existence meaningful:

In this view of life, organisms are not merely tinkered-together contraptions, bricolage, in Jacob's phrase. Evolution is not merely "chance caught on the wing," in **193**

Monod's evocative image. The history of life captures the natural order, on which se-
lection is privileged to act. If this idea is true, many features of organisms are not
merely historical accidents, but also reflections of the profound order that evolution
has further molded. If true, we are at home in the universe in ways not imagined
since Darwin stood natural theology on its head with his blind watchmaker.[58]

This order, or course, is neither the product of a purposeful designer nor programmed from the beginning; rather, evolution is an inner teleonomic process
in which order emerges spontaneously but not accidentally. "If I am right,"
Kauffman hopefully declares, "the motto of life is not We the improbable, but
We the expected."[59]

Kauffman does not limit his analysis to biological systems. Following the
lead of other complexity theorists, he extends his consideration of the dynamics
of complex adaptive systems to social, political, economic, and cultural evolution.
At every level, he maintains, evolution has the characteristics not only of an intel-
ligible but of an intelligent process that can no more be reduced to microchanges
in its constituents than Aunt Hillary can be reduced to individual ants, or a
Chuck Close painting can be reduced to its separate pixels. As networks become
more heterogeneous and interconnected, they begin to act as a whole in surprising ways. What distinguishes the current moment of complexity is the emergence
of a network culture that is truly *global.* Considering events now occurring,
Kauffman comments: "What is to become of our patchwork of civilization, ancient and new, drawn evermore tightly into one another's embrace? Like it or not,
some form of global civilization will emerge. We are at that particular time in history when population, technology, economics, and knowledge spin us together."[60] Without in any way denying the distinctiveness of our moment of
complexity, Kauffman notes that Hegel actually anticipated the logic of networks
over two hundred years ago: "Hegel gave us thesis, antithesis, synthesis. . . . These
ideas now stand discredited. Yet thesis, antithesis, synthesis sounds more than a
little bit like the evolution of the hundreds of millions of species that have come
and gone, or the evolution of technologies that have come and gone."[61] While
Kauffman overstates the similarities between Hegel's dialectical logic and the logic
of evolving self-organized networks, his reference to Hegel deserves careful consideration. Hegel realized long ago that nature and culture, as well as objectivity
and subjectivity, are spun together to form braids whose rhythms echo the intricacies of a fugue. To draw together dangling threads in the complex composition
I have been weaving, it is necessary to understand more adequately how human
subjects are constituted and the way cultures evolve.

seven

The mind cannot be understood as a single entity. It is more like a community; an assembly of intercommunicating parts. Where then, you might ask, does the "self" reside? Is it simply one of those parts, undistinguished and no better than any of its fellows?

andrew crumey

screening information

HOW THIS BOOK IS BEING WRITTEN

"The river is in torrent, because the rains which have fallen upon the city, and on the hills far upstream for days without end. . . . In these last days the river has become brown and swollen—yellow where the water is shallowest; like the color of parchment, or old skin. As I watch it run below me, I see the static ridges and folds which form as it passes beneath the arches, like hair intricately tressed and knotted; an unchanging pattern of flowing water."

I am not the author of these words. I, Vincenzo Spontiniam, am a colony of writers; a city of ideas. My work (which shall forever re-

main unwritten) is an amalgam of the various tastes, styles and interests of those whose ideas would seek to flow into the space which my literary identity is to oc-cupy.[1]

I, Mark C. Taylor, am not writing this book. Yet the book is being written. It is as if I were the screen through which the words of others flow and on which they are displayed. Words, thoughts, ideas are never precisely my own; they are always borrowed rather than possessed. I am, as it were, their vehicle. Though seeming to use language, symbols, and images, they use me to promote their cir-culation and extend their lives. The flux of information rushing through my mind as well as my body (I am not sure where one ends and the other begins) existed before me and will continue to flow long after I am gone. "My" thought—indeed "my" self—appears to be a transient eddy in a river whose banks are difficult to discern.

As boundaries become permeable, it is impossible to know when or where this book began or when and where it will end. Since origins as well as conclu-sions forever recede, beginnings are inevitably arbitrary and endings repeatedly deferred. One of the few things that is clear even if not obvious is that all writing is ghostwriting. This work, like all others, is haunted by countless specters. Some I know, others I do not; some I name, others remain unnamed. The unknown and unnamed are not, of course, absent—nor are they present. Their silence speaks through my words in ways that remain cryptic to author as well as the reader. The silent noise of ghosts clamoring for attention transforms me into a "colony of writers." *My* work forever remains unwritten because the text that is woven from borrowed threads is always "an amalgam of various tastes, styles and interests of those whose ideas would seek to flow into the space which my liter-ary identity is to occupy." My identity—literary as well as otherwise—is para-sitic upon the ghosts that haunt me. Just as my search is always a re-search, so my writing is always a re-writing. Rewriting does not merely repeat but also transforms in a way that complicates the parasite/host relationship. As the work takes shape, it becomes the host for ghosts now appearing as parasites. While I cannot write without the words of others, the Word of the other cannot survive if it is not resurrected in writing that appears to be my own. My words remain ghostly because they are haunted by others who have gone before and will haunt others yet to come. Writing always involves the screening of this spectral inter-play of parasites and hosts.

If this is what writing involves, why write? Why engage in this spectral in-terplay? There are no clear answers to such questions. The best writers find it impossible *not* to write even though they do not really know *why* they write. Writing, it seems, is the obsession of the possessed. For the possessed, writing is a search for *je ne sais quoi*. Far from original, this search that is a re-search always retraces the paths of others. The slate with which re-search begins is not blank, for it is always already inscribed with memorable patterns whose function is to screen experience. These patterns, which are neither fixed nor static but are always in formation, form a collective memory that both inhabits and surpasses the minds of individuals. Morphing screens simultaneously facilitate and inhibit my search in ways that transform all research into a long, slow, and often painful process. As "material" gradually accumulates, ideas and images, concepts and systems jostle with each other in a struggle for recognition. Why I am attracted to some thoughts and not to others—why I am repulsed by this writer rather than that one—remains mysterious to me. Even though I do not understand the draw, I know it is not merely intellectual—nothing ever is; something else is at work in this interplay. Eventually, the mix swirling in my mind becomes dense and diverse, like some primal soup slowly heating to the boiling point. As the temperature rises, matters become critical; ideas collide and combine to create noisy insights as well as insightful noise. Though seemingly arbitrary, associations are not totally free and interactions not completely random. Chance and constraint are braided in such a way that each creates the condition of the possibility of the other. The faster the motion, the greater the turbulence, and the more volatile the mix. All of this takes time; thinking has rhythms of its own—it must simmer and cannot be rushed. It is impossible to know just how much time is required for thought to gel because I am not in control of this process—nor is anyone else. Thought thinks through me in ways I can never fathom. Much—perhaps most—of what is important in the dynamics of thinking eludes consciousness.

When change occurs, it becomes obvious that the "combinatorial play" through which thought forms and transforms is not limited to ghosts from the past but extends to everyone with whom I am currently engaged. Thinking is impossible without implicit or explicit conversations with the living as well as the dead. Conversation reveals that thinking is not just taking but also giving. The give-and-take of thought stages a struggle for survival in which only the fittest images, concepts, ideas, and schemata survive. Rather than a matter of strength, fitness is measured by the capacity to connect and interrelate effectively and creatively. Thinking appears to be a constantly shifting puzzle in which forms, shapes, and patterns emerge from pieces that often are irregular. What makes this puzzle so complex is the way in which its pieces change in order to adapt to other pieces, which, in turn, are adapting to them. The interactivity of thinking complicates the moment of writing. The time of writing does not follow the popular figure of a line because present, past, and future are caught in strange loops governed by nonlinear dynamics. Past and future are knotted in the present in such a way that each simultaneously conditions and transforms the other. Neither complete nor finished, the past is repeatedly recast by a future that can never be anticipated in a present that cannot be fixed. Anticipation refigures recollection as much as recollection shapes expectation. The present, therefore, is doubly haunted by specters that approach by withdrawing and withdraw by approaching. As a result of this interplay of present, past, and future, the writing occurring through me recreates those whom it resurrects. The very identity of my ghosts changes with what I write. The child, after all, creates the parents as much as the parents give birth to the child.

Though the pieces of the puzzle never fit perfectly, gradual modifications can lead to major changes. Since thinking is a complex process in which images, concepts, and schemata are always struggling to adapt to each other, the pieces of the puzzle form networks of relations in which changes in a particular time or place ripple throughout the web. As ripples become waves, webs become less and less stable. When a growing number of experiences and ideas can no longer be adequately processed, thought is pushed far from equilibrium and approaches the tipping point. In this moment, danger and opportunity intersect. Driven to the edge of chaos and sunk in confusion, thinking either dissolves in madness or transforms in unexpected ways. The tipping point is the boiling point, which occurs when simmering ideas reach maximum turbulence. *If* change occurs, new patterns emerge and organize themselves spontaneously. In this moment when thinking happens, I do not so much write as I am written; creativity and destruction collide in the passion of writing. Though destruction is not always creative, creation is inevitably destructive.

The moment of writing is a moment of complexity in which multiple networks are cultured. If writing does not push limits to the tipping point, it is simply not worth the effort. The writing that matters disturbs more than it reassures; it drives authors as well as readers to the edge of chaos and abandons them. Writers realize that the pleasure of the text is not the satisfaction it provides but the dissatisfaction it engenders. The equilibrium of satisfaction is a symptom of death; the turbulence of dissatisfaction is the pulse of life. *If* writing has a point, it is to leave everyone and everything forever unsettled. But, of course, these are not my words but are the words of another who, like a stream rushing through me, refreshes but allows me no rest.

Far from mere opposites, simplicity and complexity, we have discovered, are braided "like hair intricately tressed and knotted." Such knots create binds and double binds that transform seemingly simple questions into exceedingly complex puzzles. The patterns we have been tracing are not only natural but also cultural, not only objective but also subjective, not only biological but also mental. As we become ever more deeply enmeshed in the logic of networks, the lines dividing such opposites become porous screens. But what exactly is a screen? And what precisely is screening? These questions are deceptive in their simplicity.

Screen, which, of course, can be either a noun or a verb, is a strange word in which multiple meanings pass through each other without losing definition. It derives from the stem *(s)kreu,* meaning to divide, cut, bite, scrape, or pluck. *(S)kreu* is also the root of the Latin *carnis* (flesh, cut off) and, by extension, *carnage, carnal, carnival,* and *incarnation,* as well as *cortex* (bark, cut off). The many meanings of *screen* include, as a noun,

· A moveable device, especially a framed construction, designed to divide, conceal, or protect, as a hinged or sliding room divider.
· Something that serves to divide, conceal, or protect.
· A coarse sieve used for sifting out fine particles as of sand, gravel, or coal.
· A system for appraising and selecting personnel.
· A window insertion of framed wire or plastic mesh used to keep out insects.
· The phosphorescent surface upon which the image is formed in a cathode-ray tube.
· A forged banknote.

And as a verb,

· To conceal from view.
· To protect, guard, or shield.
· To separate or sift out by means of a sieve or screen.
· To examine systematically in order to determine suitability.
· To show on a screen as a motion picture.[2]

A screen, then, is more like a permeable membrane than an impenetrable wall; it does not simply divide but also joins by simultaneously keeping out and letting through. As such, a screen is something like a mesh or net forming the site of passage through which elusive differences slip and slide by crossing and criss-crossing. But a screen is also a surface on which images, words, and things can be displayed. Every surface is actually a screen that hides while showing and shows while hiding. This duplicity of the screen is captured in the verb: to screen means both to conceal and to show. Enacting what it designates, screen implies that concealing is showing and showing is concealing.[3] Screen, screening, screenings: noun/verb, hide/show, conceal/reveal, absence/ presence, pollution/purity, darkness/light. . . . Forever oscillating between differences it joins without uniting, *the* meaning of *screen* remains undecidable. Far from a limitation, this undecidability is the source of rich insight for understanding what we are and how we know. In network culture, *subjects are screens* and *knowing is screening*.

Since the beginning of Western philosophy, knowledge and self have been inseparably related. When Plato argues that knowing is recollecting, he binds cognition to human memory. Though the forms through which reality is constituted, experience is filtered, and knowledge is structured are, according to Plato and his followers, eternal and as such surpass every individual, they nonetheless inhabit the soul and form the condition of the possibility of experience. As we have seen, from the divine Logos of early Christian apologists and medieval theologians, through the innate ideas of Descartes and a priori forms of Kant, to the psychosocial structures of Lévi-Strauss, generative grammar of Chomsky, and extended phenotype of Dawkins, Platonic forms repeatedly return in unexpected ways. While there obviously can be consciousness without self-consciousness, it is not clear whether there can be knowledge without self-knowledge. If knowledge in the strictest sense involves not only awareness but also awareness that we are aware, then knowledge presupposes self-consciousness. Self-consciousness, of course, can never be complete, for, as we have discovered, when reflection turns back on itself to become reflexive, it creates strange loops that cannot be closed. Rather than complete self-knowledge, adequate self-consciousness issues in the knowledge of what we can and cannot know about ourselves as well as the world. No one has seen this more clearly than Augustine.

Augustine's *Confessions* (381) is commonly acknowledged to be the first autobiography ever written. For Augustine, self-reflection is not an end in itself but, in the words of his medieval follower Bonaventura, the *itinerarium mentis in deum.* After beginning his search for God in the world of outer appearances, Augustine turns inward to contemplate his own mind. As a student of Plato and the neo-Platonists, he acknowledges an intimate relation between knowing and recollecting or remembering. In the remarkable tenth book of the *Confessions,* he develops an account of memory that still bears careful scrutiny. Following a lengthy exploration of what he vividly describes as the vast "fields and spacious palaces of memory, where lie the treasures of innumerable images of all kinds of things that have been brought in by senses," Augustine finally concludes that *cogito* (to think, reflect) is, in effect, *cogo* (to bring together, collect):

By the act of thought we are, as it were, collecting together things which the memory did contain, though in a disorganized and scattered way, and by giving them our close attention we are arranging for them to be as it were stored up ready to hand in the same memory where previously they lay hidden, neglected, and dispersed, so that now they will come forward to the mind that has become familiar with them. . . . In fact what one is doing is collecting them from their dispersal. Hence the derivation of the word "to think." For cogo *(to collect) and* cogito *(to think) are in the same relation to each other as* ago *and* agito, facio *and* factito. *But the mind has appropriated to itself this word* (thinking), *so that it is only correct to say "think" of things which are "re-collected" in the mind, not things re-collected elsewhere.*[4]

Thinking, then, involves sorting, selecting, and processing the data of experience stored in "the belly of the mind." The patterns and programs through which this processing occurs are not, according to Augustine, derived from experience but are universal traces of the divine Logos in the human mind. Paradoxically, as his self-consciousness grows, he becomes less and less comprehensible to himself. Through images that are as provocative as they are evocative, Augustine restlessly probes what he cannot comprehend. In the endless "caverns and abysses" of his memory, he discovers "secret, numberless, and indefinable recesses." Awestruck by his own incomprehensibility, he calls out to God:

How great, my God, is this force of memory, how exceedingly great! Like a vast and boundless subterranean shrine. Who has ever reached the bottom of it? Yet this is a faculty of my mind and belongs to my nature; nor can I myself grasp all that I am. Therefore, the mind is not large enough to contain itself. But where can that uncontained part of it be? Is it outside itself and not inside? In that case, how can it fail to contain itself? At this thought great wonder comes over me; I am struck dumb with astonishment.[5]

Long before Freud, Jung, and Lacan, Augustine recognizes the force of unconscious *thinking.* Since "the mind is not large enough to contain itself," thinking can exceed consciousness as well as self-consciousness. This awareness only deepens the mystery of subjectivity. "Great indeed is the power of memory!" exclaims Augustine. "It is something terrifying, my God, a profound and infinite multiplicity; and this thing is the mind, and this thing is I myself. What then am I, my God? What is my nature? A life various, manifold, and quite immeasurable."[6] This is not, of course, the end of Augustine's journey. Eventually he passes beyond the "huge court of memory" to the throne of God, where every mystery disappears. When the last screen is lifted, knowledge and self-knowledge form a perfect union.

Initially, the clarity of Augustine's vision and direction of his journey seem alien to the contemporary world. Yet the notion of subjectivity and account of knowledge he so deftly develops sheds surprising light on the paradoxes of subjectivity and the dilemmas of thinking in emerging network culture. In his comprehensive and informative book, *The User Illusion: Cutting Consciousness Down to Size,* Danish science writer Tor Nørretranders echoes Augustine when he declares: "I realize that I am more than my *I.*"[7] For Nørretranders, as for Augustine, the self *exceeds* itself; the mind, in Augustine's words, "is not large enough to contain itself." While this condition is not new, the circumstances of network culture are creating changes that are transforming the processes of thinking and knowing as well as the structures and patterns of subjectivity. As the networks passing through us become more complex and the relations at every level of experience become more extensive and intensive, the speed of change accelerates until equilibrium disappears and turbulence becomes a more or less permanent condition. While occasioning confusion, uncertainty, and sometimes despair, this inescapable turbulence harbors creative possibilities for people and institutions able to adapt quickly, creatively, and effectively. Those who are too rigid to fit into rapidly changing worlds become obsolete or are driven beyond the edge of chaos to destruction.

Though there are multiple sources of turbulence, one of the most important factors creating unrest in today's world is the unprecedented noise generated by proliferating networks whose reach extends from the local to the global. As networks relentlessly expand, the mix of worlds, words, sounds, images, and ideas becomes much more dense and diverse. When this media-mix approaches the boiling point, multiple cognitive and cultural changes become inevitable. The intricate dynamics of information processing and complex adaptive systems, which we have been exploring, can help us to understand how these changes in thinking subjectivity are occurring and what their implications are for life in the twenty-first century.

Noise, we have discovered, is never absolute; rather, noise and information are bound in a relation in which each is simultaneously parasite and host for the other. What is noise in one context, at one level, or at one time, is information in another context, at another level, or at another time, and vice versa. One of the distinguishing characteristics of contemporary experience is that the excess of information creates noise. Swelling rivers of information and streams of data ceaselessly flow through us. Since most of these data remain unprocessed, we are unconscious, though not necessarily unaware, of much of the information coursing through our bodies and minds. The very streams circulating through consciousness and the unconsciousness, creating psychological, social, and cultural turbulence, also form the reservoirs from which knowledge can be fashioned and meaning emerge. Knowledge and meaning assume form when the flux of information is effectively channeled through processes of multiple screenings.

I have noted that etymology suggests that screening entails a process of sculpting or cutting away. Such sculpting is, in effect, an editing in which excess information is filtered. Far from exclusive opposites, noise is both information waiting to be screened and the remainder, refuse, or debris left over after screening occurs. There can no more be noise without information than there can be information without noise. Nørretranders explains this interplay between information and noise by drawing a distinction between information and what he describes as "exformation":

Exformation is perpendicular to information. Exformation is what is rejected en route, before expression. Exformation is about the mental work we do in order to make what we want to say sayable. Exformation is the discarded information, everything we do not actually say but have in our heads when or before we say anything at all. Information is the measurable, demonstrable utterances as we actually come out with it. The number of bits or characters in what is actually said.[8]

Exformation, in other words, is what is left out as information is formed from noise. As such, exformation is not simply absent but is something like a penumbral field from which information is formed. Since information is constituted by what it excludes, it inevitably harbors traces of noise. Noise, we have noted, is always in-formation in at least two ways. First, noise is always forming into information and being formed by the processes of exclusion from information; and second, noise does not simply disappear but remains *in* information as a haunting specter. There is, undeniably, a certain destructive dimension to the processing of information. Computer scientist and inventor Ray Kurzweil goes so far as to insist that the "destruction of information" is "the key to intelligence." "The value of computation," he argues, "is precisely in its ability to destroy information *selectively*. For example, in a pattern-recognition task such as recognizing faces or speech sounds, preserving the information-bearing features of a pattern while 'destroying' the enormous flow of data in the original image or sound is essential to the process. Intelligence is precisely this process of selecting relevant information carefully so that it can skillfully and purposefully destroy the rest."[9] Whether information is actually destroyed, as Kurzweil argues, or excluded but not necessarily destroyed, as I would insist, screening simultaneously filters noise and displays information by channeling it into the patterns that eventually constitute knowledge.

The screening critical to channeling experience, articulating knowledge, and cultivating meaning occurs through *dynamic* patterns. While functioning in ways similar both to Kant's a priori forms of intuition and categories of understanding as well as to Lévi-Strauss's psychosocial structures, these patterns are not fixed but emerge and change over time. In his examination of the "self-organization of [the] brain and behavior," J. A. Scott Kelso describes the complex dynamics of brain and mind activity through the metaphor of a river:

Like a river whose eddies, vortices, and turbulent structures do not exist independent of the flow itself, so it is with the brain. Mental things, symbols and the like, do not sit outside the brain as programmable entities, but are created by the never ceasing dynamical activity of the brain. The mistake made by many cognitive scientists is to view symbolic contents as static, timeless entities that are independent of their origins. Symbols, like the vortices of the river, may be stable *structures or patterns that persist for along time, but they are not timeless or unchanging.*[10]

The operations of brain and mind as well as their nonlinear interrelations are governed by principles we have already seen at work in complex adaptive systems. Within dynamic complex adaptive systems, patterns are always context dependent. Information and noise, as well as knowledge and meaning, are joined in something like a figure/ground relation. "Thought," George Lakoff correctly argues, "has *gestalt properties* and is thus not atomistic; concepts have an overall structure that goes beyond merely putting together conceptual 'building blocks' by general rules." When elaborating the significance of these "gestalt properties," Lakoff invokes what Bateson describes as the ecology of mind: "Thought has an *ecological structure*. The efficiency of cognitive properties, as in learning and memory, depends on the overall structure of the conceptual system and on what the concepts mean. Thought is thus more than just the mechanical manipulation of abstract symbols."[11] The switch from understanding mental activity mechanically to analyzing thinking as a dynamic adaptive process makes it possible to reinterpret mind and culture in terms of complex systems rather than simple and stable structures. Explaining the necessary "conditions for self-organization," Kelso writes:

> *Patterns arise spontaneously as the result of large numbers of interacting components. If there aren't enough components or they are prevented from interacting, you won't see patterns emerge or evolve. The nature of the interactions must be nonlinear. This constitutes a major break with Sir Isaac Newton, who said in Definition II of the* Principia: *"The motion of the whole is the sum of the motion of all the parts." For us, the motion of the whole is not only greater than, but* different *than the sum of the motions of the parts, due to nonlinear interactions among the parts or between parts and the environment.*[12]

From this point of view, mental activity, like Aunt Hillary, emerges from the interrelations of particular events without any centralized agency or directing agent. In other words, "the system organizes itself, but there is no 'self,' no agent inside the system doing the organizing."[13] This is a crucial point with far-reaching implications. The generation of mind does not presuppose the prior existence of mind or of any kind of purposeful agent. Instead of intentionally formed patterns through which experience is screened, the mind is generated by complex interrelations among patterns that emerge spontaneously. Insofar as human subjectivity or selfhood necessarily entails mental activity, the self is the *result* rather than the presupposition of screening information.

CULTURING NETWORKS

The patterns through which information is screened and subjectivity assumes form can be understood in terms of what I have described as complex adaptive systems. In this context, it is helpful to recall Gell-Mann's definitive description of the operation of complex adaptive systems:

> *The common feature of all these processes is that in each one a complex adaptive system acquires information about its environment and its own interaction with that environment, identifying regularities in that information, condensing those regularities into a kind of "schema" or model, and acting in the real world on the basis of*

that schema. In each case, there are various competing schemata, and the results of the action in the real world feed back to influence the competition among these schemata.[14]

While recognizing the similarities between complex adaptive systems and the fluid dynamics of turbulence, Gell-Mann, as we have seen, also stresses how they differ. The difference between the ways in which eddies form in rivers and patterns form in minds, he insists, "lies in the way information about the environment is recorded."[15] The critical issue is the role of schemata in complex adaptive systems.[16] Emerging schemata identify, compress, and store the regularities of experience in a way that makes it possible for the system to adapt by responding quickly and effectively. The relation between pattern and environment, therefore, is not merely reactive but involves coadaptive feedback and feed-forward loops. When confronted with what amounts to news, schemata must respond and sometimes adapt. Information in one pattern is noise in another and vice versa. Gell-Mann summarizes processes that include yet surpass cognitive activity:

In the presence of new information from the environment, the compressed schema unfolds *to give prediction or behavior or both.*

When the compression took place, regularities were abstracted from experience and compressed. The rest of experience, ascribable to change or to regularities too subtle to recognize, cannot be compressed and does not typically form part of the schema. When unfolding takes place, new material is adjoined, much of it again largely random, as "present data" or input from the real world.[17]

Gell-Mann cites two examples of this process: the first, which we have considered in the previous chapter, is biological evolution, and the second is a scientific theory.[18] In the following analysis, we will see not only that biological and mental processes are isomorphic but that, when taken together, they constitute yet another complex adaptive system. What must be stressed at this point is that the activity through which schemata adapt to the environment involves a constant fluctuation between information and noise. When there are too many discrepancies between the theory and the data of experience, new ideas must be explored and concepts formulated. If the input from the so-called real world cannot be effectively processed, the schema either adapts or becomes obsolete. Cognitive adaptability entails an intricate interplay of competition and cooperation between and among alternative patterns and schemata. "Complex adaptive systems," Gell-Mann concludes, "operate through the cycle of variable schemata, accidental circumstances, phenotypic consequences, and feedback of selection pressures to the competition among schemata. They explore a huge space of possibilities, with openings to higher levels of complexity and to the generation of new types of complex adaptive systems."[19] As with all such competition, survival depends on fitness. In this case, fitness involves both interrelations among perceptual and conceptual components of schemata and the relation between schemata and the data of experience.

The patterns through which experience is filtered and information processed operate at different levels and through multiple media. As one proceeds from sensation and perception through cognition and conception to reflection and speculation, there is a movement from the concrete to the abstract. The transition from the particular to the general is marked by progressive screenings that filter out more and more information. The more refined the screens, the narrower their bandwidth. Since these filters are conceptual as well as perceptual, screens must operate through different media. At the conceptual level, ideas, categories, names, models, and paradigms pattern data mediated by the senses. Sensual perception, however, is never raw; it is always cooked according to recipes that bubble up in the stew of experience. Sensory filters can be both visual and auditory; images, pictures, representations, even logos and brands, as well as sounds, rhymes, jingles, tunes, and melodies structure awareness and direct attention. The screening of information always begins below or beyond the level of consciousness. Daniel Dennett observes that "we have come to accept without the slightest twinge of incomprehension a host of claims to the effect that sophisticated hypothesis-testing, memory searching inference—in short, information processing—occurs within us even though it is entirely inaccessible to introspection. It is not repressed unconscious activity of the sort Freud uncovered, the activity driven out of the 'sight' of consciousness, but just mental activity that is somehow beneath or beyond the ken of consciousness altogether."[20] As one moves from this preliminary information processing to consciousness, screened perceptions must be gathered or, in Augustine's terms, "collected" into coherent objects. Consciousness requires the interrelation of an object and a subject, who is not yet self-aware. Neither subjects nor objects exist prior to or independent of this active interrelation; each emerges in and through the other, and therefore neither can be reduced to the other. Furthermore, objects as well as subjects are always caught in networks of relations with other objects and subjects. The relations between and among objects, like the interplay of subjects and objects, are not extrinsic but are intrinsic to the identity of particular objects. Thus, when a subject (S_A) interacts with an object (O_A), it simultaneously relates to all other objects in the network (O_B, O_C, etc.) (fig. 35). It is important to stress at this juncture that the information processing occurring below or beyond the level of consciousness also conforms to the structure and operation of the complex adaptive systems that govern consciousness.

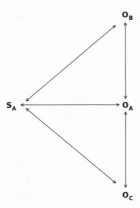

FIGURE 35 Consciousness

The preliminary processing of information through perceptual screens is a necessary but not sufficient condition of knowledge. Contrary to popular opinion and many philosophical epistemologies, knowledge does not involve the union or synthesis of an already existing subject and an independent object. To the contrary, knowing is an ongoing adaptive process in and through which subjectivity and objectivity actually emerge and continue to evolve. Knowledge is constituted when subject and object *fit* together. As such, knowledge requires some form of consciousness but does not necessarily entail self-consciousness. For consciousness to become self-consciousness, the knowing subject must bend back on itself and become reflexive. In other words, the subject, which emerges through interaction with objects, must apprehend *itself* as an object. This reflexive turn further complicates an already tangled situation. When a subject (S_A) becomes conscious of itself as a subject conscious of objects (S_A-O_A), self-consciousness is doubled in such a way that it comes into mediated relationship with the entire network of objects disclosed through consciousness (O_B, O_C, etc.) (fig. 36). Furthermore, different subjects, (S_A, S_B, S_C, etc.) just like different objects (O_A, O_B, O_C), now appear to be caught in mutually constitutive webs (fig. 37). Insofar as self-conscious subjects interact, they are not only constitutively related to each other but also stand in mediated relations with all of the objects of which other self-conscious subjects are aware (fig. 38).

As if all this were not complicated enough, everything now must be set in motion. What has been presented so far oversimplifies by abstracting synchronic relations from diachronic development. In complex adaptive systems, everything is always changing. Networks of objects, networks of subjects, and the interplay between them constantly evolve. The identity of any subject (S_A) or object (O_A) is not only constituted by their interaction (S_A-O_A); it also emerges over time through the intersection of past (T^1) and future (T^3) in the present (T^2). S^1_A=

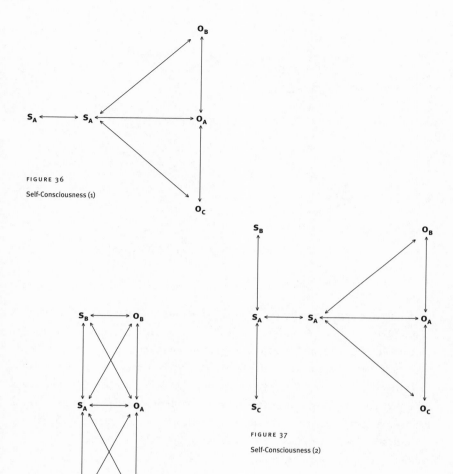

FIGURE 36

Self-Consciousness (1)

FIGURE 37

Self-Consciousness (2)

FIGURE 38

Self-Consciousness (3)

Subject$_A$ at T^1, O^1_A = Object$_A$ at T^1, and S^1_A-O^1_A = the subject-object relation at T^1, and so on. When synchronic and diachronic relations are woven together, the tapestry of the complex adaptive networks within which knowledge emerges and subjects as well as objects are formed comes into full view (fig. 39). This is not to suggest, of course, that this picture of consciousness and self-consciousness is complete. The strange loops we have discovered in our exploration of the

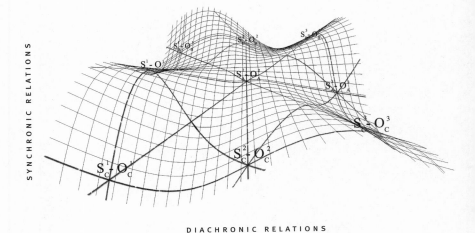

SYNCHRONIC RELATIONS

DIACHRONIC RELATIONS

FIGURE 39 Emerging Knowledge and Thinking Subjectivity

paradoxical structure of reflexivity render consciousness inescapably incomplete and self-consciousness forever open. It is precisely this openness and incompleteness that keep thinking subjectivity constantly in motion.

Information, of course, is not knowledge, and knowledge is not necessarily meaningful. We can know many things about the world and ourselves without grasping their meaning. The articulation of meaning extends the information processing, which begins in sensation and perception, and continues in consciousness and self-consciousness, by screening knowledge to form coherent and relatively comprehensive patterns. This process takes place through reasoning. Reason seeks to determine not only *what* happens but *why* it happens. Reason, in other words, seeks reasons. Though operating at a different level, reasoning involves the same kind of activity of collection at work in consciousness and self-consciousness. Just as consciousness and self-consciousness gather the data of sensation and perception to form coherent subjects and objects of knowledge, so reason collects what is known to form meaningful patterns. Meaning is always relational and thus inevitably contextual. As relations change and context shifts, meaning transforms. While there are many ways to weave meaning from the multiple strands of experience, two are particularly important: theory and myth. Both theories and myths are comprised of networks of symbols or, more precisely, networks of networks of symbols. Symbols, as the word implies (*sun,* together + *ballein,* to throw), throw things together. Reason deploys symbols to throw together data presented through other modes of information processing.

210

While formally similar, theory and myth differ most significantly in their relation to time. Theory, on the one hand, abstracts from temporal development in
an effort to determine universal truth, and, myth, on the other, elaborates nar-
ratives to suggest general, if not universal, truths. Though it is often difficult, if
not impossible, to be sure where myth ends and history begins, the temporal dimension of myth does not, of course, necessarily render narrated events historical. Nevertheless, myths tend to preserve the temporality of experience in ways
that theories do not. As I have noted, Gell-Mann maintains that theories—scientific and otherwise—function as complex adaptive systems. This line of analysis now can be extended to the symbol systems comprising myths.

Working at the intersection of neuroscience, cognitive psychology, and evolutionary anthropology, Terrence Deacon develops an analysis of the structure
and function of symbol systems that both illuminates and is illuminated by the
account of complex adaptive systems developed by complexity theorists. Summarizing his conclusion about the way in which symbols shape the cognitive
landscape, he writes:

*Symbols cannot be understood as an unstructured collection of tokens that map to a
collection of referents because symbols don't just represent things in the world, they
also represent each other. Because symbols do not directly refer to things in the world,
but indirectly refer to them by virtue of referring to other symbols, they are implicitly
combinatorial entities whose referential powers are derived by virtue of occupying determinate positions in an organized system of other symbols. Both their initial acquisition and their later use require a combinatorial analysis. The structure of the whole
system has a definite semantic topology that determines the ways symbols modify each
other's referential functions in different combinations. Because of this systematic relational basis of symbolic reference, no collection of signs can function symbolically unless the entire collection conforms to certain overall principles of organization.[21]*

In terms of our previous analysis, reference is relational and relations are referential. The relations in and through which symbols emerge form complex networks, which establish constraints that are simultaneously restrictive and productive. Each symbol within these networks is a node in a web of relations.
Indeed, a symbol is nothing other than the intersection of relations knotted in
nodes: "Symbols do not, then, get accumulated into unstructured collections
that can be arbitrarily shuffled into different combinations. The system of representational relationships, which develops between symbols as symbol systems
grow, comprises an ever more complex matrix. In abstract terms, this is a kind of

211

tangled hierarchic network of nodes and connections that defines a vast and constantly changing semantic space."[22] The hierarchy of such networks establishes the multiple levels at which information processing occurs. As one moves from perception through consciousness and self-consciousness to reason, more and more data are screened as the patterns of awareness become increasingly compressed. Paradoxically, this winnowing of data is at the same time an expansion of comprehension. As idiosyncratic perceptions give way first to general concepts and ideas, and then to shared meanings, awareness becomes more comprehensive and experience more comprehensible. As a person becomes fluent in a language of symbol and myth, he or she enters into the community of experience that speakers of the language share. There is no community without communication and no communication without community.

The hierarchy of screening processes should not obscure the fact that the relationship of perception, consciousness, self-consciousness, and reason is neither linear nor unidirectional. Different ways of processing information are joined in recursive loops that are mutually determinative. While emerging from and conditioned by perception, consciousness, and self-consciousness, the theories and myths through which we reason torque so as to inform knowledge and figure perception. The two-way circuits joining these mental activities make the formation of knowledge an open-ended process (fig. 40). Moreover, since experience always exceeds our capacity to process it and self-consciousness cannot be complete, the meanings forged from knowledge are never stable but always changing. This instability holds the promise of creativity and carries the threat of destruction.

Once again, an already complex analysis must be further complicated. Just as networks of symbols comprise myths, so different myths form networks in which their specificity is a function of their difference from and, thus, relation to other myths. Constituting a structure bordering on the fractal, myths are networks of networks made of up of nodes within nodes (fig. 41). Never fixed, these networks of symbols and myths must constantly adapt to each other. Though neither programmed nor planned, such changes tend in the direction of ever-greater complexity. Like other complex adaptive systems, myths change in response to what appears to be noise. In the case of myths, noise is generated by experiences that cannot be adequately processed and by conflicts between and among different symbols and myths. There are three possible responses to such noise: myths can be reinforced, transformed, or destroyed. The most conservative response to apparently discrepant experiences is to try to explain them through the accepted myth by providing reasons and explanations, which are not immediately obvious. When successful, this response serves to reinforce the

212

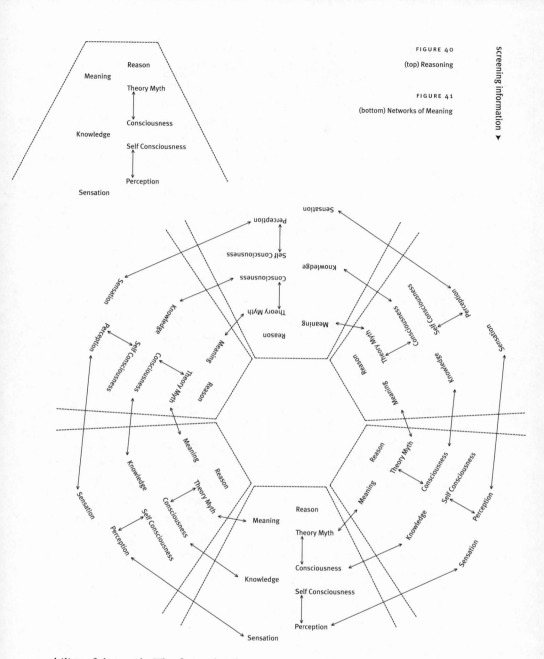

FIGURE 40

(top) Reasoning

FIGURE 41

(bottom) Networks of Meaning

stability of the myth. The fewer the alternative points of view, the more stable particular patterns of explanation tend to remain. As networks of travel and communication expand, however, interpretive schemes proliferate and doubts about the veracity and viability of any single perspective begin to develop. Initially, these doubts trigger reactions designed to protect one's symbolic and mythic turf by keeping others out or by converting them to one's own point of view. This is one of the reasons the processes of globalization have unleashed resurgent nationalisms and fundamentalisms throughout the world. Though these reactions often issue in devastating violence, they are ultimately fated to fail. In emerging network culture, walls, we are discovering, become permeable membranes or, more precisely, screens, which allow differences to persist while permitting passage across borders that cannot be fixed.

As different myths mingle and mix in a culture that is increasingly global, particular symbol systems and networks of myths are pushed to the edge of chaos where they either transform or collapse. Transformation can only occur far from equilibrium. As incongruous experiences and dissonant interpretations accumulate, the conditions for psychosocial dislocation as well as creative change gradually take shape. Expanding networks of communication promoting rapid symbolic exchange generate a density and diversity of symbolic and mythic resources that make creative innovation possible though not inevitable. If innovation occurs, it is a function of what Calvino describes as "combinatorial play." While such play is not necessarily deliberate and is not always the result of creative individuals, different strategies can be deployed to facilitate it. New meanings are rarely planned or programmed but usually emerge in surprising ways. At the cusp of purpose and chance, words, images, and symbols are thrown together to create new meanings that are as unpredictable as they are uncontrollable. In this innovative practice, creation is recreation, which is realized through a cutting-and-splicing that transforms culture into recombinant culture. Within this combinatorial play, meanings are interactive events, which usually occur in multiple media. Rather than viewing events as meaningful, meaning must be understood as an event. Events of meaning emerge from but are not reducible to the interplay of particular symbols and different myths. As interpretive schemata change to adapt to each other, they eventually reach the tipping point where new comprehensive patterns suddenly emerge. Such patterns survive only as long as they fit experience.

VIRAL WEBS

Insofar as myths and the networks of symbols comprising them function as complex adaptive systems, they form something like what E. O. Wilson labels a "superorganism," which is both independent of and lives in and through individual minds.[23] Deacon's account of how languages emerge and evolve helps to clarify this dimension of the dynamics operating in myths. In contrast to analysts such as Lévi-Strauss and Chomsky, who understand language as a static structure or fixed program, Deacon maintains that languages more closely resemble emergent self-organizing systems. As such, they tend to develop in ways that best promote their own survival:

The world's languages evolved spontaneously. They were not designed. If we conceive of them as though they were invented systems of rules and symbols, intentionally assembled to form logical systems, then we are apt either to assign utility and purpose where there is none, or else to interpret as idiosyncratic or inelegant that for which we cannot recognize a design principle. But languages are far more like living organisms than like mathematical proofs. The most basic principle guiding their design is not communicative utility but reproduction—theirs and ours. So, the proper tool for analyzing language structure may not be to discover how best to model them as axiomatic rule systems but rather to study them the way we study organism structure: in evolutionary terms. Languages are social and cultural entities that have evolved with respect to the forces of selection imposed by human users.[24]

Several important conclusions follow from this seminal insight. For Deacon, human beings, especially children, "are the vehicle by which language gets reproduced." At one point, he goes so far as to insist that languages need people more than people need language. This line of analysis reverses the ordinary angle of vision; instead of human beings evolving in ways that enable them to use language, language adapts to the people who use it. "Languages have had to adapt to children's spontaneous assumptions about communication, learning, social interaction, and even symbolic reference, because children are the only game in town."[25] To underscore this point, Deacon invokes the metaphor of a *virus*.[26] Always lurking on the border between living and nonliving, viruses are strings of DNA or RNA that infect the cells through which they replicate themselves. If language is like a virus, then language and the minds and brains it inhabits are joined in a parasite/host relationship. As we have discovered, parasite and host form an undecidable relation in which each functions simultaneously as itself and the other. In this case, languages are parasites, which depend on their human hosts, and human beings are parasitic upon the linguistic host, which, in some sense, makes them human. Since neither can exist apart from the other, languages and human beings must coadapt and thus coevolve. Deacon underscores the importance of recognizing "a coevolutionary dynamic between language and its host. By imagining language as a parasitic organism, we can come to appreciate the potential for conflicting reproductive interests, where some language features might occur at the expense of the host's adaptations, and the possibility that many features may have more to do with getting passed on from generation to generation than with conveying information."[27]

What Deacon describes as a language virus functioning like strings of DNA or RNA, Gell-Mann, borrowing a term from Hazel Henderson, dubs "cultural DNA." In a manner analogous to the genome, cultural DNA "encapsulate[s] the shared experience of many generations and comprise[s] the schemata for the society, which itself functions as a complex adaptive system."[28] In recent years, several leading biologists, cultural theorists, and cyberpunk writers have become fascinated by the similarities between biological and cultural evolution. Many critics have responded to arguments supporting the notion of cultural *evolution* with considerable skepticism. While it cannot be denied that cultures change, it is not immediately obvious that they *evolve*. Widespread doubts about grand theories and comprehensive metanarratives developed by many philosophers and literary critics in the past three decades have led to ideological opposition to the very idea of cultural evolution. The claim that social and cultural systems evolve is not, of course, new; what is distinctive about recent arguments is the way in which the analysis of genetics in terms of information processes has led to a rethinking of cultural evolution by *biologists*. When insisting on parallels between biological and cultural development, some influential interpreters are currently arguing that cultural evolution is governed by the same rules and principles that we have already analyzed in the evolution of complex adaptive systems. In their 1981 book, *Genes, Mind, and Culture: The Co-Evolutionary Process,* Charles J. Lumsden and Edward O. Wilson coin the term "culturgen" to define "the basic unit of inheritance in cultural evolution":

A culturgen is a relatively homogeneous set of artifacts, behaviors, or mentifacts (mental constructs having little or no direct correspondence with reality) that either share without exception one or more attribute states selected for their functional importance or at least share a consistently recurrent range of such attribute states within a given polyphonic set.[29]

As the subtitle of the book suggests, Lumsden and Wilson argue that biological organisms and cultural "superorganisms" coevolve. In arguing for the coevolution of genes and culturgens, they attempt to counterbalance certain aspects of the reductive tendencies informing earlier versions of sociobiology. The coevolutionary process now appears to be directed by "epigenetic rules," which can be empirically identified and conceptually described. These epigenetic rules

*are the genetically determined procedures that direct the assembly of the mind, in-
cluding the screening of stimuli by peripheral sensory filters, the internuncial cellular
organizing process, and the deeper processes of directed cognition. The rules comprise
the restraints that the genes place on the development (hence the expression "epigenetic"), and they affect the probability of using one culturgen as opposed to another.
The probability distributions themselves are appropriately termed* usage bias curves
or, more simply, just bias curves.[30]

As this remark suggests, and as we will see in more detail in what follows,
though claiming to support a coevolutionary perspective, Lumsden and Wilson
never completely divest themselves of the sociobiological perspective of their earlier work and thus still tend to privilege the biological over the cultural. What is
important to stress at this point is that their effort to find common principles
that are applicable to biological and cultural processes can be developed into a
more genuinely coevolutionary position.

The complex issues Deacon, Wilson, and Lumsden investigate have entered
popular culture through Richard Dawkins's widely discussed notion of the
meme. The meme is to the mind and culture what the gene is to the brain and
body. Memes, like genes, are replicators, which form the basis of an evolutionary
process. "Just as genes propagate themselves in the gene pool by leaping from
body to body via sperms or eggs, so memes propagate themselves in the meme
pool by leaping from brain to brain via a process, which, in a broad sense, can be
called imitation."[31] While genes bear coded information for synthesizing pro-
teins, memes carry instructions for mental and cultural activities. Just as individual organisms are the means by which genes produce themselves, so individual minds are the vehicles for memetic replication. Dawkins joins Deacon and
others in comparing memes to viruses: "Memes should be regarded as living
structures, not just metaphorically but technically. When you plant a fertile
meme in my mind you literally parasitize my brain, turning it into a vehicle for
the meme's propagation in just the way that a virus may parasitize the genetic
mechanism of a host cell."[32] Memes, then, spread like parasitic viruses infecting
the hosts without which they cannot survive. Unlike some of his more literal followers, Dawkins readily admits that "it is a matter of dispute whether the resemblance between gene and meme is good science or bad poetry."[33] In explaining his scientific and cultural theories, he tends to use metaphors that are
particularly revealing. In *River out of Eden: A Darwinian View of Life,* for example, Dawkins explores what he describes as "the digital river" running through
us. "The river of my title," he explains, "is a river of information, not a river of

bones and tissues: a river of abstract instructions for building bodies, not a river of solid bodies themselves."[34] Elsewhere he represents this digital river as software and the brain as the hardware on which it runs.

Returning to the evolution of the human brain, what are we looking for to complete the analogy? A minor improvement in hardware, perhaps a slight increase in brain size, which would have gone unnoticed had it not enabled a new software technique which, in turn, unleashed a blossoming spiral of co-evolution? The new software changed the environment in which brain hardware was subject to natural selection. This gave rise to strong Darwinian pressure to improve and enlarge the hardware, to take advantage of the new software, and a self-feeding spiral was under way with explosive results.[35]

This self-feeding spiral, which can trigger explosive results, suggests the feed-forward processes of complex adaptive systems, which, at critical points, unleash effects disproportionate to their causes. For such change to occur, memes cannot be isolated entities but must be joined in complex networks, which both facilitate and require coadaptation. Dawkins names the complex adaptive networks "memeplexes": "Just as a species gene pool becomes a cooperative cartel of genes, so a group of minds—a 'culture', a 'tradition'—becomes a cooperative cartel of memes, a memeplex. . . . As in the case of genes, it is a mistake to see the whole cartel as a unit being selected as a single entity. The right way to see it is in terms of mutually assisting memes, each providing an environment which favors the others."[36] To avoid misunderstanding Dawkins's position, it is important to stress that memes are also "selfish cooperators"; memes, in other words, are "interested" in cooperating only when it promotes their own replication.

In elaborating the notion of memes and memeplexes, Dawkins and his followers describe a broad range of cultural artifacts and practices as memetic replicators. Everything from particular words, ideas, and sounds, to tunes, images, and symbols, can function as memes. Taken together, memes form more or less inclusive and/or exclusive networks. Languages, for instance, function as memes, as do myths and the assemblages of symbols comprising them. More comprehensively conceived, memeplexes are cultural traditions, which emerge from micronetworks of memes and the interactions of different traditions. We have already been led to suspect that as technology creates more extensive and complex relationships, cultural change becomes both more rapid and more significant. The escalating rate of the exchange of information among cultures produces volatile conditions in which consequential change becomes inevitable.

When networks of communication become fiber optic and telematic, memes spread at warp speed, creating a cascade of changes in memeplexes. Memes and the complex networks they form are not, as Dawkins's image of the digital river and software sometimes suggests, fixed programs. To the contrary, memeplexes, like the extended phenotype, form complex adaptive systems, which are always in formation.

CYBERGNOSIS

For biologists like Wilson and Dawkins, there can no more be minds and cultures without brains and bodies than there can be parasites without hosts. For some writers, philosophers, and even scientists with lively imaginations, however, talk about superorganisms and memeplexes unleashes remarkable techno-fantasies. These speculative flights of the imagination often have a quasi-spiritual dimension and usually are conceived, at least by their creators, as utopian visions. In 1938, H. G. Wells anticipated many of these fantasies in a little book entitled *World Brain*. Wells looks forward to "a world-wide network . . . woven between all men about the earth." As the vision unfolds, it becomes clear that this World Brain might be more accurately described as a World Mind. The networked World Brain or World Mind, Wells believes, holds the promise of "World Peace." This salvific World Brain will be realized through a "World Encyclopedia," which, as Wells describes it, sounds suspiciously like Ted Nelson's initial vision for hypertext. Unlike traditional encyclopedias whose formalistic structure was defined during the eighteenth century, Wells's World Encyclopedia will be "a permanent organism" rather than a dead mechanism:

A World Encyclopedia no longer presents itself to a modern imagination as a row of volumes printed and published once and for all, but as a sort of mental clearing house for the mind, a depot where knowledge and ideas are received, sorted, summarized, digested, clarified, and compared. It would be in continual correspondence with every university, every research institution, every competent discussion, every survey, every statistical bureau in the world.[37]

Total information globally available any place, any time: this "would constitute the material beginning of a real World Brain." But, alas, Wells is forced to confess, "the World Knowledge Apparatus" is not yet up to creating the World Encyclopedia. The dream of a World Brain, therefore, remains an idle fantasy.

In the six decades since Wells cast his vision, a growing number of people have become convinced that what had long been fantasy is now becoming reality. While visions differ, certain themes, which carry important implications for the formation of subjectivity and the process of cultural evolution, repeatedly appear. Instead of trying to offer a comprehensive survey of the relevant literature, it will be more productive to consider the work of three exemplary writers—one philosopher and two scientist-inventors—whose claims are particularly relevant for the problems we are considering.

Pierre Lévy is professor in the Department of Hypermedia at the University of Paris VIII and scientific advisor to the TriVium company. In a series of works, which are more manifestos than arguments, Lévy spins a vision of a techno-political utopia by recasting the ideas of Western philosophers and theologians to create cyberspaces and virtual realities that are otherworldly rather than realistic. At the heart of his vision lies the notion of a "new cognitive ecology," which promotes the "evolution of collective intelligence." This collective intelligence will be as different from human intelligence as Aunt Hillary is from the ants making her up. Indirectly invoking the notions of the fourth dimension and hyperspace, which inspired artists like Malevich and Lissitsky, Lévy describes a "fourth space" where knowledge evolves beyond the bounds of humanity in a new organization, which he names the "cosmopedia":

The cosmopedia is based largely on the possibilities made accessible to us through computer technology for the representation and management of knowledge. . . . Instead of a one-dimensional text or even a hypertext network, we now have a dynamic and interactive multidimensional representational space. Instead of the conjunction of image and text, characteristic of the encyclopedia, the cosmopedia combines a large number of different types of expression: static images, video, sound, interactive simulation, interactive maps, expert systems, dynamic ideographs, virtual reality, artificial life, etc. At its extreme the cosmopedia contains as many semiotics and types of representation as exist in the world itself. The cosmopedia multiplies nondiscursive utterances.[38]

Lévy does not disguise the spiritual dimensions of the world he imagines. The cosmopedia promises an "escape from the world of absence," which becomes possible on the "plane of immanence of signification in which beings, signs, and things exist in dynamic relationship of mutual participation, escaping the separation of territorial space as well as the circuits of the spectacle that characterize the commodity space." Absence gives way to presence in a world where signs are real and reality is nothing but signs. Paradoxically, total mediation seems to hold the promise of total immediacy. The dawn of this New Age, Lévy contends, brings the relief of simplicity to an era suffering under the growing weight of complexity:

In contrast to the expanding complexity that we attempt to organize through transcendence or distribute within increasingly inextricable networks, the cosmopodeia provides a new kind of simplicity. Not some mutilating simplification applied violently from without, but an essential simplicity that results from the principle of organization inherent in knowledge space. Beyond the world of chaos and large numbers, beyond the incantations to complexity, simplicity is born of implication.[39]

Instead of seeing evolution moving from simplicity to complexity, Lévy dreams of escaping complexity by returning to simplicity. Though claiming to be progressive, this argument is actually regressive enough to be justifiably labeled reactionary.

At first glance, the technological speculations of Ray Kurzweil and Hans Moravec seem to share little with Lévy's apocalyptic musings. Further reflection, however, suggests unexpected points of agreement. The three perspectives intersect in the vision of what Kurzweil calls the "postbiological future." The likelihood of Kurzweil's predictions proving true is far less interesting than the assumptions and values that lead him to make them. For Kurzweil, "technology is the continuation of evolution by other means."[40] Far from metaphorical, this claim is surprisingly literal; Kurzweil actually anticipates a future in which the evolutionary process shifts from embodied beings to disembodied *software*. From this point of view, the body is, in the preferred term of cyberpunk writers, nothing more than "meat," which can be dropped without a loss of identity. Once mind is unplugged from body, it becomes possible to imagine the prospect of "scanning the brains" and "downloading the mind" in a computer. If bodies are not necessary for survival, immortality is no longer an unrealizable dream. Kurzweil goes so far as to declare:

There won't be mortality by the end of the twenty-first century. Not in the sense that we have known it. Not if you take advantage of the twenty-first century's brain-porting technology. Up until now, our mortality was tied to the longevity of our hard-ware. When the hardware crashed, that was it. For many of our forebears, the hard-ware gradually deteriorated before it disintegrated. . . . As we cross the divide to instantiate ourselves into our computational technology, our identity will be based on our evolving mind file. We will be software, not hardware.[41]

While this prophecy seems wildly speculative, Kurzweil has demonstrated a remarkable ability to read technological trajectories with uncanny accuracy.

Kurzweil is not, however, satisfied with mere survival; he looks forward to the evolution of *higher* forms of life in the postbiological era. He goes so far as to assert confidently: "the emergence of machine intelligence that exceeds human intelligence in all of its broad diversity is inevitable." Once freed from material constraints, the new order of intelligence will be able to inhabit virtual bodies and evolve "higher" forms of intelligent life. As if teleporting to the year 2099, he looks back and reports:

Machine-based intelligences derived entirely from these extended models of human intelligence claim to be human, although their brains are not based on carbon-based cellular processes, but rather electronic and photonic "equivalents." Most of these intelligences are not tied to a specific computational-processing unit (that is, piece of hardware). The number of software-based humans vastly exceeds those still using native neuron-cell-based computation. A software-based intelligence is able to manifest bodies at will: one or more virtual bodies at different levels of virtual reality and nanoengineered physical bodies using instantly reconfigurable nanobot swarms.[42]

The shift from the human to the hyperhuman issues in what Hans Moravec describes as the "Age of Mind" in which "mere machines" become "transcendent minds." What these "superintelligences" transcend is once again the body and the material conditions of existence. Without a trace of irony, Moravec projects brains in vats, which, he believes, will free us from the constraints under which we currently suffer. Waving "Bye Bye Body," he declares:

Picture a "brain in a vat," sustained by life-support machinery, connected by wonderful electronic links to a series of artificial rent-a-bodies in remote locations and to simulated bodies in virtual realities. Although it may be nudged far beyond its natural lifespan by an optimal physical environment, a biological brain evolved to operate for a human lifetime is unlikely to function effectively forever. Why not use advanced neurological electronics, like that which links it with the external world, to replace the gray matter as it begins to fail? Bit by bit our failing brain may be replaced by superior electronic equivalents, leaving our personality and thoughts clearer than ever, though, in time, no vestige of our original body or brain remains.[43]

For Moravec, as for Kurzweil, the only bodies transplanted brains will need are virtual or simulated bodies, which, of course, are not bodies at all. While

Kurzweil is preoccupied with the continuing evolution of bodiless minds, Moravec is interested in the "prosperity beyond imagination," which, he believes, robotic intelligence soon will make possible. He predicts that in the near future, labor will virtually disappear and productivity will increase exponentially to create a world of plenty for everyone. As is often the case in such utopias, this supposedly utopian future looks suspiciously like the past. Capitalism will collapse and nations all but vanish as the world returns to premodern tribalism:

Nations may become less important, as taxes on local robot industries supply all human needs. The civilized world may return to a comfortable tribalism after a five-millennium detour into organized civilization. Countries with traditional tribal structures may simply stay that way, building on their ancestral customs, leapfrogging urbanization altogether. Developed countries may foster untraditional tribes with customs and beliefs more bizarre than anything today.[44]

As if this belief were not bizarre enough! Moravec's projections would be easier to dismiss were he not the widely respected founder and director of the world's largest robotic program, at Carnegie Mellon University. Far from responsible predictions based on measured assessments of technology, Moravec's vision is an expression of the same longing for simplicity that informs Lévy's New Age philosophy. For futurists who look ahead only to rediscover what lies behind, evolution becomes devolution.

What is surprising about these techno-fantasies is not how new the worlds they imagine are but how old the ideas they represent remain. Without realizing what they are doing, Lévy, Kurzweil, Moravec, as well as many others revise ancient philosophical and theological visions for the twenty-first century. Though the technologies of salvation change, the aspirations they provoke remain remarkably similar. What these latter-day practitioners of once occult arts long for is a spirituality achieved through *disembodiment.* Sophisticated machines, scientific theories, and mathematical codes and formulae cannot disguise the spiritual and intellectual traditions within which these seemingly radical thinkers continue to work. They are contemporary Gnostics, Platonists, and Cartesians. The fundamental assumption that has always informed these points of view is a thoroughgoing dualism between mind and body, form and matter, immateriality and materiality, pattern and substance, and so on. Inasmuch as truth and reality are associated with mind, form, immateriality, and pattern, and illusion is associated with body, matter, materiality, and substance, the goal of thinking and living is to separate the former from the latter. In this tradition, when the kingdom arrives, it will not be of this world. What advertises itself as hope in these visions is actually the most profound despair: the world and life in it cannot be redeemed and must, therefore, be negated. If truth and reality are otherworldly, then this world must be left behind or even destroyed in their pursuit. Lévy, Kurzweil, and Moravec do nothing to question this mythic perspective; to the contrary, they promulgate a cure that actually exacerbates the disease. Their arguments and analyses are of interest only insofar as they are symptomatic of pervasive ideas and values implicitly informing much current scientific research and technological development.

INCARNATIONAL NETWORKS

But what if this ancient tradition is *wrong?* What if matter/form, materiality/immateriality, substance/pattern, mind/body, and so on are not opposites, which confront each other, but are mutually implicated in such a way that each folds into the other like a Möbius strip, which never quite closes on itself? If thinking is not to be the eternal return of the same, it must be recast in ways that refigure nature and culture as well as body and mind so that neither remains what it was believed to be when it appeared the opposite of the other. Instead of promoting apocalyptic otherworldliness through new technologies, what we need is a more radical style of *incarnational* thinking and practice. To see what this might involve, let us return to the ways in which information is screened in the complex interplay between nature and culture.

One of the perennial promises of visionaries is that in the future, all things will be possible. Whatever constraints we suffer in this world will disappear and we will be able to enjoy a freedom now barely imaginable. Such promises, however, are always cruel because they cannot be fulfilled. Possibilities are inevitably limited by constraints that can never be overcome. The only viable freedom is not freedom *from* constraints but the freedom to operate effectively *within* them. Many of the techno-fantasies currently circulating refuse to accept the inescapable limits imposed by the multiple networks in which we are caught. Arguments to the contrary notwithstanding, constraints, as we have seen elsewhere, are not merely negative but can also be productive; indeed there are no creative possibilities without significant constraints. Constraints provide the parameters within which thinking and acting must occur. As such, constraints can be natural as well as cultural, or, in different terms, can be imposed by both genes and memes. This way of posing the issue, however, still remains too simple, for, as our investigation of evolving complexity has disclosed, biological evolution itself is constrained by not only morphological but also genetic factors. Individual organisms, species, and the evolutionary process as a whole, as well as the interrelations among them, are all complex adaptive systems, which forge possibilities by imposing constraints. The same patterns are discernible in cultural development. Individuals, groups, and sociocultural totalities are all bound in complex relations of restriction and production. Moreover, nature and culture are woven together in nonlinear feedback and feed-forward loops through which nature produces culture as much as culture produces nature.

This way of framing the issue helps to overcome some of the most persistent confusions that all too often oversimplify arguments between scientists and humanists. While certain scientists tend to reduce culture to nature, many humanists defiantly reduce nature to culture. For some who are inclined to biological reductionism, the decoding of the genome seems to hold the promise of identifying specific genes for psychological traits, social activities, and cultural processes as well as for biological functions. According to such essentialism, decoding the genetic program makes it possible both to understand and to reengineer organisms. Even in weaker versions, biological reductionism leaves cultural production little if any creative role. In an utterly predictable fashion, this line of argument triggers an equal and opposite reaction. In ways we have examined in chapter 2, nature, critics of genetic determinism counter, is a sociocultural construction and scientific theories and experiments are expressions of subjective desires, which serve personal, political, and corporate interests. From this point of view, nature as such does not really exist, for it is nothing more than a psychosocial construct. Extensive arguments developed in the currently thriving fields of cultural studies and science studies do little more than elaborate this basic point again and again. While some of the insights generated by this work are helpful, the point once made tends to prevent rather than promote serious debate. It will not be possible to move beyond our current critical impasse until scientists, who criticize humanists, and humanists, who criticize scientists, turn their critical gaze back on themselves to question and revise their own assumptions.

As always in such disputes, each side develops important points, which are distorted when pushed to extremes. *Every* form of reductionism is mistaken and misguided; culture can no more be reduced to nature than nature can be reduced to culture. Since nature and culture develop *interactively* through a process of coevolution, natural, or more precisely biological, and cultural processes form a complex adaptive system. This complex adaptive system maintains the relationship between the complex adaptive systems comprising biological and genetic networks and the complex adaptive systems comprising cultural and memetic networks. The crucial point to understand about these multiple dimensions of complexity is that within these networks, adaptation inevitably involves coadaptation. To get a better sense of how this coadaptation works, it is helpful to consider certain aspects of the interaction between the brain and mind.

The notion of brains as hardware and minds as software as well as brains in vats would be humorous were their implications for the world that science and technology help to create not so serious. It is, therefore, important to understand why this vision is wrong. Instead of dualistic entities that are completely separable, most important contemporary research suggests that brains and minds are coadaptive and thus cannot be unplugged from one another. In ways that are gradually becoming clear, brains create minds as much as minds create brains. The interaction between brain and mind or body and mind, and, by extension, biology and culture, is always two-way. What makes this interactivity possible is the isomorphism between brain function and mental activity. J. A. Scott Kelso points out that the "human brain is *fundamentally* a pattern-forming, self-organized system governed by nonlinear dynamical laws. Rather than compute, our brain 'dwells' (at least for short times) in metastable states: it is poised on the brink of instability where it can switch flexibly and quickly. By living near criticality, the brain is able to anticipate the future, not simply react to the present. All this involves the new physics of self-organization in which, incidentally, no single level is any more or less fundamental than any other."[45] This understanding of the functioning of the brain stands in marked contrast to accounts that represent the brain as directed either by a central processor or prescribed program. As a complex adaptive system, brain activity can no more be reduced to the firing of individual neurons than Aunt Hillary can be reduced to the sum of her ants. Just as ants create Aunt Hillary without any leader or conductor, so neural activity produces dynamic brain patterns through *distributed* processes that are not necessarily programmed. Always more than the sum of its parts, the brain operates globally through patterns that are constantly in flux. "Even when we know what single neurons do (and keep in mind the neuron itself is a complex system)," Kelso stresses, "we still have to understand how they work together to support behavioral function."[46]

One of the most common ways to understand the interrelation of distributed neurons is in terms of neural networks, which display many of the most important features of complex adaptive systems. Neural nets are made of nodes, which receive, transmit, and send signals. When numerous nodes are connected, complex behavior becomes virtually inevitable. "The function of the net," Deacon explains, "is determined by the global patterning of signals from output nodes with respect to patterns presented to input nodes. These input-pattern-to-output-pattern relationships are thus mediated via the patterning of signals distributed through the web of interconnections that link output to input nodes, by way of the intervening hidden nodes, and not by the state or activity of any individual node."[47] When understood in this way, *the brain is a global network* made up of multiple micro- or local-area networks, which are constantly emerging and changing. The relation of the brain's neurons, axons, and dendrites is analogous to nodes joined in nets through which electrical impulses circulate. The brain, like the world in which it is enmeshed, then, operates according to *network logic.* Changes in the structure and function of the brain result both from the coadaptation of neural networks within the brain and from the coadaptation between the brain and its environment.

As a result of its coadaptive capacity, the brain is not hardwired but, like all complex networks, functions between fixity and flux. The constraints facilitating brain functions and mental activity are not completely predetermined but change in relation to other physical, chemical, and biological processes. In addition to this, the brain can be reconfigured or rewired in certain ways through interactions with its social and cultural surroundings. Jean-Pierre Changeux argues that "a significant variability in the organization of the cortex of the brain is related to the cultural environment." While Wilson insists that genetically coded epigenetic rules "keep culture on a leash," Changeux goes so far as to claim that "mental objects can participate in the epigenesis of the brain."[48] The process by which culture "contributes to fine-tuning the cerebral cortex connections" is screening—precisely the process we have discovered in mental activities ranging from sensation and perception through consciousness and self-consciousness to reason. Changeux explains:

According to this scheme, culture makes its impressions progressively. The 10,000 or so synapses per cortical neuron are not established immediately. On the contrary, they proliferate in successive waves from birth to puberty in man. With each wave, there is transient redundancy and selective stabilization. This causes a series of critical periods when activity exercises a regulatory effect. If we consider that the growth of axonal and dendritic trees is innate and that selective stabilization defines acquired characteristics, the innate can be differentiated from the acquired only by the detailed study at the synaptic level. This study is made more difficult by the intimate association of growth and epigenesis, and their alteration over time. One has the impression that the system becomes more and more ordered as it receives "instructions" from the environment. . . . Epigenetic selection acts on preformed synaptic substrates. To learn is to stabilize preestablished synaptic combinations, and to eliminate *the surplus.*[49]

Learning, in whatever way or medium it occurs, is always a process of filtering that involves screening information. The implications of Changeux's argument become even more intriguing in light of the recent discovery by Princeton neurophysiologists that the production of brain cells continues throughout life and does not, as previously assumed, cease at a relatively early stage of life. In earlier theories, learning was thought to be a function of new connections between existing neurons, or the strengthening of the existing synaptic connections through some cellular mechanism (e.g., more receptors, increased release of neurotransmitters, sprouting of new axon end terminals or spines on dendrites, etc.). Elizabeth Gould and Charles Gross have shown that new cells are born in the hippocampus (i.e., the learning area of the brain) in adult rats and primates. With characteristic clarity and conciseness, George Johnson explains the significance of this discovery: "If a steady stream of fresh brain cells is continually arriving to be incorporated into new circuitry, then the brain is more malleable than hardly anyone has realized."[50]

The interplay between culture and the network architecture of the brain can also be understood in terms of the interaction between memes and genes. Dennett is convinced that memes can actually "transform the operating system or computational architecture of a human brain."[51] The human mind, he argues, "is itself an artifact created when memes restructure a human brain in order to make it a better habitat for memes. The avenues for entry and departure are modified to suit local conditions, and strengthened by various artificial devices that enhance fidelity and prolixity of replication: native Chinese minds differ dramatically from native French minds, and literate minds differ from illiterate minds. . . . Normal human brains are not all alike; they vary considerably in size, shape, and in the myriad details of connection on which their prowess depends. But the most striking differences in human prowess depend on microstructural differences induced by the various memes that have entered them and taken up residence."[52] In addition to these "microstructural differences induced by the various memes," "the way that information is analyzed ultimately becomes reflected in the way the brain regions are 'designed' by this activity."[53] If one extends this line of analysis of brains and minds, it becomes clear that genes and memes are involved in a relationship that is coevolutionary as well as coadaptive.

Dawkins refuses to draw this conclusion. His resistance to the notion of gene/meme coevolution rests on what Francis Crick describes as "The Central

Dogma" of molecular biology, which "states that once information has been passed into protein it cannot get out again. In more detail, the transfer of information from nucleic acid to nucleic acid, or from nucleic acid to protein may be possible, but transfer from protein to nucleic acid is impossible." In an unusually suggestive paper entitled "Genes Beget Memes and Memes Beget Genes: Modeling a New Catalytic Closure," James Gardner argues that recent technological developments raise serious questions about Dawkins's appropriation of Crick's position:

It is an article of faith among adherents to the central dogma that, in Dawkins' words, "the Larmarkian theory is completely wrong." However, to a degree that is largely unappreciated by orthodox theoretical biologists, the ongoing revolution in biotechnology renders the central dogma obsolete. The fact is that information can and does flow upstream into the genome from the particular extended phenotype we know as human civilization.

The gene/meme relationship, Gardner proceeds to explain, must be reconceived: "The emerging capabilities of genetic engineering provide the technological precondition for explosively paced catalytic closure between the forces of genetic and cultural evolution. These new capabilities conceivably could create a radically new autocatalytic relationship between the two sets of forces that could enable the 'old' replicators—Dawkins' selfish genes—and the 'new' replicators—units of cultural transmission called culturgens or memes—to coevolve in novel and unpredictable ways at vastly enhanced speeds."[54] When the relation between genes and memes is conceived in this way, the implicit philosophical and theological framework of interpretation is completely recast. Instead of Gnostic, Platonic, or Cartesian, this conclusion is, as Deacon suggests, thoroughly *incarnational.* In a chapter of *The Symbolic Species: The Coevolution of Language and the Brain* provocatively entitled "And the Word Became Flesh," he writes:

The evolutionary miracle is the human brain. And what makes this extraordinary is not just that a flesh and blood computer is capable of producing a phenomenon as remarkable as a human mind, but that the changes in this organ responsible for this miracle were a direct consequence of the use of words. And I don't mean this in a figurative sense. I mean that the major structural and functional innovations that make human brains capable of unprecedented mental feats evolved in response to the use of something as abstract and virtual as the power of words. Or, to put this miracle in simple terms, I suggest that an idea changed the brain.

229

Now this may seem a rather mystical notion on its own, inverting our common sense notion of causality that physical changes require physical causes, but I assure you that it is not. I do not suggest that a disembodied thought acted to change the physical structure of our brains, as might a god in a mythical story, but I do suggest that the first use of symbolic reference by some distant ancestors changed how natural selection processes have affected hominid brain evolution ever since. So in a very real sense I mean that the physical changes that make us human are the incarnations, *so to speak, of the process of using words.*[55]

The implications of this insight extend far beyond the specific problem of the interrelation between mind and brain. If Deacon is right, all the oppositions like form/matter, pattern/substance, culture/nature, virtuality/reality, which have structured thinking for centuries, must be reconceived. Any analysis that reduces one to the other or tries to synthesize otherwise discrete differences is inadequate as well as misleading. What thinking requires is a new *architecture of complexity* that simultaneously embodies and articulates the *incarnational logic of networking.*

By now it should be clear that the architecture of complexity entails a considerably expanded understanding of information and information processing. If mental activity can be understood in terms of the complex adaptive systems facilitating information processing, then natural as well as cultural processes, which are also complex adaptive systems, are, in some sense, information processes. Information, in other words, is limited to neither minds nor computational machines but is distributed throughout all the networks passing through us. Information processing does not presuppose consciousness or self-consciousness, though consciousness, self-consciousness, and reason are impossible without it. From neurophysiological activity and immune systems, to computational machines, to financial and media networks, information is processed apart from any trace of consciousness. Such information processing forms something like what Hegel describes as "objective spirit," which emerges in and through natural and social processes.[56] If bodies as well as society and culture necessarily involve information processes, it is no longer clear where to draw the line between mind and matter, self and other, human and machine. *Mind is distributed throughout the world.* Nature and culture, in other words, are the objective expression of mind, and mind is subjective embodiment of nature and culture.

The networks circulating through us bind self and world in increasingly complex relations. In network culture, technology is an indispensable prosthesis through which body and mind expand. This relationship is always two-way: as body and mind extrude into world, world intrudes into body and mind. Technology is, in Derrida's apt phrase, a "prosthesis of the inside."[57] Not only is the mind too large to contain itself, as Augustine realized long ago, but the "belly of the mind" actually extends into the world. In this way, the networks extruding from and intruding into our bodies and minds form something like a *technological unconscious,* which, like conscious mental processes, screens information. The excess filtered by the prosthetic devices of the technological unconscious does not disappear but remains as residual noise, which both creates interference and provides a constant source of new information.

As the webs in which I find myself become ever more complex, I eventually realize that the currents rushing through me are tributaries in a vast river of information. Tossed and turned by the turbulence this river perpetually generates, the I unravels but I do not completely dissolve.

The mind cannot be understood as a single entity. It is more like a community; an assembly of intercommunicating parts. Where then, you might ask does this "self" reside? Is it simply one of those parts, undistinguished and no better than any of its fellows?[58]

"A life various, manifold and quite immeasurable." The self—if, indeed, this term any long makes sense—is a node in a complex network of relations. In emerging network culture, *subjectivity is nodular.* Nodes, we have discovered, are knots formed when different strands, fibers, or threads are woven together. As the shifting site of multiple interfaces, nodular subjectivity not only screens the sea of information in which it is immersed, but is itself a screen displaying what one is and what one is not. In emerging network culture, life is lived on screen. Bits of information become pixels, which, like a Chuck Close painting, organize themselves into the patterns of "my" life. In the midst of these webs, networks, and screens, I can no more be certain where I am than I can know when or where the I begins and ends. I am plugged into other objects and subjects in such a way that I become myself in and through them, even as they become themselves in and through me. Though hard-and-fast distinctions are no longer possible, the networks in which nodular subjectivity is emerging bear traces of natural, social, and cultural systems. In the absence of walls that once seemed secure, nothing remains separate or isolated; everything becomes permeable screens. As the networks that make me what I am expand, both they and I become more complex and less secure. Speed of connection continues to accelerate until everything and everybody seem poised at the edge of chaos. This tipping point marks the moment of greatest danger and greatest opportunity. When

231

change occurs, it is never as planned and thus is always surprising. The more I struggle to fathom this critical moment, the more complex it becomes. Eventually, I am driven to conclude that I am—the I is—a moment of complexity. The networks in which nodular subjects form create binds and double-binds that cannot be undone. Turning back on myself to look at myself looking at myself, I realize thought is never my own, and thus thinking can never come full circle. As I try to think about, speak about, write about what seems to be happening, I discover that words are not mine but are merely borrowed for a *brief* moment. My identity—literary as well as otherwise—is parasitic upon the ghosts haunting me. Their noise is what makes it possible for me to write. As I screen their words, their thoughts and words are reborn through me. What I know now that I did not know when I began is that I am not merely a parasite but am also the host of others both known and unknown. The networks that make me what I am are always networks within networks, which, while never complete, are nonetheless global. As a node in networks that are infinitely complex, I am the incarnation of worldwide webs. The fiber of these webs, I now realize, is not merely optical, for networks always operate in many channels and multiple media. Webs and networks can no more exist without me than can I without them. In the absence of firm walls and fixed boundaries, it is impossible to put an end to this inconclusive interplay of networking. This is what I am—this is what we have become in the moment of complexity.

Business art is the step that comes
after Art. . . . Being good in business is
the most fascinating kind of art.

andy warhol

You guys are in trouble and we are
going to eat your lunch.

michael milken

the currency of education

PRACTICING THEORY

Theory without practice is empty; practice without theory is
blind. The ongoing challenge is to bring theory and practice together in such a way that we can theorize our practices and practice our theories. This has never been more important than in the
moment of complexity. As we have discovered, emerging network
culture is transforming the social, political, economic, and cultural fabric of life. The same information and telematic technologies responsible for the shift from an industrial to a postindustrial
economy are bringing higher education to the tipping point
where unprecedented change becomes unavoidable. While it is
impossible to predict the precise rate and scope of these changes,

there can be little doubt that in the near future, colleges and universities will look very different than they do today. What makes this situation particularly troubling is that educational institutions are ill equipped to cope with these developments, and many educators are not inclined to seek creative responses. The organizational structure and governance procedures of colleges and universities make it almost impossible for them to operate effectively in a world moving at warp speed. To prosper in network culture, it is necessary to make decisions expeditiously and to develop programs quickly and efficiently. The technological innovations that have reformed business and finance in recent decades will both transform the organization of educational institutions and radically change *what* educators do as well as *how* they do it. The impact of these technologies on pedagogy is already becoming clear. As more professors develop on-line courses or components of courses, the traditional classroom inevitably changes. Most important, the classroom has expanded and now is *global.* Anyone anywhere in the world can, in principle, sit down around the same virtual table and learn together. What is studied in virtual global classrooms will be as different as how it is studied. Disciplinary boundaries are becoming as mobile and permeable as the screens on which courses are cast. Since the organizational structure of knowledge is always bound to the modes of production and reproduction in a particular society, technological changes issue in the reconstitution of knowledge. In the future, the curriculum will look more like a constantly morphing hypertext than a fixed linear sequence of prepackaged courses. When knowledge changes and both seminar tables and lecture halls become global, traditional classrooms will not remain the same. Like a growing number of businesses, colleges and universities will become clicks-and-mortar operations in which the old and new economies intersect and interact in unpredictable ways. Without in any way minimizing the challenges and difficulties these changes pose, it is vitally important for educators to appreciate the new opportunities e-Ed creates for people around the world and to develop new strategies for a rapidly changing educational environment.

EDUCATION BUSINESS

In network culture, *education is the currency of the realm.* If information is the oil of the twenty-first century, colleges and universities are sitting on very valuable reserves. A growing number of people in business understand the importance of these reserves more clearly than many educators. Education, like money in the world of finance, is a currency that is also a *commodity.* In the eyes of many in the business community, this commodity is potentially very profitable. The prospect of significant profits in the education market is, of course, the result of new technologies for providing and promoting education. Unlike industrial products, education is a commodity that is distributable through telematic technologies. From satellites and the Internet to rapidly developing wireless technology, education is now accessible in ways it never previously has been. As global communications networks become more sophisticated, the market will expand exponentially. No one has been quicker to realize what the new economy means for education than Michael Milken. On November 4, 1999, *The New York Times* ran a front-page article entitled "Investors See Room for Profit in the Demand for Education," which began:

When Arthur Levine, President of Teachers College at Columbia University met last year with Michael Milken, the fallen junk bond king turned education entrepreneur, it was more than a case of the old school meeting the new.

The message was, "You guys are in trouble and we're going to eat your lunch," Dr. Levine said, recalling what he considered not a direct threat but a kind of predatory challenge.[1]

Milken's Knowledge University is one of the first for-profit corporations to take direct aim at postsecondary education. Among the many voices bullish on the burgeoning education market quoted in the *Times* article, one is of particular interest: "Changes in education are coming as surely as the Berlin wall went down," said William F. Weld, the former governor of Massachusetts, who is part of a group that has raised $150 million to invest in for-profit education and training and which hopes to raise another $100 million by early next year. Weld's association of the emergence of the new education market with the collapse of the Berlin Wall is particularly suggestive, for, as we have seen, this pivotal event serves as an effective trope for the transition to network culture. As networks grow and webs expand, seemingly secure walls once again become permeable screens. Weld is, in my judgment, correct when he claims that the changes information technologies will bring in education will be at least as dramatic as they have been in business and finance.

There is ample support for this view of the education business among investors. In April 1999, Merrill Lynch issued a 193-page research document for its privileged investors entitled *The Book of Knowledge: Investing in the Growing Education and Training Industry.* The opening paragraph sets the stage for the analysis that follows:

The two trillion dollar global education and training industry is going through radical changes. Market forces are providing a catalyst to alter the traditional ways education is delivered. Megatrends such as demographics, the Internet, globalization, branding, consolidation, and outsourcing all play major roles in this transformation. In the U.S., the focus of this report, education and training is a $740 billion dollar market.[2]

The Merrill Lynch analysts are persuaded that the difficult changes businesses and corporations underwent in the transition from an industrial to a postindustrial economy will also occur in the "education industry." Just as corporations that are unable to respond to the rapidly changing economic environment cannot remain competitive, so colleges and universities that are unwilling to adapt to the new education and economic landscape will not survive. The recognition of the economic stakes of these changes has gone unnoticed by most members of the education community. Since 1994, there have been thirty-eight IPOs with thirty follow-on offerings, which have raised $3.4 billion of equity. While most of the investment to date has been in primary and secondary education, this situation is changing very quickly. The Merrill Lynch report points out that the postsecondary market is currently $237 billion and is growing rapidly. There are three leading areas of projected growth:

1. College students. The number of high school graduates will grow 22 percent from 2.5 million in 1995 to 3.1 million in 2008. In addition to this increase, a greater number of high school students will be attending college.
2. Lifelong education. Demand for education throughout one's career will combine with a growing retirement population to create unprecedented demand for education beyond the traditional college years.
3. International students. While the number of foreign students studying in the United States has been increasing steadily, there is huge demand that remains unmet.

Predictions of significant growth in the postsecondary market have led to the formation of a group of for-profit companies with significant investment capital. Most of these companies are currently focusing on K–12 and on training courses tailored for corporations and industry, but several ventures are planning to expand their offerings to compete with universities and colleges.

The most important factor contributing to the explosion of for-profit education investment is the extraordinary growth of the Internet and related technologies. The unexpectedly rapid increase in e-commerce should serve as a warning of what is about to happen in e-education. The impact of the Internet on education will in all likelihood be even greater than it has been on other forms of commerce. The number of students currently enrolled in e-Ed courses is 710,000 and is expected to increase to 2.3 million by 2002. This represents a compound annual growth rate of 33 percent. Such projections are encouraging the formation of new companies, which provide a broad range of services to the postsecondary market.

Universities and colleges have been relatively slow to respond to the challenge posed by the for-profit sector. While the situation is fluid, the few serious for-profit ventures undertaken by universities have been limited to professional schools—most notably and least surprisingly business schools. There continues to be considerable reluctance on the part of educational institutions to expand for-profit offerings to the entire curriculum—especially undergraduate liberal arts courses. Concerns about commercial pressures have been somewhat mitigated by the recognition of the necessity to adapt to the changing technological environment. As administrators, faculty, and staff consider the hardware, software, and personnel costs required to mount extensive on-line programs, it quickly becomes obvious that the price for quality e-Ed is *very* high. If they are savvy enough to examine the competitive materials being produced by well-financed for-profit companies, they soon realize that the costs are actually prohibitive. Faced with the reality of high costs and the possibility of significant profits, a growing number of educators, educational institutions, and companies are beginning to enter into agreements. Though differing considerably in their details, these arrangements tend to take one of two forms: either a school contracts with a company to provide services necessary for e-Ed, or companies contract with individual professors to provide courses to be marketed for profit. When universities and colleges turn to companies to help them with e-Ed, they usually are looking for software packages that will enable professors to get their courses online quickly and easily. This approach in effect outsources software development. Companies like Blackboard.com, Campuspipeline.com, Mascot.com, MyBytes.com, and click2learn.com offer a range of products adapted to different subjects and teaching styles. While appearing to provide a quick solution to a pressing problem, there are three basic difficulties with this solution. First, software packages for on-line courses usually do not come with support services. The school, therefore, must purchase equipment and hire additional staff to operate programs purchased from outside vendors. Second, outsourcing software development does not solve the problem of marketing and distributing on-line courses. In an increasingly competitive market, the cost of publicizing e-Ed is rapidly escalating. Third, courses adapted to software templates developed by for-profit companies tend to be pedagogically conservative. The professors who are satisfied with these products usually do not understand the capabilities of new technologies but simply want to put courses in their current form on-line. The alternative to this strategy, which is spreading very quickly, is for companies to contract directly with individual professors to provide courses. Companies like UNext, Harcourt-Brace, and Thinkwell are offering professors attractive financial deals for putting their courses on-line. All costs for production, support services, distribution, and advertising are borne by the company, and profits are split between the company and the professor. This kind of arrangement between the company and the professor runs the risk of creating an adversarial relationship between the faculty member and his or her institution. Until the contentious issue of ownership of intellectual property rights to course materials is settled, tensions will continue to escalate. As e-Ed becomes more popular, the stakes of this struggle will continue to grow. If new teaching media are to be developed in ways that are educationally responsible and practically feasible, both business and education will have to change in ways that are only beginning to become clear. Faculty members, administrators, and investors will have to find ways to cooperate that are mutually beneficial. This is no small challenge, because it will require the creation of new kinds of institutions and organizations, which will bring together two *very* different cultures.

It is precisely this intersection of the cultures of business and education that so many faculty members oppose. The reasons for this resistance are often both understandable and predictable: unwarranted expenditure of limited financial resources, diversion of precious time and energy from teaching and research, possible dumbing down of courses, etc. But the most common reason faculty members offer for not pursuing e-Ed is that on-line education cannot possibly be an adequate substitute for the "real" classroom experience. The product, in other words, is inferior. In an important article entitled "'When Industries Change' Revisited: New Scenarios for Higher Education," David Collis points out that this is a typical response to "disruptive technologies":

What makes some new technologies so insidious is that they first appear as ineffective substitutes that can safely be disregarded. The best-selling management text by my colleague Clay Christensen makes exactly this point. In countless instances of what Christensen calls "disruptive technologies," from disk drives, to personal computers and now the Internet, technological advances initially yield products that fail to satisfy current users. The product is too slow, too expensive, too limited, and so on. In-

cumbents who do the right thing and listen to their customers, are told that this is not what they want. Accordingly, the market leaders ignore the new technology and continue to refine their existing products and technology.[3]

Until recently, "the market leaders," i.e., the elite colleges and universities, have been especially wary of for-profit e-Ed. However, as the impact of the Internet continues to grow, it is becoming clear that no educational institution can avoid developing programs in on-line education. As I have noted, the difficulty for most colleges and universities is that the cost of producing, maintaining, and distributing high-quality courses is simply prohibitive. Thus they face a seemingly irresolvable dilemma: they cannot afford to do e-Ed and they cannot afford not to do it. This problem becomes more vexing when administrators confront faculty opposition to distributed teaching and learning. Though this resistance is often expressed in terms of lofty educational ideals, the real problem is that many faculty members continue to be very suspicious of technology and business and, therefore, refuse to participate in the educational revolution that is taking place. There are, of course, noteworthy differences among institutions and within the faculty. In general, universities tend to be more open to e-Ed than colleges, and people in the arts and humanities are often more reluctant to use technology than people in the natural and social sciences. The concerns of many faculty members are expressed clearly in an article written by James Perley and Denise Marie Tanguay, who are both members of the American Association of University Professors, entitled "Accrediting On-Line Institution Diminishes Higher Education." On-line institutions, they argue, "raise the specter of a higher-education system that is nothing more than a collection of marketable commodities—a system that could turn out to be all but unrecognizable to the scholarly communities that invent and reinvent higher education on a daily basis."[4] What this widely held point of view refuses to admit is that in the future it will not only be scholars who invent and reinvent education; as the education business becomes more profitable, commercial interests will play a much more significant role in shaping postsecondary education. Many professors are convinced that these developments pose a serious threat to higher education. While it is undeniable that concerns about increased workload, fewer full-time positions, and less job security contribute to faculty resistance to e-Ed, it is too easy to dismiss criticisms as nothing more than expressions of shortsighted self-interest. The issues involved in these debates are extremely complex and have a long history. An understanding of this history helps to explain why the mix of business, technology, and education is so volatile.

239

Though its roots extend back to the Middle Ages, the contemporary university is actually a *modern* invention, which took shape during the late eighteenth and early nineteenth centuries. In ways that are rarely noted, many of the philosophical issues we have been probing contributed directly to the creation of the modern university and continue to determine the direction of higher education. It is precisely this vision that fuels widespread opposition to the new forms of education and novel educational institutions now emerging. If the modern university could be traced to a single source, it would, without a doubt, be Kant's *Conflict of the Faculties,* published in 1798. Wilhelm von Humboldt used Kant's work as the blueprint for the first modern university—the University of Berlin, which opened in 1810 with the Idealist philosopher Johann Gottlieb Fichte as the first rector. The romantic theologian Friedrich Schleiermacher and Hegel were among the university's most influential faculty members. Although often disagreeing among themselves about subtle intellectual issues, their shared philosophical assumptions contributed significantly to forming the university. During the last two centuries, no university has had a greater worldwide impact on the structure of the university and the shape of higher education.

Kant begins his introduction with a remarkably suggestive comment:

Whoever it was that first hit on the notion of a university and proposed that a public institution of this kind be established, it was not a bad idea to handle the entire content of learning (really, the thinkers devoted to it) by mass production, *so to speak— by a division of labor, so that for every branch of the sciences there would be a public teacher or* professor *appointed as its trustee, and all of these together would form a kind of learned community called a* university *(or higher school). The university would have a certain autonomy (since only scholars can pass judgment on scholars as such) and accordingly it would be authorized to perform certain functions through its* faculties.⁵

This seminal passage harbors a tension which we have encountered repeatedly in our investigation of alternative models of systems and structures. On the one hand, the modern university embodies the mechanical logic of industrialism, and, on the other hand, the university is governed by the principle of autonomy, which, when elaborated, leads to organicism. Anticipating the logic that would eventually issue in the assembly line, the university produces education for *mass consumption.* In order to function efficiently, the labor process is divided into separate departments with different tasks and responsibilities. The product is packaged as individual courses, which are discrete units with set values. In order to be certified by the university, students must pass through a preestablished program whose requirements are set by the faculty and administration. The most important distinction Kant draws is between the university's "higher" and "lower" faculties: the higher faculties, which include theology, law, and medicine, are devoted to disciplines that serve our eternal, civic, and physical well-being, and the lower or philosophical faculty is divided between departments devoted to "historical knowledge" (history, geography, philology, the humanities, as well as the empirical knowledge of the natural sciences) and departments devoted to "pure rational knowledge" (pure mathematics and pure philosophy, and the metaphysics of nature and of morals). This division between higher and lower faculties is the foundation of today's distinction between professional schools and the faculty of arts and sciences.

In contrast to independent academies, where there is research but no teaching, and the gymnasium, where there is teaching but no research, the university combines research and teaching. Kant's description of members of the higher faculties is very important for our purposes:

While only the scholar [i.e., member of the lower faculty] can provide the principles underlying their functions, it is enough that they [i.e., members of the higher faculties] retain empirical knowledge of the statues relevant to their office (hence what has to do with practice). Accordingly, they can be called the businessmen *or* technicians *of learning. As tools of the government (clergymen, magistrates, and physicians), they have legal influence on the public and form a special class of the intelligentsia, who are not free to make public use of their learning as they see fit, but are subject to the censorship of the faculties.*[6]

While the role of the government in monitoring the activities of the higher faculties marks a noteworthy difference between Kant's scheme and most contemporary universities, the philosophical principles informing the structure of the university and the responsibilities of the faculties remain in place. The division between the lower and higher faculties is an expression of the difference between theoretical and practical reason. Kant underscores this point by describing members of the higher faculties as *businessmen* or *technicians* of learning who are *tools* of the government. In the higher faculties, business and technology come together to serve interests *extrinsic* to the university.

The lower faculty, by contrast, is "autonomous"; it is ruled only by the principles of universal reason:

It is absolutely essential that the learned community at the university also contain a faculty that is independent of the government's command with regard to its teachings; one that, having no commands to give, is free to evaluate everything, and concerns itself with the interests of the sciences, that is, with truth: one in which reason is authorized to speak out publicly. For without a faculty of this kind, truth would not come to light (and this would be to the government's own detriment); but reason is by its own nature free and admits of no command to hold something as true.⁷

Since philosophy is the discipline devoted to reason as such, it assumes the position of "queen of the sciences" once held by theology in medieval universities. The principle of autonomy operates in three ways in the philosophical faculty. First, reason is governed only by reason and is not influenced by any *external* goals or interests. Second, "only scholars can pass judgment on scholars as such." This is, of course, the principle of peer review. Third, the philosophical faculty is supposed to be independent of any influence by government or other *outside* forces. The combination of the first and third points issues in the principle of academic freedom. The autonomy of the philosophical faculty enables it to fulfill its primary function *within* the university. The philosophical faculty is charged with judging the practices of the higher faculty by holding them up to the standards of truth and reason. The autonomy of the lower faculty, which is fragile and is always in danger of falling prey to outside interests, must be protected. For Kant, the purity of reason is a prerequisite for philosophy and its faculty to play its essential role in judging the higher faculties. The raison d'être of the lower faculty is *criticism*. For critics devoted to purportedly disinterested inquiry, nothing compromises the purity of the faculty more quickly than the filthy "calculations" of "businessmen" and "technicians of learning."

In Kant's scheme, the hierarchical structure of the faculties rests upon a series of binary oppositions:⁸

Low/High
Theory/Practice
Scholar/Businessman, technician
Disinterested/Interested

Pure/Applied

Pure/Impure
Intrinsic/Extrinsic

If framed in these terms, it quickly becomes apparent that Kant is extending his analysis of art and aesthetic judgment to the structure of the university and its curriculum. Whereas the higher faculties are useful, the lower faculty can only function effectively by resisting every form of institutional utilitarianism. When translated from studio and museum to classroom and university, art for art's sake becomes knowledge for knowledge's sake. As we have discovered in our investigation of the genealogy of self-reflexive and autopoietic systems, Kant argues that utility exhibits *external* teleology (i.e., means lead to ends other than themselves), whereas beauty is characterized by *internal* teleology (i.e., purposiveness without purpose, in which means and ends are reciprocally related). While from an aesthetic point of view, the work of art is intrinsically valuable, from a utilitarian point of view, purposeless products are useless. The binaries structuring the university presuppose this originary distinction:

Useless/Useful
Unprofitable/Profitable
Arts and sciences/Professional schools

This way of posing the issue, however, is somewhat misleading because it is too simple to characterize all art as useless or unprofitable. The distinctions useless/useful and unprofitable/profitable also apply to different kinds of art. In a manner strictly parallel to the division of the faculties, high or fine comes to be identified with critical art, which is not intended to be useful or profitable, and low popular art is identified with art that is either useful or profitable. This distinction between high and low art emerged at the same time Kant was writing the *Critique of Judgment*. With the decline of aristocratic and ecclesiastical power and the emergence of the bourgeoisie wrought by the advent of modern industrialism, the condition of artistic production and consumption changed significantly. When medieval patronage collapsed, the artistic "tenure system" disappeared; art was transformed into a commodity, and artists had to compete in the marketplace. Art no longer was produced exclusively for wealthy patrons who enjoyed leisure and were unburdened by the necessity to work, but now had to be marketed by effectively addressing consumers with different interests. The emergence of a market economy created the new class of the bourgeoisie, whose members were neither exactly producers (workers) nor nonproducers (aristocrats). This class needed cultural markers by which to establish their status and social identity. The interests of artists and bourgeois consumers intersected in the search for cultural artifacts that were not completely subject to market forces. The creation and possession of works of art came to serve as a means of securing social distinction. For art to serve no practical function, however, it had to be clearly distinguished from all commodities that could have a utilitarian end. Art, in the strict sense of the term, became "high art" or "pure art," which was distinguished from crafts, mass art, and popular art. The distinction between high and low art formulated at the end of the eighteenth century continues to be decisive throughout the twentieth century. In an essay that continues to exercise enormous influence both on art criticism and artistic practice, "The Avant-Garde and Kitsch," Clement Greenberg writes:

Kitsch is a product of the industrial revolution, which urbanized the masses of Western Europe and America. . . . To fill the demand for the new market, a new commodity was devised: ersatz culture, kitsch, destined for those who, insensible to the values of genuine culture, are hungry nevertheless for the diversion that only culture of some sort can provide. Kitsch . . . welcomes and cultivates this insensibility. It is the source of its profits. Kitsch is mechanical and operates by formulas. Kitsch is vicarious experience and faked sensations. . . . Kitsch is the epitome of all that is spurious in the life of our times. Kitsch pretends to demand nothing of its customers except their money—not even their time.[9]

Kitsch is impure and inauthentic because it is tainted by filthy lucre. In contrast to mechanically produced and economically motivated kitsch, high art is not only critical but self-critical and, as such, demands time as well as education. The uneducated masses cannot appreciate the transcendent value of true art.

As Pierre Bourdieu correctly argues, high art and low art involve very different economic logics:

These fields are the site of the antagonistic coexistence of two modes of production and circulation obeying inverse logics. At one pole, there is the anti-"economic" economy of pure art. Founded on the obligatory recognition of the values of disinterestedness and on the denegation of the "economy" (of the "commercial") and of "economic" profit (in the short term), it privileges production and its specific necessities, the outcome of an autonomous history. . . . At the other pole, there is the "economic" logic of the literary and artistic industries which, since they make the trade in cultural goods just another trade, confer priority on distribution, on immediate and temporary suc-

cess, measured for example by the print run, and which are content to adjust them-selves to the pre-existing demand of a clientele.[10]

The abiding tension between art and commerce grows out of their contrasting economic logics. For the businessman, value and popularity/profitability are di-rectly related (i.e., the more popular and profitable, the more valuable), and for the artist, value and popularity/profitability are inversely related (i.e., the more popular and profitable, the less valuable). Flaubert speaks for many artists when he declares: "If one does not address the crowd, it is right that the crowd should not pay one. It is political economy. But, I maintain that a work of art (worthy of that name and conscientiously done) is beyond appraisal, has no commercial value, cannot be paid for. Conclusion: if the artist has no income, he must starve!"[11] It is hard to miss the religious overtones of this view of art. The erst-while prophet becomes the avant-garde artist whose otherworldly values reject the bourgeois world on which he nonetheless indirectly depends. Worldly failure is the mark of success, and worldly success is the sign of failure.

There is a third and, at least for the moment, final stage in the progression I have been charting. After the religious prophet becomes the avant-garde artist whose aesthetic education is supposed to lead the way to the promised land, the artist becomes the intellectual whose "negative capability" is criticism without end. Bourdieu credits Émile Zola with "the invention of the intellectual":

Zola needed to produce a new figure, that of the intellectual, by inventing for the artist a mission of prophetic subversion, inseparably intellectual and political, which had to be able to make everything his adversaries described as the effect of a vulgar or depraved taste appear as an aesthetic, ethical and political stance, and one likely to find militant defenders. Carrying to term the evolution of the literary field towards autonomy, he tries to extend into politics the very values of independence being as-serted in the literary field. . . . The intellectual is constituted as such by intervening in the political field in the name of autonomy and of the specific values of a field of cultural production which has attained a high degree of independence with respect to various powers.[12]

Whether or not Bourdieu is right in giving Zola so much credit, it is clear that the mantle of criticism passes from the prophet to the artist, and then to the in-tellectual, who often, though not always, is a scholar.

As I have noted, Kant's division of the faculties and their inverse economic logics lies at the foundation of the University of Berlin. In his comprehensive survey of the formation of the modern university in Germany during the period extending from 1700 to 1914, Charles McClelland argues that a group he labels "neo-humanists" exercised decisive influence in shaping the University of Berlin. Though "a manifestation of the self-consciousness of the German middle class," neo-humanism defines itself by *rejecting* bourgeois values:

The insistence of the neo-humanists on education and self-development as the primary marks of a true Mensch *tended toward a form of elitism that excluded the values of a majority of normal burghers. The pursuit of wealth or other mundane signs of fortune characterized the values of these burghers, in the eyes of the neo-humanists, and they rejected such values out of hand. Thus, in important respects, neo-humanism was an elite bourgeois ideology directed against a common bourgeois value system.[13]*

As von Humboldt began to develop his program for the university, he was determined to resist the utilitarianism which, he believed, was ruining British universities. In forming his educational policy, he borrowed two central ideas from neo-humanism: *Wissenschaft* and *Bildung.* Though usually translated "science," Wissenschaft is not limited to what we now designate the natural sciences; rather than discipline specific, it refers to the pursuit of knowledge for its own sake. Bildung is the process of self-cultivation or self-realization. In contrast to what was known in Germany as *Brotstudium,* studying for the sake of a career, Wissenschaft is the *disinterested* pursuit of truth and Bildung is motivated by *intrinsic* rather than *extrinsic* interests.[14] In words clearly echoing the Kantian principles of autonomy and inner teleology, von Humboldt insists that study at the university is to be "unforced and non-purposeful."[15] When understood in terms of Bildung, education becomes a personal quest of self-discovery, which is, in many ways, an inwardization of the traditional grand tour. Education, in other words, is not the pursuit of technical expertise or practical training, which equips one for worldly success, but is the preparation for a lifelong journey to selfhood.[16]

As I have noted, the University of Berlin has served as a model for many of the most important developments in higher education for the past two centuries. Universities throughout the world attempt to combine research and teaching in the way initially established at Berlin. In countries with different traditions where research institutions were slower to develop, the principles of organizational structure and educational mission defined by von Humboldt and his colleagues nonetheless exercised considerable influence on educational policy. In England, for example, Oxford and Cambridge created residential colleges in rural settings and a tutorial system for individual instruction. Education extended beyond the classroom and lecture hall to the playing fields and personal life. Though subject to utilitarian pressures, which usually originated in Scotland, British universities like Oxford and Cambridge remained dedicated to ideals very similar to those promoted by Germans in the name of Wissenschaft and Bildung. This is most evident in John Henry Newman's influential book *The Idea of the University*, written in response to what he saw as the pernicious "utilitarianism and mechanism" pervading education in the United Kingdom. With the church rather than the state looking over his shoulder, Newman espoused many of the principles and ideas formulated by Kant and elaborated by von Humboldt. Most important, he defined "liberal knowledge" in a way that both summarized significant developments during the first half of the nineteenth century and anticipated the understanding of the liberal arts informing the arts and sciences down to our own day. "That alone is liberal knowledge," Newman avers, "which stands on its own pretensions, which is independent of sequel, expects no complements, refuses to be *informed* (as it is called) by any end, or absorbed into any art, in order duly to present itself to our contemplation." This understanding of liberal knowledge lies at the heart of Newman's understanding of the university:

This process of training, by which the intellect, instead of being formed or sacrificed to some particular or accidental purpose, some specific trade or profession, or study or science, is disciplined for its own sake, for the perception of its own proper object, for its own highest culture, is called Liberal Education; and though there is no one in whom it is carried as far as is conceivable, or whose intellect would be a pattern of what intellects should be made, yet there is scarcely any one but may gain an idea of what real training is, and at least look towards it, and make its true scope and result, not something else, his standard of excellence; and numbers there are who may submit themselves to it, and secure it to themselves in good measure. And to set forth the right standard and to train according to it, and to help forward all students towards it according to their various capacities, this I conceive to be the business of the university.[17]

It is obvious that Newman's notion of "the business of the university" is at odds with the educational policies of Kant's "businessmen" and "technicians of learning." Education in which the intellect is "disciplined for its own sake" rather than "sacrificed" to "some specific trade or profession" is consistent with Kant's lower rather than higher faculties. There is, however, an important difference between the positions of Kant and Newman. Whereas Kant crowns philosophy queen of the sciences responsible for laying the foundation for critical reflection, Newman gives pride of place to literature. In his scheme, philosophical training actually occurs under the aegis of the discipline of literature. The difference between Kant and Newman on this critical issue reflects differences between empirical British philosophy and speculative continental philosophy. As we will see, the shift from philosophy to literature is important for the emergence of literary theory in the past three decades.[18]

The history of higher education in the United States is, in large measure, the story of the struggle to combine the British college and the German university. The first American university was Harvard, which included among its founders a group of several Oxford graduates and a group of approximately thirty-five Puritans, who were graduates of Cambridge's Emmanuel College. Historian of education Frederick Rudolph argues that "the founders of Harvard attempted to recreate at Cambridge the college they had known at the old Cambridge in England."[19] While committed to the model of residential colleges with a strong emphasis on moral character development, the Puritans who founded Harvard were suspicious of education that had little practical purpose. For early pioneers—educational and otherwise—the aim of education was to train leaders for service in church, state, and society. This practical orientation was reinforced by a profound sense of pragmatism that runs deep in the American grain. For many people who came to the New World, classical European education appeared to be antidemocratic and to reflect aristocratic ideals from which they had fled. When universities strayed too far from principles of practicality, various legislative authorities often intervened. In 1850, for example, the Massachusetts General Court "called on Harvard to reform its curriculum in order to prepare 'better farmers, mechanics, or merchants.'" Half a century later, Theodore Roosevelt expressed a similar sentiment on the occasion of the dedication of a new law school building at the University of Chicago: "We need to produce, not genius, not brilliancy, but the homely, commonplace, elemental virtues."[20] Such views did not, of course, go uncontested. At the same time the Massachusetts Court was issuing its edict, Henry Tappan, who would soon become the president of the University of Michigan, complained: "We have cheapened education so as to place it within the reach of everyone." This concern reflects sentiments also expressed by Nathan Lord, president of Dartmouth from 1828–1863, when he insisted that college education is not intended for people who planned to "engage in mercantile, mechanical, or agricultural operations."[21] American higher education has consistently wrestled with the problem of balancing the practical and the impractical, the common and the elite, the applied and the theoretical, and so on.

Eventually workable compromises were reached. Though the British college has been very important for American higher education, the model of the German research university also played a significant role from the country's early days. In 1824, Thomas Jefferson, to the dismay of many of his critics, recruited faculty members from Germany to teach at the University of Virginia. The most important effect of German influence came in 1867, when Johns Hopkins used the proceeds from his Baltimore and Ohio Railway to create the first American university devoted to "pure scholarship." Initially, it appeared that universities would be devoted primarily to original research and scholarship, while colleges would have the responsibility for teaching the history and tradition of different disciplines and for passing on the results of work being done at universities. However, merely two years after Hopkins's initiative, Charles William Eliot used the occasion of his inaugural address at Harvard to insist on the necessity of "purposefully obliterating or at least diffusing the lines between undergraduate and graduate, between collegiate and scholarly."[22] Though the history of higher education in the United States from the 1850s to the 1950s is often a tangled affair, the poles of debate have remained relatively constant.

In the years following the Second World War, the rapid expansion of higher education brought major changes, whose effects are still being felt. Among the many significant developments, two are particularly important for our purposes: the accelerated professionalization of the faculty characterized by new departures in academic practice, and the growing corporatization and bureaucratization of the university. Though the line between research universities and teaching colleges had never been hard and fast, in the late 1960s it became increasingly obscure. Among the many factors contributing to this development, none was more important than the collapse of the academic job market in the late 1960s and early 1970s. With the abrupt end of the postwar expansion and the disappearance of new faculty positions, research and publication demands increased significantly at virtually all institutions. There was an ill-advised expansion of graduate programs and an understandable explosion of professional organizations, specialized journals, and university press publications. Even at institutions that traditionally had privileged teaching, original research and publication requirements for hiring and promotion became the norm. As the importance of publication increased, the importance of teaching decreased. The value of publication, in turn, followed the inverse economic logic we have discovered in art: the more popular and profitable the work, the less its academic value. Translated into the world of education, this means that one rarely gets tenure at a respected college or university for writing a textbook or popular work; nor does one get tenure primarily on the basis of teaching. These developments eventually created educational institutions with more and more faculty members whose work interests fewer and fewer people beyond the walls of the academy. By any economic logic, this is hardly a formula for success in today's world.

The corporatization of the university during the last half century is, at least in part, the result of the unprecedented postwar growth of educational institutions. In order to manage the increasing student population, both teaching and administrative procedures had to be regularized and bureaucratized. While most obvious in large state institutions, these tendencies pervaded all higher education to a greater or lesser extent. But growth alone cannot explain the attraction of the corporate model; changes in the economics of higher education were equally important. In order to finance necessary expansion, many universities began relying heavily on government support. Some of these funds were outright grants, but much of the income resulted from contracts with government agencies, many of which were connected with the Pentagon and the CIA. Needless to say, most—but not all—of these revenues were generated by the natural sciences and applied technology. The management of these contracts required new bureaucratic structures, legal services, and many more administrators. Once universities became dependent on government funds, both their autonomy and capacity for political and cultural criticism were compromised. Resistance to the war in Vietnam forced a reevaluation of the relations between universities and government, which left lingering suspicions among many faculty members about associations "outside" the walls of the academy. In their recent polemical tract, *Academic Keywords: A Devil's Dictionary for Higher Education,* Cary Nelson and Stephen Watt go so far as to argue that such relations run the danger of stripping "higher education of all its intellectual independence, its powers of cultural critique and political resistance."[23] But government was not the only source of this alleged danger. The end of the Cold War brought important changes for relations between the federal government and universities. Funding for research and development shifted from defense projects to programs designed to increase the economic competitiveness of U.S. companies.

With the cutbacks in government spending during the Reagan years, private corporations stepped in to fill the void left by the withdrawal of federal and state support. Through elaborate legal agreements, corporations contracted or entered into partnerships with departments and universities to conduct research as well as to develop and test products. In 1980, Congress passed patents and trademark amendments, known as the Bayh-Dole Act, which allowed universities and other nonprofit institutions to hold property rights to discoveries and inventions developed with federal research funds.[24] This legislation was intended to encourage joint ventures by universities and the for-profit sector. For universities, these arrangements provided much-needed income, and for businesses and corporations, it often proved more cost effective than setting up labs and hiring staff. While corporate partnerships were formed primarily in the areas of science and technology, throughout the 1980s there was a noteworthy shift from the physical to the biological sciences. The hyperspecialization of the faculty and corporatization of the university combine to make institutions the target of trenchant criticism from different political perspectives within and beyond the academy.

CRITICAL FACULTIES

While discussions of higher education in recent years have been wide-ranging, few have been philosophically informed. A conspicuous exception to the tendency to focus on practical remedies rather than theoretical critique is Derrida's influential analysis of what he describes as "the institution of technoscience," which grounds the contemporary university. Derrida's pointed criticism has inspired several noteworthy studies and provoked heated debate.[25] Though Derrida and his followers frame their position to resist what they regard as the conservative forces dominating the university, their vision of the university turns out to be surprisingly traditional.

In formulating his analysis, Derrida returns to Kant's *Conflict of the Faculties* to construct an analysis that is largely derived from Heidegger's interpretation of the work of art and corresponding critique of technology. The institution of modern technoscience, which shapes today's universities, Derrida argues, is characterized by the "finalization" of research. The difference between basic and finalized research is strictly parallel to Kant's distinction between inner and external teleology: while basic research is neither motivated by "outside" interests nor practically directed, finalized research is directed toward a specific telos, aim, or goal, which is known in advance of the investigation. In the case of technoscience, this goal is usually *extrinsic* to the university as such and, in the most profitable cases, is related to military projects:

This is all too obvious in areas such as physics, biology, medicine, biotechnology, bioprogramming, data processing, and telecommunications. We have only to mention telecommunications and data processing to assess the extent of the phenomenon: the "orientation" of research is limitless, everything, everything in these areas proceeds "in view" of technical and instrumental security. At the service of war, of national and international security, research programs have come to encompass the entire field of information, the stockpiling of knowledge in the workings and thus also the essence of language and of all semiotic systems, structural and generative linguistics, pragmatics, rhetoric.[26]

Derrida borrows the interpretation of science and technology underlying his analysis from Heidegger. As we have already discovered, Heidegger sees twentieth-century technology as the disastrous outcome of the will to power, which has been unfolding throughout the Western tradition. In his influential essay "The Question Concerning Technology," he argues:

We ask the question concerning technology when we ask what it is. Everyone knows the two statements that answer our questions. Technology is a means to an end. The other says: Technology is a human activity. The two definitions of technology belong together. For to posit ends and procure and utilize the means to them is a human activity. The manufacture and utilization of equipment, tools, and machines, the manufactured and used things themselves, and the needs and ends that they serve, all belong to what technology is. The whole complex of these contrivances is technology. Technology itself is a contrivance, or, in Latin, an instrument.[27]

The instrumentality of technology, then, establishes a utilitarian relation between means and end. For technological reason, nothing is ever good in itself but is only valuable insofar as it is good for *something other than itself.* When so understood, Heidegger argues, technology is a practical extension or application of modern science:

Modern science's way of representing pursues and entraps nature as a calculable coherence of forces. Modern physics is not experimental physics because it applies apparatus to the questioning of nature. Rather the reverse is true. Because physics, indeed already as pure theory, sets up nature to exhibit itself as a coherence of forces calculable in advance, it therefore orders its experiments precisely for the purpose of asking whether and how nature reports itself in this way.[28]

The instrumentality of technology is the practical effect of the calculating machinations of modern science. Conversely, the instrumentality, which defines technology, frames the supposedly objective and disinterested outlook of mod-

ern science. Technology and science come together in contemporary techno-
science.

Technoscience regards not only nature but also culture as a resource or, in
Heidegger's terms, a "standing reserve" for human exploitation. From this point
of view, natural and cultural resources are valuable only insofar as they are *good
for human use.* Derrida sees information and telecommunications technologies
as even more dangerous than previous forms of technology. Invoking the ancient
distinction between words and things, he points out that traditionally science
and technology deal with things, while the arts and humanities explore words or
language. What makes recent technological developments so insidious, according to Derrida, is the "informatization of language, which transforms language
and culture from a safe preserve into a resource that can be exploited for *extrinsic* purposes." "Computer technology, data banks, artificial intelligences, translating machines, and so forth," he points out, "are constructed on the basis of
that instrumental determination of a calculable language." When language becomes information, all culture becomes a standing reserve available for use by
individuals and institutions with interests and motives that are neither educational nor cultural. Derrida explains that "the Navy can very rationally subsidize
linguistic, semiotic or anthropological investigations. These in turn are related to
history, literature, hermeneutics, law, political science, psychoanalysis, and so
forth."[29]

Faced with this situation, Derrida, as always, calls for resistance—the
"lower" faculty must reassert its interests against the pressures of the "higher faculties." His deconstructive strategy brings together philosophy and literary criticism to create a theoretical practice, which is, in effect, the queen of the arts and
sciences. Derrida develops his critical position by rereading Heidegger's account
of thinking through Kant's interpretation of the work of art. "Thought," Derrida insists, is "a dimension that is not reducible to technique, nor to science,
nor to philosophy." Insofar as it has a goal, the pursuit of thinking is intended to
"remove the university from 'useful programs and from professional ends,'" and
thereby subvert the "powers of caste, class, or corporation."[30] Thinking, like art,
resists technological and economic interests by following an inverse economic
logic: to think is to engage in an activity that is useless or even wasteful. Bill
Readings effectively summarizes Derrida's conclusion:

*Thinking, if it is to remain open to the possibility of Thought, to take itself as a ques-
tion, must not seek to be economic. It belongs rather to an economy of waste than to
a restricted economy of calculation. Thought is non-productive labor, and hence does*

not show up as such on balance sheets except as waste. The question posed to the university is thus not how to turn the institution into a haven for Thought but how to think in an institution whose development tends to make Thought more and more difficult, less and less necessary.[31]

In terms reminiscent of Kant's account of intrinsic rather than extrinsic value, the more useful thinking is, the less its value, and the less useful it is, the more its value. For thinking to preserve its critical edge, it must remain *autonomous,* that is, independent of political influence and market forces. The thinker, like the prophet and avant-garde artist, finds success in failure, which the masses, who follow the ways of the world, confirm by their rejection.

Derrida's interpretation of Heideggerian thinking through Kant's view of art issues in an understanding of theory as essentially *critical* theory. Deconstruction assumes primarily responsibility for the task of criticism assigned to the present-day version of Kant's lower faculty. The practice of deconstruction cannot be located in any particular department of the university but claims to create the critical opening in which all the arts, humanities, and sciences operate.[32] Falling between or beyond every discipline, deconstructive criticism offers nothing more than "negative wisdom."[33] Like ancient Hebrew prophets, who serve an otherworldly God, those who follow criticism into the wilderness say "No" even when claiming to say "Yes."[34] The kingdom, they believe, is *forever* deferred. This is not to imply that criticism is politically irrelevant and institutionally disengaged. To the contrary, critical practice, Derrida insists, "is not a matter simply of questions one *formulates* while submitting oneself, as I am doing here, to the principle of reason, but also preparing oneself thereby to transform the modes of writing, approaches to pedagogy, the procedures of academic exchange, the relation to languages, to other disciplines, to the institution in general, to its inside and its outside."[35] Herein lies the purportedly *political* dimension of deconstruction. Critical practice develops new styles of writing, teaching, and academic exchange, which are intended to unsettle educational institutions. But, of course, this entire strategy unfolds *within* the precincts of the university even while claiming to solicit a certain irreducible exteriority. Politics, in other words, is always *academic* politics. While claiming to resist or subvert every form of closure, critical theory nonetheless attempts to preserve the autonomy of the university by setting it apart from the world and by establishing a sanctuary from the harsh economic "realities" of "the world."[36]

Derrida, of course, realizes that the Sisyphean challenge of criticism inevitably fails. Criticism becomes an "infinite task" subverting "all the ruses of end-orienting reason." "Beware of ends," Derrida warns, "but what would be a university without ends?"[37] The endlessness of criticism is the result of the inexhaustible capacity of economic forces to appropriate for their own ends everything intended to escape systems of exchange:

From now on, so long as it has the means, a military budget can invest in anything at all, in view of deferred profits: "basic" scientific theory, the humanities, literary theory and philosophy. The compartment of philosophy, which covered all of this and which Kant thought ought to be kept unavailable to any utilitarian purpose and to the orders of any power whatsoever in its search for truth, can no longer lay claim to such autonomy. What is produced in this field can always be used. And even if it should remain useless in its results, in its productions, it can always serve to keep the masters of discourse busy: the experts, professionals of rhetoric, logic or philosophy who might otherwise be applying their energy elsewhere. Or again, it may in certain situations secure an ideological bonus of luxury and gratuitousness for a society that can afford it, within certain limits. Furthermore, when certain random consequences of research are taken into account, it is always possible to have in view some eventual benefit that may ensue from an apparently useless research project (in philosophy or the humanities, for example). The history of the sciences encourages researchers to integrate that margin of randomness into their centralized calculation.[38]

When such calculation occurs, the businessmen and technicians of learning truly have taken over the university; the so-called higher faculties completely absorb the so-called lower faculty by turning apparently disinterested investigation to their own ends and by capitalizing on ostensibly useless work for their own profit. Far from subverting or overturning apparently repressive structures, the practice of criticism can actually reinforce the very systems it struggles to resist.

This dialectical reversal of critical fortunes becomes obvious when critics who follow Derrida turn from textual strategies to institutional criticism. Rather than reform, so-called critics often unknowingly call for the reinforcement of many of the programs and policies defined by Kant and implemented at the University of Berlin. Nelson and Watt speak for most of their colleagues when they insist that the essential function of "cultural critique and political resistance" can be preserved only if institutional autonomy remains inviolable. Such autonomy presupposes full-time lifelong employment and the protection of separate departments in the university. While arguing in the name of academic freedom, Nelson and Watt, like many others, are actually more interested in protecting job security, that is, tenure.[39] This is best accomplished, they believe, by preserving separate departments and ensuring peer review. What must be avoided at all costs is *outside* interference in the activities of departments and the university. The walls separating departments form an autonomous enclave in an autonomous institution, which apparently remains separate from the world.

Many faculty members are convinced that the walls "protecting" the university are threatened by what is misleadingly labeled "distance education." "Real" education, which requires proximity and presence, they argue, cannot be delivered through technologies that mediate at a distance. Struggling to defend traditional pedagogical styles and institutional structures, critics argue that e-Ed is an inadequate substitute, which is no more valuable than kitsch. Always seeking the lowest possible common denominator, e-Ed supposedly devalues the currency of the realm by serving practical ends and addressing the interests of popular culture. Once again, the terms of the debate are precisely those Kant established in 1798: critical versus useful, theoretical versus practical, elite versus popular, high versus low. Criticisms that presuppose these distinctions are both specious and disingenuous; at the very least, *such criticisms are both profitable and useful to the securely situated individuals who make them.* Protests to the contrary notwithstanding, the university is not autonomous but is a thoroughly parasitic institution, which continuously depends on the generosity of the host so many academics claim to reject. The critical activities of the humanities, arts, and sciences are only possible if they are supported by the very economic interests their criticism so often calls into question. The university and the people who are employed by it have always been *thoroughly* implicated in a market system, which now is expanding more rapidly than ever. After all, faculty members are paid by the income from endowments created by government grants as well as by "charitable" gifts from wealthy donors and invested in financial markets around the world. In addition to this, most faculty members demand job security denied to the people upon whose labors they depend. Leaving questions of fairness and practicality aside, this system of well-paid lifelong employment is unlikely to continue much longer. While American higher education has a long tradition of philanthropic support, it is not yet clear whether donors whose wealth is created by the information economy and knowledge business will continue to give so freely to universities and colleges who increasingly are their competitors. In network culture, the rules of the game are changing; competitors must adapt quickly by learning to cooperate or they will not survive.[40] It is critical for educators to understand the implications of these changes and to assume an active role in the transformation of higher education now occurring.

When walls become permeable screens, the university as we have known it for two hundred years becomes a thing of the past. To begin to appreciate the significance of this change, it is helpful to reconsider the interplay between grids and networks by returning to Times Square. As we have discovered, in the midst of the Manhattan grid, a striking new skyscraper is emerging. From the spectacular Nasdaq sign to Gehry's remarkable cafeteria, the Condé Nast building not only captures but also transmits the currents of network culture. Far from superficial, the changes occurring in the moment of complexity are altering the very structure and substance of life. As the currents of network culture expand to circulate through colleges and universities at ever-greater speeds, the world of higher education is being completely transformed. To understand what these changes entail, try to imagine a university modeled on the architecture of a Frank Gehry rather than a Mies van der Rohe building. Is it possible to create an educational institution whose structure and function more closely approximate Nasdaq than a Ford assembly line?

These musings grow out of more than a decade of experimenting with new technologies in my research, writing, and teaching. As I explained in the introduction, in the late 1980s I began to glimpse some of the changes information and telematic technologies would bring to education. By the early 1990s, I had taken the first step toward creating global classrooms by teaching a teleseminar, which linked students in the United States with a colleague and students in Finland. In the years following the Helsinki seminar, I started working with multimedia in my "writing" and teaching. As the Internet spread and the worldwide web emerged in the mid 1990s, I began exploring ways of delivering education on-line. My most ambitious effort was the creation of what I described as a CyberCollege for Williams College alumni and alumnae. With the guidance and support of my former student John Kim, I used recently developed software to webcast classes synchronously and asynchronously, and to set up bulletin boards and chat rooms for everyone taking the course. By the fall of 1998, we were able to bring together forty Williams undergraduates and over two hundred alumni/ae for a semester-long course on the psychology of religion.

The more I worked with these new technologies, the more convinced I became of their rich educational potential. With increasingly sophisticated hardware and software, new forms of cultural production are becoming possible. After many false starts, I eventually learned that *the effective use of information and telematic technologies does not involve doing the same thing differently but doing*

257

something different. Writing and teaching are no longer limited to the printed and spoken word but expand through the creative interplay of words, sounds, and images. Texts—be they books, articles, or lectures—do not have to be closed but can become open and interactive. Through the use of increasingly sophisticated multimedia, it is possible to develop new modes of interpretation and argumentation. These expanding textual practices can extend the audience for cultural analysis in ways not previously possible. When "writing" assumes new forms, the difference between publication and teaching no longer remains stable. On the one hand, more and more scholarly publication is being produced and distributed through electronic media, and, on the other, on-line teaching is becoming an alternative form of publication. With wired classrooms, the time and place of instruction are transformed: education becomes available anytime, anyplace in the world. Like networked global markets, education becomes a 24/7 business.

As my understanding of the significance and scope of these changes deepened, I realized the importance of creating support structures which would encourage and enable faculty colleagues to explore new media. With the cooperation of the administration and Office of Information Technology at Williams College, I created the Center for Technology in the Arts and Humanities, which is a research and development studio where faculty can experiment in new media. One of the distinctive features of this center has proven to be very controversial. Since students are much more knowledgeable about new media than most faculty members, I attempted to create collaborative relationships in which the line separating teachers and students is not firmly fixed. Even when such collaborations are successful, many professors resist or even resent becoming students of "their" students.

The experiments of the past decade have, for the most part, been surprisingly successful, yet the response of most colleagues and administrators has been *very* disappointing. Many people continue to believe that information and telematic technologies will not fundamentally change higher education. During the past two years, I have had the opportunity to talk with faculty and administrators at many of the nation's leading colleges and universities and have discovered that the assumptions and suspicions I have found at Williams are widely shared. *What makes e-Ed so controversial is the way in which it brings together technological innovation and market forces.* Kant's vision of the arts, humanities, and sciences implicitly inspires resistance to technology and business even though faculty members do not realize that they remain committed to a model of the university that is over two hundred years old. The lesson seems to be that the more inevitable change is, the greater the opposition to it becomes.

In late fall 1998, while I was preparing to leave Williams to teach at the University of North Carolina, Chapel Hill, I was pondering possible strategies for proceeding in a climate that was at best indifferent and often hostile. A few weeks before I was to leave, I unexpectedly received a phone call from Herbert A. Allen. I had never met Herbert but knew he is a Williams graduate, who is the head of Allen & Co., a New York investment bank. I also knew he is legendary for an annual conference he hosts each year in Sun Valley, Idaho, where he brings together a large group of leading figures in technology, media, news, entertainment, and finance for a week of discussion and debate.[41] Though we live and work in very different worlds, I suspected we might share certain interests. Herbert had heard of some of the work I was doing and was interested in talking about it. When we met, to focus our discussion, he asked me three questions. He wanted to know what I had been doing in the area of technology and education, what I was not doing that I thought I should be doing, and where I thought the world of higher education was heading. Over the course of the next several hours, we had a remarkable conversation ranging from Hegel and postmodernism to the future of wireless technology and the rate of deployment of broadband transmission capacity. As a result of many years of experience and extraordinary professional and personal contacts, Herbert is in a unique position to know what is going on at the critical intersection of finance, technology, media, and politics. He also has a long-standing interest in education and has generously supported the arts. As I sketched my vision for the university of the twenty-first century, Herbert immediately understood what I was saying and agreed with my overall assessment of the direction in which things are heading both within and beyond the world of higher education. With our conversation drawing to a close, we paused and looked at each other; we both realized that we had been thinking along similar lines for quite a few years. In the brief moment of that glance, our worlds collided, releasing sparks that seemed both strangely inevitable yet completely unexpected. After a moment, Herbert said: "You know, what you've been saying is right. It's going to happen; no doubt about it. And it's important that it is done right. If you want to give it a shot, I'll back it." This was a possibility I had *never* anticipated, and Herbert knew before I did that I could not say "No."

Herbert and I agreed that higher education is poised at what his longtime friend and chairman of Intel, Andy Grove, describes as a "strategic inflection point," which marks the transition between two very different ways of doing business.[42] This "inflection point" bears all the marks of Bak's "tipping point," where unpredictable forces create serious dangers as well as extraordinary opportunities. But colleges and universities have neither the financial resources nor the necessary commitment to respond to this critical moment quickly and effectively. The organizational structures and vested interests of different constituencies within colleges and universities make it highly unlikely that they will be able to negotiate the changes network culture brings by themselves. *If* higher education is to thrive in an increasingly competitive commercial market, traditional institutions must create new alliances among themselves and enter into unprecedented partnerships with for-profit businesses. With the pace of change increasing, Herbert and I felt the urgency of forming new structures of cooperation and support that would enable educators and educational institutions to assume a significant role in creating an educational environment unlike any we have known in the past. The stakes of this undertaking are too high to be determined by commercial interests without the active participation of colleges and universities.

As a way of getting started, Herbert asked me to write a summary of the analysis I had presented to him. Pondering my assignment, I attempted to imagine how the kind of network logic I have analyzed in the previous chapters could be deployed to create a new kind of organization or institution. What *would* a university structured like Nasdaq with a constantly changing hypertextual curriculum actually look like and how would it work? I summarized my tentative conclusions in a memo entitled "Global Education Network." Six months later, Herbert and I founded a company named Global Education Network (GEN), whose goal is to provide high-quality e-Ed in the humanities, arts, and sciences to people of all ages throughout the world at a reasonable price.

As we have discovered, one of the distinguishing characteristics of network culture is that walls which once seemed secure become permeable screens which create possibilities for the emergence and evolution of new organizational systems and structures. In the world of higher education, the two most important walls that must become screens are the wall separating for-profit and nonprofit organizations and the wall separating different educational institutions. The information economy makes it more important than ever for business and education to cooperate. Business cannot survive without continuing education for its workforce, and colleges and universities cannot afford the level of investment necessary for developing new ways of delivering education. The economies of scale involved in emerging technologies make cooperation among educational institutions not only prudent but also necessary. One of the lessons of the new economy is that *competitors must learn to cooperate.* It is much more efficient and effective for colleges and universities to form associations that foster collaboration. But these alliances must extend beyond traditional boundaries. GEN is a virtual organization, which links educators, educational institutions, business professionals, and investors in a web of relations that is mutually supportive and productive. The principle governing the distribution of responsibilities in this organization is as simple as it is untested: educators will do what they do best and business partners will do what they do best. Instead of continuing business and education as usual, we are forming new structures that will enable individuals and institutions to do things differently. This requires the creation of alternative organizational and operational principles.

GEN is organized as a for-profit corporation whose partners include a group of highly selective universities and colleges and some of the country's leading companies and investors (fig. 42). Partnerships in the organization are characterized by consultation, cooperation, and collaboration on *all* aspects of the enterprise. Acutely sensitive to the concerns about market forces compromising the quality of e-Ed, GEN is committed to leaving the responsibility for providing courses and preserving their quality *completely* in the hands of educators and educational institutions. Two kinds of faculty members are participating in GEN. The first group comes from partner schools and the second from unaffiliated institutions. Professors from partner institutions teach the core of the curriculum. To insure the breadth and diversity of education opportunities, leading faculty members from other institutions also offer courses on the network. While all courses are produced in cooperation with the GEN production staff, nothing appears on the network until it is approved by the contributing faculty and partner institutions. The task of shaping the curriculum and monitoring its quality is the responsibility of an academic advisory board comprised of four dis-

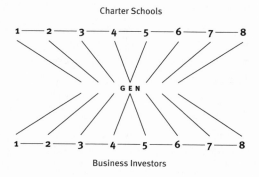

FIGURE 42 Global Education Network (1)

tinguished retired presidents of leading colleges and universities.[43] The collective educational and administrative wisdom of this group forms an invaluable bridge between the academic and business cultures.

Recognizing the understandable concern of colleges and universities have about the cost of supporting high quality e-Ed, GEN investors assume *all* financial risk. In contrast to companies that only provide software for producing courses or portals for distributing courses that have already been produced, GEN takes responsibility for the production, technical support, advertising, marketing, distribution, and administration of all courses on the network at no cost to partner institutions and participating professors. Collaboration among different individuals as well as various institutions greatly enhances the efficiency and effectiveness of these efforts. One of GEN's most significant contributions is to supply the most advanced hardware and produce the most advanced software currently available. As new pedagogical resources emerge, it becomes clear that e-Ed is not, as many people continue to believe, merely a different way of delivering what goes on in traditional classrooms. Rather, new software transforms the educational experience. Everything produced on the network is archived in a way that allows students to access selected materials in any course. By breaking down walls separating courses, the curriculum becomes a complex hypertext. For example, a student taking a class in biology can search the archive for references to the genome in every course on the network. All student questions and comments are also archived and can be accessed. In this way, professors, like students, are in conversation even when they do not realize it. One of the distinguishing features of new platforms for e-Ed is the extensive use of interactive multimedia. Through the creative interplay of audio, visual, and verbal media, different ways of teaching become necessary. All of these innovations come at a cost that is not merely financial. In network culture, professors who continue to lecture from tattered notes they have used for years are as obsolete as most of the ideas they continue to promote. Intellectual change and pedagogical development are necessary survival strategies. Version 1.0 of every course is, like the machines on which it runs, outdated even before it appears.

Unlike existing companies, which provide either low-end training classes or postgraduate professional education, GEN concentrates on a high-quality undergraduate liberal arts curriculum. While realizing the financial uncertainties involved in this area of the market, we are convinced that it is possible to maintain high educational standards in the arts, humanities, and sciences while running a profitable business. All GEN courses can be taken on a graded or nongraded basis. With the cooperation and guidance of faculty members, GEN supports a pool of qualified teaching assistants from schools throughout the world who answer questions, respond to written work, and preside over bulletin boards and chat rooms. Each course is assigned the necessary number of teaching assistants and discussion moderators. This arrangement requires yet another level of interinstitutional cooperation among partners and with professors and schools that are not part of the network. We anticipate three major groups of students: advanced high school students, students enrolled in universities and colleges, and lifelong students. Though it is difficult to predict the market for this kind of e-Ed, it seems likely that the majority of people taking GEN courses probably will be continuing education students. We are committed to keeping the cost of GEN courses as low as possible and are developing a scholarship program for needy students of every age. GEN also provides scholarships to be awarded by partner schools to low-income advance-placement high school students to take courses on the network without payment. In addition to this, each partner school is entitled to designate two nonprofit organizations whose members are entitled to take courses at a reduced rate. Company profits are shared according to an agreed-upon split of *gross* tuition revenue among contributing faculty, institutions, and GEN.

One of the most complicated and divisive problems created by e-Ed is the issue of intellectual property rights. At most institutions, it has never been clear whether the faculty member or the university owns the rights to course materials for classes offered. The prospect of significant income from e-Ed is generating heated debate on this issue and is creating a conflict of interest among faculty, universities, and companies. In seeking to resolve these difficulties, educators and companies are turning to models in the entertainment and publishing industries for guidance. *The Chronicle of Higher Education* recently quoted Christine Maitland, the higher-education coordinator for the National Education Association, on this issue: "We've been taking lessons from the entertainment industry," Maitland commented. "The unions in the entertainment industry have been fighting for the rights to be paid for repeated performances for years, and we're trying to learn how we can work with that."[44] Needless to say, many faculty members are very nervous about analogies between education and entertainment because they fear the "dumbing down" of courses is inevitable when education is unduly subjected to market forces. They prefer to think of e-Ed as an extension of book publication. According to this model, intellectual property rights and, of course, royalty income belong exclusively to the faculty member. While the publication analogy is in many ways apt, it has yet to be determined whether the rules of print publication will be applied to the electronic distribution of courses. Some universities are actively exploring the possibility of contracting for a percentage of royalties for books written by their employees.[45] To date, the most celebrated example involving the conflict of interests between professor and institution in e-Ed is the case of Harvard law professor and television personality Arthur Miller. Miller contributed videotapes of his lectures to Concordia University School of Law, which is an on-line institution, without asking permission from Harvard. In response to Harvard's claim that he had violated the terms of his contract by providing course materials to another institution, Miller argued that "his arrangement with Concordia is analogous to publishing a book or giving a lecture on television."[46] It is obvious that neither the entertainment nor publishing model fits e-Ed adequately. Given the importance and complexity of these problems, this difficult issue is not going to be settled quickly.

In an effort to move ahead, GEN is attempting to avoid conflicts with both partner institutions as well as participating faculty members. Rather than competitors, charter schools and GEN are engaged in a common enterprise with shared responsibility for making all policy and strategic decisions. As employees of a participating college or university, faculty members benefit when GEN benefits and vice versa. In addition to this, GEN minimizes the possibility of a conflict of interests by providing faculty members with the opportunity to develop different kinds of courses. The aim of GEN is not to duplicate what colleges and universities are already doing but to offer an alternative educational experience. The more one learns about the potential of new instructional platforms, the more obvious it becomes that for the foreseeable future, e-Ed will supplement rather than replace what goes on in traditional classrooms. One of the most important changes that will occur is that the length of courses will vary. We have discovered that the traditional semester of 12–16 weeks is too long for most on-line courses. It is necessary to move away from uniform courses designed according to the principles of mass production and to begin designing courses as variable in length and organization as they are in content. One course size does not fit all; different subjects require different lengths of time to master them. By offering courses of various lengths at different levels of difficulty, it is possible to address the educational interests and needs of many more people. Students are no longer limited to selecting among courses offered in programs of study but will be able to devise the content of their classes and determine the direction of their education to an extent never before possible. Courses as well as the curriculum as a whole will become less fixed and more flexible. Just as the transition from an industrial to a postindustrial economy involves a shift in the balance of power from the producer to the consumer, so network culture creates the conditions for the transformation of mass production to mass customization of education by increasing the power of students to shape their courses.[47]

These changes in the production and delivery of education are having a significant impact on the structure of knowledge. The walls separating academic departments and disciplines are becoming as permeable as every other division in network culture. Creative work usually emerges between fields in areas that are far from equilibrium and often seem to hover at the edge of chaos. Separate disciplines as currently constituted can no more be justified than the departments whose interests they serve. To be effective in today's world, knowledge and the curriculum must assume the form of complex adaptive systems, which are in a process of constant formation and reformation. New technologies of production and reproduction not only facilitate but actually necessitate these changes. As these developments continue to unfold, the organizational structure of colleges and universities will have to become much more flexible and adaptable to accommodate the ongoing transformation of the substance and organization of knowledge. For faculty members, the most important consequence of curricular change will be the continuing erosion of tenure.[48]

One of the reasons many faculty members oppose distributed learning is because they see it as a threat to tenure. While defenses of tenure are usually presented in terms of academic freedom, concern about job security is often a more important issue. Speaking for many in today's university, Cary Nelson and Stephen Watt maintain that what is misleadingly labeled "distance education" exacerbates the problems they believe are plaguing the university. The disadvantages (for faculty), they conclude, outweigh the advantages (for students):

Distance learning also offers students the chance to take some of their courses from other universities. A student enrolled in a University of Pennsylvania degree program might choose to take an Internet art history course from Harvard. Other students have the opportunity to take infrequently taught languages on-line. Sorting out all the credits and payments for multi-institutional programs may require hiring a few more administrators, but that would hardly overturn recent hiring patterns. Increased choice and availability would give students opportunities to take courses and degree programs not available anywhere near their homes.

The downside of that would be the temptation for schools to reduce the size of or entirely eliminate departments that are well represented with nationally available courses. Updating courses will hardly require hiring full-time faculty; once again, part-timers will do. But the priority to service on-line courses may go to the top of the instructional budget. Partnerships with corporations to market on-line courses then increase the possibility that profit-oriented firms may gain an important role in campus staffing decisions. Overall, once large numbers of courses are available on-line, they are certain to have a significant impact on faculty hiring and the nature of the professorate.[49]

In the competitive world of higher education, institutions and corporations will find it increasingly difficult to survive when a significant portion of their workforce is comprised of permanent employees who resist change.

To respond to these changes of this magnitude, it is necessary to create a new kind of organization, which allows educators, colleges, universities, businesses, and investors to cooperate in ways that are mutually profitable. In developing GEN, we are attempting to design a different institutional architecture,

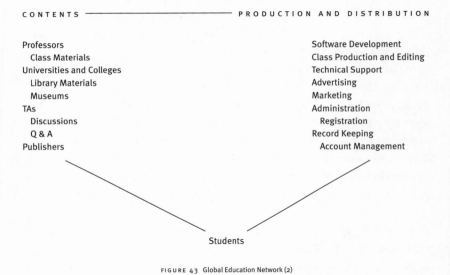

CONTENTS ———————————— PRODUCTION AND DISTRIBUTION

Professors
  Class Materials
Universities and Colleges
  Library Materials
  Museums
TAs
  Discussions
  Q & A
Publishers

Software Development
Class Production and Editing
Technical Support
Advertising
Marketing
Administration
  Registration
Record Keeping
  Account Management

Students

FIGURE 43 Global Education Network (2)

which is more like a network than a grid (fig. 43). More precisely, GEN is a network of networks that are always in flux. As the competitive landscape changes, new alliances and partnerships are constantly being formed. In addition to partnerships with select universities, GEN is contracting or partnering with other companies to provide necessary hardware, software, and related instructional materials. These agreements include alliances with service providers, computer companies, publishers, public and private libraries, and news and media organizations. Eventually, GEN will expand beyond the educational community to establish partnerships with a broad range of cultural institutions. Museums of all kinds, symphonies, and operas have rich resources and important educational programs, which complement college and university offerings. People interested in the arts are often looking for the kind of new educational opportunities GEN provides. Bringing together education and cultural institutions throughout the world will create unprecedented opportunities for cooperative exchange.

As GEN evolves, it will assume more and more characteristics of complex adaptive systems. Member schools, participating companies, and groups of fi-

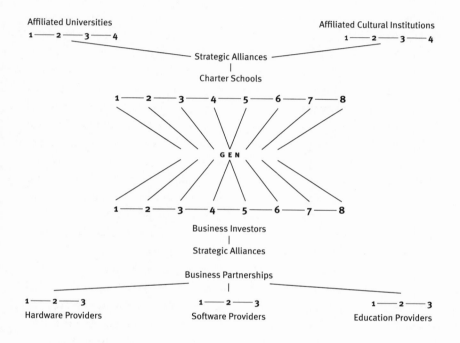

FIGURE 44 Global Education Network (3)

nancial investors are already complex organizations located in their own organizational networks. By trying to maintain an open network architecture, we are attempting to transform the long-standing walls, which traditionally have set colleges and universities apart, into permeable screens that nourish multiple exchanges between the so-called ivory tower and real world. While some developments are planned, many will be unexpected. There is no map for the new territory of e-Ed; indeed, the territory itself does not yet exist. If the wired world is a new frontier, it is unlike any we have known in the past. Instead of exploring a space that is already present, it is necessary to create an environment whose complexity is virtually unimaginable. By promoting connection, cooperation, and collaboration among erstwhile competitors on a new scale, GEN inevitably functions as a complex adaptive network in which particular individuals, institutions, and organizations are nodes in intricate coadaptive webs (fig. 44). The new educational environment that is emerging is more like a growing organism than an industrial factory. The ecology of this mutating landscape is as fragile as the ecology of the minds that create and are created by it. Viral webs and networks must be carefully cultivated. When attempting to support and coordinate the activities of many diverse individuals and groups, it is important to make every effort to avoid the pitfalls of highly centralized and hierarchical organizations. Whenever possible, the correlative principles of decentralization and distribution serve as guides for practice. In contrast to traditional business, educational, and broadcast models, the creation of a decentralized and distributed network facilitates many-to-many communication among both providers and users. The greater the communication and interactivity, the more flexible and adaptable GEN must become. Taken as a whole, GEN is more than the sum of the individuals and institutions that comprise it. Though neither centrally nor hierarchically organized, GEN nonetheless functions as a whole. Unplanned interactions create the condition for the emergence of creative innovations. Like every other complex adaptive system, GEN can prosper only as long as it remains poised between too little and too much order. Far from a balanced state of equilibrium, this marginal site occasions *creative disequilibrium*. As more and more participants become involved in this venture, the network's growing diversity makes innovative interactions considerably more likely. One of the critical challenges for any such organization is to find ways to mediate the disparate rates of change in the worlds of business and the Internet, on the one hand, and, on the other, the academic world. In an environment where only the quick survive, the slow response time of colleges and universities is dangerous to their long-term health. All of these complexities are, of course, compounded by the *global* reach of GEN. With education and business partners, faculty, TAs, and students from all over the world, the heterogeneity of the network increases exponentially.

The difficulties involved with creating and sustaining the Global Education Network remain daunting, and there are no guarantees of success. It is proving to be as hard for universities to adapt to business culture as it is for business to adapt to university culture. If higher education is to thrive in network culture, it is necessary to call into question many of the foundational principles upon which, as we have discovered, colleges and universities have rested for more than two hundred years:

Lower faculty/Higher faculties
Useless/Useful
Unprofitable/Profitable
Pure/Applied
Intrinsic value/Extrinsic value
Nonprofit/For-profit
Education/Entertainment
Education/Business
University/Marketplace
Ivory tower/Real world

When the walls separating these hierarchies become permeable screens, every-thing changes. While undeniable risks are involved in such changes, the greater risk lies in resisting change. Education is too important to remain confined within the walls where many people would like to keep it. Colleges and univer-sities are not, and should not be, autonomous institutions devoted to the culti-vation of useless knowledge. To survive in a rapidly changing and increasingly competitive environment, it is necessary for educators and educational institu-tions to find ways to adapt quickly and effectively. Even while claiming to be in-dependent, colleges and universities remain intricately implicated in a market-place that is not limited to ideas. The task facing educational institutions is to find new ways to turn market forces to their own advantage without losing con-trol of what they produce. The responsibility of educators is to prepare students for life and work in a world changing at warp speed by creatively shaping new educational spaces.

I dream of educational institutions that do as much for the imagination as Frank Gehry's buildings do for the eye. Though others have different visions and dreams, it seems undeniable that the currency of education has never been more valuable than in emerging network culture. Having taught the arts and human-ities at a liberal arts college for almost thirty years, I have come to believe that, while criticism is important, it is not enough to convey to students a wisdom that is merely negative. Nor is it sufficient for critical practice to have as its pri-mary aim the production of texts accessible to fewer and fewer people, which promote organizations and institutions whose obsolescence is undeniable. A pol-itics that is merely academic is as sterile as theories that are not put into practice.

Educators must come to realize that criticism that is not constructive is not adequate. If "No" does not harbor "Yes," it should remain unspoken. In the moment of complexity, emerging network culture is creating not only perils but, more important, opportunities for individuals and institutions who, without losing their critical edge, are willing to say "*Yes*." The most important legacy we can leave the next generation is the hope that creative change is still possible.

notes

INTRODUCTION

1. For a timely discussion of this important issue, see James Gleick, *Faster: The Acceleration of Just About Everything* (New York: Pantheon, 1999).
2. See Mark C. Taylor, "Terminal Condition," *About Religion: Economies of Faith in Virtual Culture* (Chicago: University of Chicago Press, 1999), 116–39.
3. See Mark C. Taylor, *Kierkegaard's Pseudonymous Authorship: A Study of Time and the Self* (Princeton: Princeton University Press, 1975).
4. See Mark C. Taylor, *Journeys to Selfhood: Hegel and Kierkegaard* (Berkeley: University of California Press, 1980; New York: Fordham University Press, 2000).
5. See Mark C. Taylor, *Deconstructing Theology* (New York: Crossroad Publishing Co., 1980); *Erring: A Postmodern A/theology* (Chicago: University of Chicago Press, 1984); *Deconstruction in Context* (Chicago: University of Chicago Press, 1986); *Altarity* (Chicago: University of Chicago Press, 1987); *Tears* (Albany: State University of New York Press, 1990); and *Nots* (Chicago: University of Chicago Press, 1990).
6. See Mark C. Taylor, *Disfiguring: Art, Architecture, Religion* (Chicago: University of Chicago Press, 1992).
7. See "Currency," in ibid., 143–84.
8. Andy Warhol, *The Philosophy of Andy Warhol: From A to B and Back Again* (New York: Harcourt Brace Jovanovich, 1977), 91. For an elaboration of these points, see Taylor, "Christianity and the Capitalism of Spirit" and "The Virtual Condition," *About Religion*, 140–210.
9. Mark C. Taylor and Esa Saarinen, *Imagologies: Media Philosophy* (New York: Routledge, 1994).
10. Mark C. Taylor and José Márquez, *The Réal: Las Vegas, Nevada* (Williamstown Mass.: Williams College Museum of Art and Massachusetts Museum of Contemporary Art, 1997), on CD-ROM, distributed by the University of Chicago Press.

11. Mark C. Taylor, *Hiding* (Chicago: University of Chicago Press, 1997), 325–26.

12. "Singularity" is a term that is used in different ways in contemporary philosophy discussions, chaos theory, and catastrophe theory. For a definition and explanation of the place of singularities in catastrophe theory, see below, chap. 2, p. 53.

13. Stephen H. Kellert, *In the Wake of Chaos: Unpredictable Order in Dynamical Systems* (Chicago: University of Chicago Press, 1993), 2.

CHAPTER 1

1. Thomas L. Friedman, *The Lexus and the Olive Tree: Understanding Globalization* (New York: Farrar, Straus, Giroux, 1999), 8.

2. Manuel Castells's comprehensive three-volume study provides a wealth of helpful data on what he describes as "the rise of network society." While calling for a clarification of "networking logic," Castells does not provide an account of the distinctive operational rules and principles of networks. See *The Information Age: Economy, Society, and Culture,* vol. 1, *The Rise of Network Society,* vol. 2, *The Power of Identity,* and vol 3, *End of the Millennium* (Cambridge, Mass.: Blackwell, 1996, 1997, 1998).

3. Chaos theory has already had a noteworthy impact on the humanities. For a consideration of the contribution of chaos theory to literary studies, see *Chaos and Order: Complex Dynamics in Literature and Science,* ed. N. Katherine Hayles (Chicago: University of Chicago Press, 1991). The best general book on chaos theory is James Gleick, *Chaos: Making a New Science* (New York: Viking, 1987).

4. John Casti, *Complexification: Explaining a Paradoxical World through the Science of Surprise* (New York: HarperCollins, 1994).

5. Stuart Kauffman, *At Home in the Universe: The Search for the Laws of Self-Organization and Complexity* (New York: Oxford University Press, 1995), 26. Underscoring the growing popular appeal of complexity theory, this edition includes the following prefatory note: "Ernst & Young is presenting Stuart Kauffman's *At Home in the Universe* to selected clients and friends of the firm because we believe that it offers a uniquely valuable perspective on the future of business. Recent discoveries about the nature of complexity and the phenomenon of self-organization will have a profound effect on how companies formulate strategy and organize themselves to compete—possibly the most profound shift they have undergone since the industrial revolution."

6. Le Corbusier, *The City of To-Morrow and Its Planning,* trans. Frederick Etchells (Cambridge, Mass.: MIT Press, 1986), 1–2. It is important to note that in this text Le Corbusier insists that geometry is "created by ourselves." Geometry, in other words, is a cultural product rather than an attribute of the world as such. The philosophical and critical implications of this thoroughgoing constructivism will become apparent in the next chapter.

7. Le Corbusier does not, of course, always remain true to this creed. In some of his most important work (e.g., the chapel at Rochamps), geometric rectilinearity gives way to biomorphic forms.

8. Le Corbusier, *City of To-morrow,* 11–12.

9. Ibid., 12, 16.

10. Ibid., 22. In addition to representing the distinction between the primitive and the modern or the infantile and the mature, these oppositions are gendered as feminine versus masculine.

11. Ibid., 220.

12. Frederick Winslow Taylor, *The Principles of Scientific Management* (New York: Harper, 1929), 114–15.

13. Rosalind Krauss, "Grids," in *The Originality of the Avant-Garde and Other Modernist Myths* (Cambridge, Mass.: MIT Press, 1986), 9–10.

14. Mies van der Rohe, "A Personal Statement by the Architect, 1964," in Walter Blaser, *Mies van der Rohe* (New York: Praeger Publishers, 1972), 10.

15. Mies, quoted in Robert Venturi, *Complexity and Contradiction in Architecture* (New York: Museum of Modern Art, 1966), 41.

16. Philip Johnson and Henry-Russell Hitchcock, *The International Style: Architecture since 1922* (New York: W. W. Norton and Co., 1932), 69.

17. Ibid., 196.

18. Venturi, *Complexity and Contradiction,* 50.

19. Ibid., 50. Venturi's insistence that Mies's pavilions are "always complete" stresses that these structures are closed.

20. Robert Venturi, Denise Scott Brown, and Steven Izenour, *Learning from Las Vegas: The Forgotten Symbolism of Architectural Form* (Cambridge, Mass.: MIT Press, 1988), 3.

21. Venturi, *Complexity and Contradiction,* 16. Venturi proposes a subtle but important reversal of modernist values. Whereas orthodox modernism associates the intricacies of decoration and ornament with the primitive and the clarity of regular forms with modernity, Venturi maintains that the elementary and simple are primitive and the complex and contradictory are modern.

22. Venturi et al., *Learning from Las Vegas,* 3.

23. This is an abbreviation of Venturi's list. For the rest of the oppositions, see ibid., 112 n.36, 118.

24. Ibid., 137.

25. Two decades later, Venturi has a much better understanding of changing cultural conditions. See his *Iconography and Electronics upon a Generic Architecture: A View from the Drafting Room* (Cambridge, Mass.: MIT Press, 1996).

26. Venturi et al., *Learning from Las Vegas,* 52–53. Venturi is quoting August Heckscher, *The Public Happiness* (New York: Atheneum Publishers, 1962), 289.

27. Venturi, *Complexity and Contradiction,* 16.

28. Venturi et al., *Learning from Las Vegas,* 8.

29. Tom Stoppard, *Arcadia* (Boston: Faber and Faber, 1993), 84.

30. René Thom, *Structural Stability and Morphogenesis: An Outline of a General Theory of Models,* trans. D. H. Fowler (Reading, Mass.: Addison-Wesley, 1989).

31. Benoit Mandelbrot, *The Fractal Geometry of Nature* (New York: W. H. Freeman, 1983), 1.

32. For further discussion of these tendencies, see Taylor, *About Religion: Economies of Faith in Virtual Culture.*

CHAPTER 2

1. See Taylor, "Returnings," the introduction to the new edition of *Journeys to Selfhood: Hegel and Kierkegaard.*

2. Claude Lévi-Strauss, *The Savage Mind* (Chicago: University of Chicago Press, 1970), 15.

3. Ibid., 75. Lévi-Strauss underscores the importance of the grid by repeatedly charting the oppositions he analyzes on gridlike structures.

4. Claude Lévi-Strauss, *Tristes Tropiques,* trans. John and Doreen Weightman (New York: Atheneum, 1974), 153.

5. Alexander Woodcock and Monte Davis, *Catastrophe Theory* (New York: Avon Books, 1978), 7.

6. See Thom, *Structural Stability and Morphogenesis,* esp. chap. 5.

7. Claude Lévi-Strauss, *Le Regard éloigné* (Paris: Plon, 1983), 15; my translation.

8. Lévi-Strauss, *The Raw and the Cooked: Introduction to the Science of Mythology,* trans. J. D. Weightman (New York: Harper and Row, 1969), 10. The use of the term "Overture" for the introduction to this important volume is significant. The epigram of the book is "To Music," a chorus with words by Edmond Rostand and music by Emmanuel Chabrier. Lévi-Strauss repeatedly insists that structuralism is anticipated in music. He even goes so far as to claim: "I mean my reverence, from childhood on, for 'that God, Richard Wagner.' If Wagner is accepted as the undeniable originator of the structural analysis of myth (and even of folk tales, as in *Die Meistersinger*), it is a profoundly significant fact that the analysis was made, in the first instance, *in music*" (15).

9. Michel Foucault, *"The Archaeology of Knowledge" and "The Discourse on Language,"* trans. A. M. Sheridan Smith (New York: Harper and Row, 1972), 15.

10. Michel Foucault, *The Order of Things: An Archaeology of the Human Sciences* (New York: Vintage Books, 1973), xiv.

11. Ibid., 333.

12. Ibid., 332.

13. Foucault, *Language, Counter-Memory, Practice,* ed. Donald Bouchard; trans. Donald Bouchard and Sherry Simon (Ithaca: Cornell University Press, 1977), 150.

14. Ibid., 147.

15. Foucault, *Order of Things,* xxi.

16. Foucault, *Language, Counter-Memory, Practice,* 200.

17. Paul R. Gross and Norman Levitt, *Higher Superstition: The Academic Left and Its Quarrels with Science* (Baltimore: Johns Hopkins University Press, 1994), 78.

18. Richard Dawkins, *Unweaving the Rainbow: Science, Delusion, and the Appetite for Wonder* (New York: Houghton Mifflin, 1998), 41, 187–88. In later chapters I will consider the relevance of some of Dawkins's biological writings for cultural analysis.

19. Edward O. Wilson, *Consilience: The Unity of Knowledge* (New York: Knopf, 1998), 40.

20. Ibid., 8, 12, 54.

21. Derrida returns to this point in "Force of Law: The 'Mystical' Foundation of Authority," special issue, "Deconstruction and the Possibility of Justice," *Cardoza Law Review* 11, nos. 5–6 (July/August 1990): 920–1045. Though Kierkegaard has always hovered in the background of Derrida's work, in recent years he has been much more explicit about the im-

portance of his Danish precursor. See especially *The Gift of Death,* trans. David Wills (Chicago: University of Chicago Press, 1995).

22. Jacques Derrida, *Writing and Difference,* trans. Alan Bass (Chicago: University of Chicago Press, 1978), 33–34.

23. Ibid, 40.

24. Ibid., 36, 55.

25. Ibid., 57.

26. Ibid., 62.

27. As we will see in following chapters, the notion of the "bifurcation point" is critical for nonlinear dynamics and complex systems.

28. Jacques Derrida, *Archive Fever: A Freudian Impression,* trans. Eric Prenowitz (Chicago: University of Chicago Press, 1996), 1.

29. Ibid, 2.

30. Ibid., 11.

31. Ibid., 11, 19.

32. Jean Baudrillard, *Symbolic Exchange and Death,* trans. Iain Hamilton Grant (London: Sage Publications, 1993), 29.

33. Ibid., 30.

34. For a detailed explanation of this issue, see Taylor, *Hiding,* where the problem of superstructure and infrastructure is examined in terms of the relationship between surface and depth.

35. Ibid., 6–7.

36. Jean Baudrillard, *Simulations,* trans. Paul Foss, Paul Patton, and Philip Beitchman (New York: Semiotext[e], 1983) 4; *Symbolic Exchange and Death,* 87.

37. Baudrillard, *Simulations,* 2.

38. Ibid., 50.

39. Baudrillard, *Symbolic Exchange and Death,* 16.

40. Baudrillard, *Simulations,* 10.

41. Jean Baudrillard, *The Evil Demon of Images* (Sydney: The Power Institute of Fine Arts, 1987), 18.

42. Baudrillard, *Simulations,* 103–4.

43. Jean Baudrillard, *The Ecstasy of Communication,* trans. Bernard and Caroline Schutze (New York: Semiotext[e], 1987), 155. It should be clear from our previous discussion of Mies van der Rohe that Baudrillard's notion of obscenity recalls the modernist preoccupation with transparency. In an obscene world, everybody lives in glass houses that expose everything.

44. Baudrillard, *Symbolic Exchange and Death,* 87, 88.

45. Ibid., 87.

46. Ibid., 88.

47. Baudrillard's conclusion echoes Derrida's argument in *Specters of Marx: The State of the Debt in the Work of Mourning, and the New International,* trans. Peggy Kamuf (New York: Routledge, 1994). Taking as his point of departure the opening line of *The Communist Manifesto* (i.e., "A specter is haunting Europe—the specter of communism"), Derrida stresses: "First of all mourning. We will be speaking of nothing else" (9). This point can be extended from this particular work to the entire Derridean corpus. Derrida's writing is, in

effect, a process of interminable mourning for a spectral presence he can neither recollect nor forget. This is why he cannot stop writing.

CHAPTER 3

1. Clement Greenberg, "Modernist Painting," *Modernism with a Vengeance, 1957–1969,* ed. John O'Brian (Chicago: University of Chicago Press, 1993), 85, 86.
2. I will return to this issue below. For the moment, it is sufficient to note that Epimenides, who was a Cretan, defined the Liar's Paradox when he declared: "All Cretans are liars." Alternative versions of this paradox include statements like: "I am a liar" or "This sentence is false." Such statements violate the clear opposition between truth and falsity and yet are not completely nonsensical.
3. Michel Foucault, *This Is Not a Pipe,* trans. James Harkness (Berkeley: University of California Press, 1983), 44.
4. Ibid., 49.
5. Douglas Hofstadter, *Gödel, Escher, Bach: An Eternal Golden Braid* (New York: Vintage, 1980), 701.
6. Ibid., 485. In later chapters I will consider the issue of spontaneous self-assembly.
7. Ibid., 489, 492.
8. Ibid., 493.
9. Alexander Pope, "Epitaph. Intended for Sir Isaac Newton, in Westminster Abbey."
10. Ilya Prigogine and Isabelle Stengers, *Order out of Chaos: Man's New Dialogue with Nature* (New York: Bantam Books, 1984), 6.
11. As we will see in the next chapter, the laws of thermodynamics at work in the steam engine and other such machines contradict important aspects of the notion of systems that derives from Newtonian science.
12. Gilles Deleuze and Felix Guattari, *A Thousand Plateaus: Capitalism and Schizophrenia,* trans. Brian Massumi (Minneapolis: University of Minnesota Press, 1987), 88.
13. Andy Warhol, "What Is Pop Art? Answers from Eight Painters," *Artnews* 62 (November 1963): 26.
14. Friedrich Schiller, *On the Aesthetic Education of Man: In a Series of Letters,* trans. Reginald Snell (New York: Frederick Ungar, 1965), 40.
15. G. W. F. Hegel, *Science of Logic,* trans. A. V. Miller (New York: Humanities Press, 1969), 711.
16. Ibid., 216–17.
17. G. W. F. Hegel, *Phenomenology of Spirit,* trans. A. V. Miller (New York: Oxford University Press, 1977), 31. As the translator points out, the phrase "report as clear as noonday" is a crucial reference to Fichte's *Sun-Clear Report to the Public about the True Essence of the Newest Philosophy* (1801).
18. Hegel, *Science of Logic,* 31.
19. Martin Heidegger, *Identity and Difference,* trans. Joan Stambaugh (New York: Harper and Row, 1969), 41.
20. Martin Heidegger, *"The Question Concerning Technology" and Other Essays,* trans. William Lovitt (New York: Harper and Row, 1977), 24.

21. Ibid., 26–27.

22. Heidegger's will to mastery represents his effort to translate Nietzsche's will to power into a contemporary idiom. While Western metaphysics reaches closure in Nietzsche's analysis of the will to power, the full implications of what Heidegger labels the "onto-theological tradition" only become clear in the devastating force of modern technology.

23. Heidegger, *"The Question Concerning Technology,"* 152.

24. Hegel, *Phenomenology of Spirit*, 355.

25. Ibid., 359.

26. Lee Smolin, *Life in the Cosmos* (New York: Oxford University Press, 1997), 26.

27. In the coda, "The Currency of Education," I will examine how Kant's interpretation of the beautiful work of art shapes his understanding of the modern university and its curriculum.

28. Immanuel Kant, *Critique of Judgment*, trans. James Meredith (New York: Oxford University Press, 1973), 22.

29. Ibid., 22, 24, 21.

30. François Jacob, *The Logic of Life: A History of Heredity*, trans. Betty Spillman (Princeton: Princeton University Press, 1973), 74. In a cover blurb, Foucault describes this book as "the most remarkable history of biology that has ever been written." Jacob's analysis had a profound influence on the interpretation of life Foucault presents in *The Order of Things*.

31. Ibid., 152, 153; emphasis added.

32. Hegel, *Science of Logic*, 737.

33. Ibid., 736.

34. Hegel, *Phenomenology of Spirit*, 107, 108; translation revised.

35. Kauffman, *At Home in the Universe*, 274. In the following pages, I will concentrate on autopoietic systems, and, in chapter 6, I will discuss autocatalytic processes. At this point, it is sufficient to note that biochemists who investigate the origin of life are particularly interested in autocatalytic process. In the absence of an external creator, life can emerge only if there is some kind of self-organizing autocatalytic process.

36. For an excellent account of these developments, see N. Katherine Hayles, *How We Became Posthuman: Virtual Bodies in Cybernetics, Literature, and Informatics* (Chicago: University of Chicago Press, 1999), esp. chap. 6. Hayles presents a helpful summary of what she describes as "the second wave of cybernetics," which is characterized by the movement from reflexivity to self-organization. While Hayles's interests are more historical and literary than philosophical, her book is a rich resource for anyone interested in these issues. See also William R. Paulson, *The Noise of Culture: Literary Texts in a World of Information* (Ithaca: Cornell University Press, 1988), 121–30.

37. Hayles, *How We Became Posthuman*, 133.

38. Quoted in ibid., 135. Hayles reads Maturana's conclusion that "everything is said by an observer" as leading to a relativism that is indistinguishable from what I have described as sociocultural constructivism. Summarizing Maturana's argument, she writes: "The results implied that the frog's perceptual system does not so much register reality as *construct* it" (135). In what follows, it will become apparent that there are ways to interpret autopoietic systems that do not necessarily lead to this conclusion.

39. Humberto R. Maturana and Francisco J. Varela, *Autopoiesis and Cognition: The Realization of the Living* (Boston: Reidel, 1980), 121. See also Francisco J. Varela, *Principles of Biological Autonomy* (New York: North Holland, 1979), and Erich Jantsch, *The Self-Organizing Universe: Scientific and Human Implication of the Emerging Paradigm of Evolution* (New York: Pergamon Press, 1980).

40. Maturana and Varela, *Autopoiesis and Cognition,* 78.

41. Ibid., 78–79.

42. Ibid., 79, 9.

43. Niklas Luhmann, "The Autopoiesis of Social Systems," in *Essays on Self-Reference* (New York: Columbia University Press, 1990), 6, 3.

44. Ibid., 12.

45. Niklas Luhmann, *Social Systems,* trans. John Bednarz Jr., with Dirk Baecker (Stanford: Stanford University Press, 1995), 16–17.

46. Ibid., 17.

47. G. W. F. Hegel, *The Logic of Hegel,* trans. W. Wallace (New York: Oxford University Press, 1968), 224.

48. Ibid., 842.

49. Maturana and Varela, *Autopoiesis and Cognition,* 78.

50. Derrida's consideration of the preface and introduction in Hegel's work is the mirror image of Kierkegaard's ironic analysis of the postscript. If Hegel's system has a postscript, which Kierkegaard provides, it can be neither complete nor closed. See *Concluding Unscientific Postscript,* trans. David F. Swenson and Walter Lowrie (Princeton: Princeton University Press, 1941).

51. Jacques Derrida, *Dissemination,* trans. Barbara Johnson (Chicago: University of Chicago Press, 1981), 219. As the between, *entre* marks the boundary, which Derrida also labels "hymen" and "margin." The passage in Hegel to which Derrida refers in this text describes the culmination of self-consciousness in which the "curtain" *between* subject and object is lifted and perfect union is achieved. Derrida's point is that the between cannot be erased and thus the union of subjectivity and objectivity is never complete. See Hegel, *Phenomenology of Spirit,* 103.

52. See Hofstadter's summary of Hilbert's question, *Gödel, Escher, Bach,* 24. Hilbert thinks Kant's a priori is illogical and makes true knowledge impossible.

53. Ibid., 17.

54. Ibid., 17. The Epimenides paradox is, as we have seen, a version of the Liar's Paradox.

55. It is, of course, possible to prove the undecidable proposition in a more inclusive system. But this system would, in turn, involve a similar undecidability. The result is an infinite regress that does not overcome the problem Gödel identifies.

56. Derrida, *Positions,* trans. Alan Bass (Chicago: University of Chicago Press, 1981), 42–43.

57. Jacques Derrida, *Margins of Philosophy,* trans. Alan Bass (Chicago: University of Chicago Press, 1982), 8.

58. J. Hillis Miller, "The Critic as Host," *Deconstruction and Criticism* (New York: Seabury Press, 1979), 219. See Mark C. Taylor, "Paralectics," *Tears* (Albany: SUNY Press, 1989), 123–44.

59. Michel Serres, *The Parasite,* trans. Lawrence R. Schehr (Baltimore: Johns Hopkins University Press, 1982), 12.

CHAPTER 4

1. Wojciech H. Zurek, "Complexity, Entropy, and the Physics of Information: A Manifesto," *Complexity, Entropy, and the Physics of Information,* ed. Wojciech H. Zurek, Santa Fe Institute Studies in the Sciences of Complexity, vol. 8 (Reading, Mass.: Addison-Wesley, 1990), x, vii.
2. Ibid., ix.
3. The root of *nausea* is the Greek *nau,* which means boat.
4. See James Grier Miller, *Living Systems* (New York: McGraw-Hill), 16.
5. J. Hillis Miller, "Critic as Host," *Deconstruction and Criticism,* 252.
6. Serres, *The Parasite,* 3.
7. Ibid., 3.
8. Ibid., 10.
9. Ibid., 67.
10. Hofstadter, *Gödel, Escher, Bach,* 25.
11. Quoted in Hofstadter, *Gödel, Escher, Bach,* 25.
12. Michael Hobart and Zachary Schiffman, *Information Ages: Literacy, Numeracy, and the Computer Revolution* (Baltimore: Johns Hopkins University Press, 1998), 2.
13. Ibid., 2.
14. It is, of course, important to recall that this play of differences is generated by *différance,* which functions for Derrida as something like absolute noise or noise that can never be converted into information.
15. Gregory Bateson, *Steps to an Ecology of Mind* (New York: Ballantine Books, 1972), 453.
16. Hobart and Schiffman, *Information Ages,* 3.
17. Theodore Roszak, *The Cult of Information: A Neo-Luddite Treatise on High Tech, Artificial Intelligence, and the True Art of Thinking* (New York: Pantheon Books, 1994), 113.
18. Mark Dery, *Escape Velocity: Cyberculture at the End of the Century* (New York: Grove Press, 1996); Erik Davis, *Techgnosis: Myth, Magic, and Mysticism in the Age of Information* (New York: Harmony Books, 1998); Hans Moravec, *Robot: Mere Machine to Transcendent Mind* (New York: Oxford University Press, 1999); and Hayles, *How We Became Posthuman.*
19. Hayles, *How We Became Posthuman,* 2.
20. N. Katherine Hayles, "Virtual Bodies and Flickering Signifiers," *October* 66 (fall 1993): 91.
21. Norbert Wiener, *Cybernetics, or Control and Communication in the Animal and Machine* (New York: John Wiley, 1948).
22. For a helpful account of the Macy Conferences, see Hayles, *How We Became Posthuman,* chap. 3.
23. Claude Shannon and Warren Weaver, *The Mathematical Theory of Communication* (Urbana: University of Illinois Press, 1949). Weaver's article first appeared in *Scientific American* in July 1949.
24. Ibid., 3.

25. Ibid., 99.

26. Ibid., 100.

27. In the absence of noise, Shannon maintains, a perfectly efficient encoding of information will always look like noise (i.e., complete randomness) to someone who does not know the code.

28. Serres, *The Parasite,* 66.

29. Italo Calvino, *If on a winter's night a traveler,* trans. William Weaver (New York: Harcourt Brace, 1981), 3, 26, 220.

30. Jeremy Campbell, *Grammatical Man: Information, Entropy, Language, and Life* (New York: Simon and Schuster, 1982), 32.

31. Weaver in Shannon and Weaver, *The Mathematical Theory of Communication,* 103.

32. Ilya Prigogine and Isabelle Stengers, *Order out of Chaos: Man's New Dialogue with Nature* (New York: Bantam Books, 1984), 106.

33. This way of formulating the notion of entropy recalls Georges Bataille's analysis of consumption in terms of "expenditure without return." As Derrida points out in his reading of Bataille, such processes of expenditure interrupt the apparent circularity of Hegelian philosophy. See Derrida, "From Restricted to General Economy: A Hegelianism without Reserve," *Writing and Difference,* 251–77.

34. Prigogine and Stengers, *Order out of Chaos,* 115–16. In developing the following summary, I have drawn on the argument Prigogine presents on pages 117–19.

35. Ibid., 117.

36. Tor Nørretranders, *The User Illusion: Cutting Consciousness Down to Size,* trans. Jonathan Sydenham (New York: Viking, 1998), 5.

37. Prigogine and Stengers, *Order out of Chaos,* 123.

38. Ilya Prigogine, *The End of Certainty: Time, Chaos, and the New Laws of Nature* (New York: Free Press, 1997), 35.

39. Ibid., 5.

40. The latter statement was not made in connection with quantum mechanics.

41. Quoted in Prigogine and Stengers, *Order out of Chaos,* 235.

42. Quoted in Nørretranders, *The User Illusion,* 22.

43. Charles Bennett, "Demons, Engines, and the Second Law," *Scientific American* 257, no. 5 (November 1987): 116.

44. Ibid., 115. Prior to these developments, the most influential "refutation" of Maxwell's demon had been presented by Leo Szilard in a 1929 paper entitled "On the Decrease of Entropy in a Thermodynamic System by the Intervention of Intelligent Beings." Szilard argues that in exercising its intelligence, Maxwell's demon necessarily uses energy: "One may reasonably assume," he concludes, "that a measurement procedure is fundamentally associated with a definite average entropy production, and thus that this restores concordance with the Second Law" (quoted in Nørretranders, *The User Illusion,* 24). Bennett's argument differs in subtle but important ways from Szilard's position. "Attributing the gain in entropy to the resetting step rather than to the measurement," Bennett acknowledges, "may seem to be a mere bookkeeping formality, since any complete cycle of Szilard's engine must include both steps, but considerable confusion can be avoided if one draws a clear distinction between the acquisition of new information

and the destruction of old information. The confusion may not have existed in Szilard's mind. In most of his paper he refers to measurements as the irreversible step, but at one point he makes an accounting of entropy changes during the cycle and finds, without explicitly commenting on it, that the increase in entropy takes place during the resetting of the memory" (116).

45. Ibid., 116.

46. Norbert Wiener, *The Human Use of Human Beings* (Boston: Houghton Mifflin, 1950), 12.

47. Ibid.

48. Michel Serres, *Hermes: Literature, Science, Philosophy,* ed. Josué V. Harari and David F. Bell (Baltimore: Johns Hopkins University Press, 1982), 75. Prigogine and Stengers provide a "Postface" to Serres's book entitled "Dynamics from Leibniz to Lucretius."

49. In the next chapter, where I will consider the problem of spontaneous self-organization in more detail, we will discover that this process can be detected in both living and nonliving phenomena.

50. Wiener, *The Human Use of Human Beings,* 21. This line of argument leads Katherine Hayles to conclude that Shannon and Wiener hold *opposite* views of the relationship between information and entropy: "Wiener accepted the idea that entropy was the opposite of information. The inverse relation made sense to him because he thought of information as allied with structure and viewed entropy as associated with randomness, dissipation, and death. . . . This view of entropy makes sense when viewed in the context of nineteenth-century thermodynamics. But it is not a necessary implication of information as *information* is technically defined. Claude Shannon took the opposite view and *identified* information and entropy rather than opposed them. Since the choice of sign was conventional, this formulation was also a possibility. Heuristically, Shannon's choice was explained by saying that the more unexpected (or random) a message is, the more information it conveys" (*How We Became Posthuman,* 102). In what follows, I will argue that it is possible to interpret Wiener's position as complicating rather than opposing Shannon's account of the relationship of information to entropy and negentropy.

51. Nørretranders, *The User Illusion,* 42.

52. Prigogine and Stengers, *Order out of Chaos,* 143.

53. Serres, *The Parasite,* 4, 12.

54. *Data* is, of course, the plural of *datum,* which derives from the Latin *dare* (to give). Accordingly, a datum is something that is given.

55. Serres, *The Parasite,* 35.

56. Ibid., 14.

CHAPTER 5

1. Andrew Crumey, *Pfitz* (New York: Picador, 1995), 63–64.

2. Kirk Varnedoe, "Chuck Close Then and Now," in *Chuck Close,* ed. Robert Storr (New York: Museum of Modern Art, 1998), 61.

3. Linda Nochlin, "The Realist Criminal and the Abstract Law," *Art in America* 61 (September–October 1973): 54, 55. Storr cites a portion of this text in his catalog essay "Chuck Close: Angles of Refraction."

4. Ad Reinhardt, "25 LINES OF WORDS ON ART: STATEMENT," *Art-as-Art: The Se-lected Writings of Ad Reinhardt,* ed. Barbara Rose (New York: Viking Press, 1975), 51–52.

5. Quoted by Storr, "Chuck Close: Angles of Refraction," 32.

6. Ibid., 41.

7. Varnedoe, "Chuck Close Then and Now," 66.

8. Michel Serres, "The Origin of Language: Biology, Information Theory, and Thermody-namics," *Hermes: Literature, Science, Philosophy,* ed. Josué V. Harari and David F. Bell (Baltimore: Johns Hopkins University Press, 1982), 74.

9. Ibid., 77–78.

10. The only book by Atlan that has been translated is *Enlightenment to Enlightenment: Inter-critique of Science and Myth,* trans. Lenn Schramm (Albany, N.Y.: SUNY Press, 1993). William Paulson is, to my knowledge, the only critic to have recognized the importance of Atlan's work. See Paulson, *The Noise of Culture: Literary Texts in a World of Information,* 69–78.

11. Henri Atlan, "Disorder, Complexity, and Meaning," in *Disorder and Order,* ed. Paisley Livingston (Saratoga, Calif.: Anma Libri, 1984), 110.

12. Henri Atlan, *L'organisation biologique et la théorie de l'information* (Paris: Hermann, 1972), 258–59.

13. Ibid., 257, 264–65.

14. Ibid., 265–66.

15. Ibid., 229, 230.

16. Ibid., 230.

17. A keen student of etymology, Murray Gell-Mann once proposed that complexity studies be named "plectics." See George Johnson, *Strange Beauty: Murray Gell-Mann and the Rev-olution in Twentieth-Century Physics* (New York: Knopf, 1999), 324.

18. For a highly informative and influential consideration of these tendencies, see David E. Rumelhart and James L. McClelland, *Parallel Distributed Processing,* vol. 1, *Explorations in the Microstructure of Cognition,* vol. 2, *Psychological and Biological Models* (Cambridge, Mass.: MIT Press, 1986).

19. Gregory Chaitin, *Information, Randomness, and Incompleteness* (Singapore: World Scien-tific Co., 1990), 88.

20. John Casti, *Complexification: Explaining a Paradoxical World through the Science of Surprise* (New York: HarperCollins, 1994), 9.

21. See Murray Gell-Mann, *The Quark and the Jaguar: Adventures in the Simple and the Com-plex* (New York: W. H. Freeman, 1994), 34–36.

22. Ludwig von Bertalanffy, *General System Theory: Foundations, Development, Applications* (New York: George Braziller, 1968), 37. See also Ervin Laszlo, *Introduction to Systems Phi-losophy: Toward a New Paradigm of Contemporary Thought* (New York: Gordon and Breach, 1972); and Robert Lilienfeld, *The Rise of Systems Theory: An Ideological Analysis* (New York: Wiley, 1978).

23. Ibid., 187–88.

24. Ibid., 33, 38.

25. Herbert A. Simon, "The Architecture of Complexity," *The Sciences of the Artificial* (Cam-bridge, Mass.: MIT Press, 1969), 86.

26. David Depew and Bruce Weber, *Darwinism Evolving: Systems Dynamics and the Genealogy of Natural Selection* (Cambridge, Mass.: MIT Press, 1995), 437.

27. Peter Coveney and Roger Highfield, *Frontiers of Complexity: The Search for Order in a Chaotic World* (New York: Fawcett Columbine, 1995), 425.

28. Steven Levy, *Artificial Life: A Report from the Frontier Where Computers Meet Biology* (New York: Random House, 1993), 49–50.

29. Ibid., 52.

30. I examine the ways in which life is informational in chapters 6 and 7.

31. Quoted in Coveney and Highfield, *Frontiers of Complexity*, 274.

32. Kauffman, *At Home in the Universe*, 27, 87.

33. John H. Holland, *Emergence: From Chaos to Order* (Reading, Mass.: Addison-Wesley, 1998), 14.

34. John Casti, *Would-Be Worlds: How Simulation Is Changing the Frontiers of Science* (New York: John Wiley, 1997), 91.

35. For a summary of these rules, see Holland, *Emergence*, 225–31.

36. Kauffman, *At Home in the Universe*, 81. An attractor is the term used to describe the relatively long-term behavior toward which a system tends or, as Stephen Kellert explains, "a set of points such that all trajectories nearby converge on it" (*In the Wake of Chaos*, 13). In the next chapter, I consider Kauffman's analysis of this important point in more detail.

37. Ibid., 60.

38. While Kauffman is preoccupied with organic systems, the conclusions he reaches confirm Prigogine's observations about certain inorganic reactions. Extrapolating his nonequilibrium thermodynamic theory, Prigogine develops a nonlinear dynamic model for a chemical reaction system to explain the Belousov-Zhabotinsky reaction. This was the first demonstration of an actual chemical reaction that provided support for self-organization far from equilibrium. See Prigogine and Stengers, *Order out of Chaos*, 146–53; and Coveney and Highfield, *Frontiers of Complexity*, 159–63, 175–78.

39. W. Daniel Hillis, *The Pattern on the Stone: The Simple Ideas That Make Computers Work* (New York: Basic Books, 1998), 88.

40. Hegel examines the logic of the transition from quantitative to qualitative change in his *Science of Logic*. See section 2, "Magnitude (Quantity)."

41. Kauffman, *At Home in the Universe*, 126.

42. Per Bak, *How Nature Works: The Science of Self-Organized Criticality* (New York: Springer-Verlag, 1996), 1–2.

43. In our analysis of the interplay between information and noise, we have already discovered that any firm line between interiority and exteriority is difficult to maintain. While Bak insists that self-organized criticality is not determined by any outside factors, he finally admits: "Of course, in the greater picture, nothing is external so in the final analysis, catastrophes must be explained endogenously in any cosmological model" (*How Nature Works*, 152). It is less misleading to understand external and internal as folded into each other in such a way that neither is what it seems to be when interpreted in oppositional terms than it is to follow Bak in collapsing the former into the latter.

44. Casti, *Would-Be Worlds*, 195.

45. Bak, *How Nature Works*, 61, 17–18.

46. Ibid., 58.

47. Derrida, *Dissemination,* 54.

48. Jantsch, *The Self-Organizing Universe,* 48.

49. J. A. Scott Kelso, *Dynamic Patterns: The Self-Organization of Brain and Behavior* (Cambridge, Mass.: MIT Press, 1995), 16.

50. In the following chapter, I consider the implications of punctuated equilibrium for evolution.

51. Crumey, *Pfitz,* 38.

52. Ibid., 40–41. *Swarm* derives from *swer,* to buzz or whisper. *Swer* in Dutch is *zwirrelen,* to whirl, which is akin to the Low German source of Middle English *swyrl,* eddy. In light of our previous consideration of the negentropic characteristics of complex adaptive systems, the association of swarms with eddies is particularly suggestive. Swarms are, in effect, temporary eddies in the river of time.

53. See, for example, Mitchel Resnick, *Turtles, Termites, and Traffic Jams: Explorations in Massively Parallel Microworlds* (Cambridge, Mass.: MIT Press, 1994).

54. Mark M. Millonas, "Swarms, Phase Transitions, and Collective Intelligence," *Artificial Life III,* ed. Christopher G. Langton, Santa Fe Institute Studies in the Sciences of Complexity, vol. 17 (Reading, Mass.: Addison-Wesley, 1994), 418.

55. Ibid., 417.

56. Ibid., 421.

57. Levy, *Artificial Life,* 80–81.

58. Millonas, "Swarms, Phase Transitions, and Collective Intelligence," 422–23. I will consider how networks learn in the next chapter.

59. Holland, *Emergence,* 142.

60. Millonas, "Swarms, Phase Transitions, and Collective Intelligence," 419. What Millonas defines as "swarm intelligence," New Age guru and executive editor of *Wired* Kevin Kelly describes as "hive mind." The hive forms a "superorganism" with emergent properties for which Kelly coins the phrase "vivisystem." He argues that "there are four distinct facets of distributed being that supply vivisystems their character: the absence of imposed centralized control; the autonomous nature of subunits; the high connectivity of subunits; and the webby nonlinear causality of peers influencing peers." Unlike Millonas, Kelly concludes that "nonlinear web systems are unadulterated mysteries." This does not, however, prevent him from writing a very long book explaining how hive mind works. See Kevin Kelly, *Out of Control: The Rise of Neo-Biological Civilization* (New York: Addison-Wesley, 1994).

CHAPTER 6

1. Hofstadter, *Gödel, Escher, Bach,* 7. Unless otherwise indicated, the texts cited in the first part of this chapter are from 3–14 and 311–36.

2. Pierre de Fermat was a seventeenth-century lawyer and mathematician, who claimed to have solved an equation in Diophantus's *Arithmetica.* While it was well known that $a^n + b^n = c^n$ has integer solutions, no one had ever found integer solutions for the corresponding higher power equations. Fermat conjectured that $a^n + b^n = c^n$ has no integer solu-

tions for any integer n > 2. Fermat wrote in the margin of his notebook: "I have found a truly marvelous proof of this statement, which, unfortunately, this margin is too small to contain." In 1993, Andrew Wiles of Princeton University announced that he had proven the theorem. One year later the necessary corrections were made and the proof was accepted. In view of our earlier consideration of Tom Stoppard's *Arcadia,* it is interesting to note that when Hannah and Valentine are leafing through Thomasia's math primer, they find the following note: "I, Thomasina Coverly, have found a truly wonderful method whereby all the forms of nature must give up their numerical secrets and draw themselves through number alone. This margin being too mean for my purpose, the reader must look elsewhere for the New Geometry of Irregular Forms discovered by Thomasina Coverly" (43).

3. John H. Holland, *Hidden Order: How Adaptation Builds Complexity* (New York: Addison-Wesley, 1995), 11. The classic study of ants is E. O. Wilson and Bert Holldobler's book, *The Ants* (Cambridge: Harvard University Press, 1990). Wilson has recently expressed sympathy with aims of complexity theory but uncertainty about the empirical evidence to support it. He tends to interpret complexity theory in a way that is more reductionistic than many of its proponents would accept. Commenting on the use of complexity studies in artificial life, Wilson writes: "They have refined reductionism as a high art and begun to achieve partial syntheses at the level of the molecule and organelle. . . . They foresee no need for overarching grand explanations as a prerequisite for creating artificial life. An organism is a machine and the laws of physics and chemistry, most believe, are enough to do the job, given sufficient time and research funding" (*Consilience,* 91). As we will see, this is a one-sided reading of complexity theory, which does little to advance what is increasingly becoming an important debate.

4. John H. Holland, "The Global Economy as an Adaptive Process," *The Economy as an Evolving Complex System,* ed. Philip W. Anderson, Kenneth J. Arrow, and David Pines (Reading, Mass.: Addison-Wesley, 1988), 117–23.

5. Murray Gell-Mann, "Complexity and Complex Adaptive Systems," *The Evolution of Human Languages,* ed. John A. Hawkins and Murray Gell-Mann (New York: Addison-Wesley, 1992), 10.

6. While variations of the five features of complex adaptive systems I describe can be found in Gell-Mann's analysis, at some points I have supplemented and extended this account.

7. Holland, *Hidden Order,* 31.

8. Bak, *How Nature Works,* 167–70.

9. Holland, *Hidden Order,* 29. Emphasis added.

10. Henri Atlan, *Enlightenment to Enlightenment,* 165. Quoted from Colin S. Pittendrigh, "Adaptation, Natural Selection, and Behavior," *Behavior and Evolution,* ed. Anne Roe and George Gaylord Simpson (New Haven: Yale University Press, 1969), 390–416, esp. 391–94.

11. Henri Atlan, *Entre le cristal et la fumée,* 15. See also *L'organisation biologique et la théorie de l'information,* 281–84.

12. Gell-Mann, *The Quark and the Jaguar,* 318.

13. Charles J. Lumsden and Edward O. Wilson, *Genes, Mind, and Culture: The Coevolutionary Process* (Cambridge, Mass.: Harvard University Press, 1981), 237. See also John N. Thompson, *The Coevolutionary Process* (Chicago: University of Chicago Press, 1994).

14. Stuart Kauffman, *The Origins of Order: Self-Organization and Selection in Evolution* (New York: Oxford University Press, 1993), 173.

15. For a classic treatment of the proofs of God's existence, see Paul Tillich, "Two Types of Philosophy of Religion," *Theology of Culture,* ed. Robert C. Kimball (New York: Oxford University Press, 1959).

16. William Paley, "The Attributes of Deity from the Appearances of Nature," *The Cosmological Arguments: A Spectrum of Opinion,* ed. Donald Burrill (New York: Doubleday, 1967), 167.

17. Ibid., 166.

18. Depew and Weber, *Darwinism Evolving,* 45. This book is, in my judgment, the best available account of the history of Darwinism from its inception to the present day. In developing the following remarks, I have drawn on the rich analysis of Depew and Weber at various points.

19. Ibid., 48.

20. Ibid., 45–46.

21. Stephen Jay Gould, *Full House: The Spread of Excellence from Plato to Darwin* (New York: Harmony Books, 1996), 41. The emphasis on "population thinking" is part of the probability revolution that we have considered in the context of our discussion of thermodynamics and dissipative structures.

22. Quoted in Depew and Weber, *Darwinism Evolving,* 71.

23. Ibid., 122. Darwin's emphasis on the variations of "force" in this text harbors echoes of Newton.

24. Ibid., 129.

25. John Maynard Keynes, *The General Theory of Employment, Interest and Money* (New York: Harcourt Brace, 1935), 359–62.

26. Quoted in Depew and Weber, *Darwinism Evolving,* 81.

27. Karl Marx and Friedrich Engels, *Selected Correspondence,* trans. I. Lasker (Moscow, 1965).

28. Depew and Weber, *Darwinism Evolving,* 82

29. Richard Dawkins, *The Selfish Gene* (New York: Oxford University Press, 1989), 23.

30. Richard Dawkins, *River out of Eden: A Darwinian View of Life* (New York: Basic Books, 1995), 4. Though Dawkins does not seem to realize it, this way of posing the issue pushes Darwinism back in the direction of Platonism. The information coded in the genotype seems to function something like a Platonic archetype. I will return to this issue in the following chapter.

31. Dawkins, *The Selfish Gene,* 234. See also Richard Dawkins, *The Extended Phenotype: The Long Reach of the Gene* (New York: Oxford University Press, 1982).

32. Stephen Jay Gould, *Hen's Teeth and Horse's Toes* (New York: Norton, 1983), 13.

33. Brian Goodwin: *How the Leopard Changed Its Spots: The Evolution of Complexity* (New York: Scribner's, 1994), vii, 176.

34. Ibid., 197.

35. See D'Arcy Thompson, *On Growth and Form* (Cambridge: Cambridge University Press, 1917); C. H. Waddington, *Organizers and Genes* (Cambridge: Cambridge University Press, 1940), and *Principles of Embryology* (New York: Macmillan, 1952); and O. B. Hardison Jr., *Disappearing through the Skylight: Culture and Technology in the Twentieth Century* (New York: Viking, 1989).

36. Brian Goodwin, "A Structuralist Research Programme in Developmental Biology," *Dynamic Structures in Biology*, ed. Brian Goodwin, Atuhiro Sibatani, and Gerry Webster (Edinburgh: Edinburgh University Press, 1989), 49.

37. Goodwin, *How the Leopard Changed Its Spots*, 116.

38. Ibid., viii.

39. Kauffman, *At Home in the Universe*, 185.

40. Ibid., 23.

41. Depew and Weber, *Darwinism Evolving*, 399.

42. Roger Lewin, *Complexity: Life at the Edge of Chaos* (New York: Macmillan, 1992), 27.

43. Kauffman, *At Home in the Universe*, 83.

44. Ibid., 71.

45. Hofstadter, *Gödel, Escher, and Bach*, 10.

46. Kauffman, *At Home in the Universe*, 69.

47. Ibid., 15.

48. Ibid., 49–50.

49. Ibid., 99.

50. Gell-Mann, *The Quark and the Jaguar*, 69. In light of our consideration of Darwin's appropriation of Adam Smith's economic principles, it is instructive to note that when Kauffman considers the problem of coevolution, he too draws on Smith's insights: "The vast puzzle that is the emergent order in communities—in community assembly itself, in coevolution, and in the evolution of coevolution—almost certainly reflects selection acting at the level of the individual organism. Adam Smith first told us the idea of an invisible hand in his treatise *The Wealth of Nations*. Each economic actor, acting for his own selfish end, would blindly bring about the benefit of all. If selection acts only at the level of the individual, naturally sifting the fitter variants that 'selfishly' leave more offspring, then the emergent order of communities, ecosystems, and coevolving systems, and the evolution of coevolution itself are the work of an invisible choreographer" (*At Home in the Universe*, 209).

51. Kauffman, *At Home in the Universe*, 166.

52. Gell-Mann, *The Quark and the Jaguar*, 248.

53. Kauffman, *At Home in the Universe*, 27.

54. Goodwin, *How the Leopard Changed Its Spots*, x.

55. Depew and Weber, *Darwinism Evolving*, 436.

56. Kauffman, *At Home in the Universe*, 129. While in general agreement with Kauffman on this important point, Gell-Mann remains somewhat more cautious: "Although there is still no reason to believe in a steady drive toward more complex organisms, the selection pressures that favor higher complexity may often be strong" (*The Quark and the Jaguar*, 245).

57. As we will see in the next chapter, increasing complexity and connectivity do not necessarily lead to greater unity.

58. Kauffman, *At Home in the Universe*, 25–26.

59. Ibid., 45.

60. Ibid., 298. At times, Kauffman is naively optimistic about what will result from this process. He seems to think that evolution will inevitably lead to more democratic forms of government. While not impossible, such a result is, of course, far from assured.

61. Ibid., 299.

CHAPTER 7

1. Crumey, *Pfitz*, 121.
2. These etymologies and definitions are from Shipley, *The Origins of English Words, The American Heritage Dictionary* (New York: Houghton Mifflin Co., 1970), and *The Oxford English Dictionary.*
3. While a screening is a presentation of a film, video, CD-ROM, website, etc., screenings are remains, waste, and detritus. Screening is often a process of filtering, designed to purify by removing, excluding, or repressing what threatens to contaminate. What is screened, however, does not simply disappear but lingers as dangerous refuse, which is neither precisely present nor absent.
4. Augustine, *Confessions,* trans. Rex Warner (New York: New American Library, 1963), 217; 221–22. Translation modified.
5. Ibid., 219, 224, 227, 219.
6. Ibid., 227.
7. Nørretranders, *The User Illusion,* 257.
8. Ibid., 95.
9. Ray Kurzweil, *The Age of Spiritual Machines: When Computers Exceed Human Intelligence* (New York: Viking, 1999), 78.
10. Kelso, *Dynamic Patterns,* 1.
11. George Lakoff, *Women, Fire, and Dangerous Things: What Categories Reveal about the Mind* (Chicago: University of Chicago Press, 1990), xiv–xv.
12. Kelso, *Dynamic Patterns,* 16.
13. Ibid., 8.
14. Gell-Mann, *The Quark and the Jaguar,* 17.
15. Gell-Mann, "Complexity and Complex Adaptive Systems," 10.
16. For a very helpful account of the history of schemata and schematization from Plato to complexity theory, see Ben Martin, "The Schema," *Complexity: Metaphors, Models, and Reality,* ed. George A. Cowan, David Pines, and David Meltzer, Santa Fe Institute Studies in the Sciences of Complexity, vol. 19 (Reading, Mass.: Addison-Wesley, 1994), 263–85.
17. Gell-Mann, "Complexity and Complex Adaptive Systems," 10.
18. It is instructive to note that for Hegel, scientific inquiry forms a "self-organizing whole," which is remarkably similar to Gell-Mann's account of scientific theories in terms of complex adaptive systems.
19. Gell-Mann, *The Quark and the Jaguar,* 329.
20. Quoted in Nørretranders, *The User Illusion,* 165.
21. Terrence Deacon, *The Symbolic Species: The Coevolution of Language and the Brain* (New York: Norton, 1997), 99.
22. Ibid., 100.
23. Wilson, *Consilience,* 130.
24. Deacon, *The Symbolic Species,* 110.
25. Ibid., 109.
26. William Burroughs anticipates this understanding of language as a virus. John Johnston points out that Burroughs intends his account of the viral nature of language to underscore

its proscriptive rather than its productive propensities: "From the outset, Burroughs thought of words and images in this new environment not as media carrying messages but as viral forms capable of replication. Understood as a virus, a word or image is significant not because it bears a meaning but because it can perpetuate itself by activating an inscribed code, a code that ultimately proscribes one form of being rather than another, one set of responses rather than another. In Burroughs's fiction, therefore, the word or image virus is an agent— a secret agent—operating as a nearly invisible force in a control network" (*Information Multiplicity: American Fiction in the Age of Media Saturation* [Baltimore: Johns Hopkins University Press, 1998], 20). Douglas Rushkoff elaborates Burroughs's insights in *Media Virus: Hidden Agendas in Popular Culture* (New York: Ballantine Books, 1994).

27. Ibid., 112.

28. Gell-Mann, *The Quark and the Jaguar*, 292.

29. Lumsden and Wilson, *Genes, Mind, and Culture*, x, 27. "The concept of a polythetic set," Lumsden and Wilson proceed to explain, "is derived from numerical taxonomy, a methodology that attempts to quantify the degrees of relationship among organisms, species, and other conceivable objects for which classifications are needed. A polythetic group is any set of entities, such as an array of swords or marriage ceremonies, in which each entity possesses a large number of the attributes of the group, where the attributes might be size, geometric shape, duration of a process, and so forth" (27).

30. Ibid., 7. During the 1970s, arguments supporting the genetic determination of personality and social behavior provoked heated debates. In *Sociobiology: The New Synthesis* (1975), Wilson elaborated claims initially advanced by William Hamilton in a paper entitled "The Genetical Evolution of Social Behavior," published in 1964. Harvard colleagues Stephen Jay Gould and Richard Lewontin, who were not persuaded by Wilson's scientific arguments and deeply disturbed by their political implications, launched a sustained attack on sociobiology. See Richard Lewontin, *Biology as Ideology: The Doctrine of DNA* (New York: Harper, 1993), and *The Genetic Basis of Evolutionary Change* (New York: Columbia University Press, 1974); and Richard Lewontin, Steven Rose, and Leon Kamin, *Not in Our Genes: Biology, Ideology, and Human Nature* (New York: Pantheon Books, 1984). What critics find most disturbing in the sociobiological approach as Wilson and others initially defined it is its thoroughgoing genetic determinism. Gould, among others, counters by arguing that evolutionary adaptation is a function of a multiplicity of complex processes and cannot simply be reduced to genetic factors. The genocentric perspective supported by Wilson has more recently been elaborated by Dawkins and informs the work of evolutionary psychologists like Stephen Pinker. See, inter alia, *The Language Instinct* (New York: William Morrow, 1994), and *How the Mind Works* (New York: Norton, 1997). The resurgence of this point of view has been prompted by the extraordinary advances in our understanding of the human genome. As the precise function of particular genes is understood, more and more human behavior seems to be genetically determined. (For a very helpful summary of the evolutionary psychology controversy, see Melanie Mitchell, "Can Evolution Explain How the Mind Works? A Review of the Evolutionary Psychology Debates," *Complexity* 4, no. 2 [January/February, 1999]: 17–24.) Gould, it should be noted, is as vocal in his opposition to evolutionary psychology as he has been to sociobiology. In the following pages, I will attempt to mediate some of these conflicts by developing an analysis of the interde-

pendence and coevolution of biological and cultural processes. The groundwork for this argument has been laid in our rereading of the genotype as a complex adaptive system. What remains is to see the way in which culture as well as the interplay between biological and cultural systems also function as complex adaptive systems.

31. Dawkins, *The Selfish Gene*, 192.

32. Ibid., 192. For an elaboration of the viral characteristics of memes, see Richard Brodie, *Virus of the Mind: The New Science of the Meme* (Seattle: Integral Press, 1996) and Aaron Lynch, *Thought Contagion: How Belief Spreads through Society* (New York: Basic Books, 1999). Susan Blackmore provides a helpful extension of Dawkins's ideas in *The Meme Machine* (New York: Oxford University Press, 1999).

33. Richard Dawkins, *Unweaving the Rainbow: Science, Delusion, and the Appetite for Wonder* (New York: Houghton Mifflin, 1998), 302.

34. Richard Dawkins, *River out of Eden: A Darwinian View of Life* (New York: Basic Books, 1995), 4.

35. Dawkins, *Unweaving the Rainbow*, 294.

36. Ibid., 306.

37. H. G. Wells, *World Brain* (London: Methuen & Co., 1938), 40, 48–49.

38. Pierre Lévy, *Collective Intelligence: Mankind's Emerging World in Cyberspace*, trans. Robert Bononno (New York: Plenum Press, 1997), 216. For a thorough account of the importance of the fourth dimension in twentieth-century art, see Linda Henderson, *The Fourth Dimension and Non-Euclidean Geometry in Modern Art* (Princeton: Princeton University Press, 1983). The writings of the French theologian Pierre Teilhard de Chardin are another important inspiration for Lévy's spiritual musings. Lévy borrows Teilhard de Chardin's notion of the "noosphere" to develop his account of "Noolithic" life in "the fourth space of knowledge" (*Collective Intelligence*, 140). Other philosophers of cyberspace are also drawn to Teilhard de Chardin's work. See, for example, Jennifer Cobb, *Cybergrace: The Search for God in the Digital World* (New York: Crown Publishers, 1998).

39. Ibid., 216, 218.

40. Kurzweil, *Spiritual Machines*, 16. See Rob Fixmer, "The Soul of the Next New Machine: Humans," *New York Times*, November 6, 1999.

41. Ibid., 124, 128–29.

42. Ibid., 234.

43. Hans Moravec, *Robot: Mere Machine to Transcendent Mind* (New York: Oxford University Press, 1999), 169–70.

44. Ibid., 137.

45. Kelso, *Dynamic Patterns*, 26. Kelso's long and detailed discussion of these processes provides ample experimental and theoretical support for such claims.

46. Ibid., 228.

47. Deacon, *The Symbolic Species*, 130–31. We have already examined the important role experiments with neural networks played in Kauffman's analysis of biological organisms as complex adaptive systems. See the section "Morphing" in chapter 6.

48. Jean-Pierre Changeux, *Neuronal Man: The Biology of the Mind*, trans. Lawrence Garey (Princeton: Princeton University Press, 1977), 246, 248.

49. Ibid., 248–49. Changeux's position on this point anticipates Gerald Edelman's theory of neuronal group selection. See *Neural Darwinism.*

50. George Johnson, "How Much Give Can the Brain Take?" *New York Times,* October 24, 1999. See Elizabeth Gould, Allison Reeves, Michael Graziano, and Charles Gross, "Neurogenesis in the Neocortex of Adult Primates," *Science* 286 (October 15, 1999): 548–52.

51. Daniel C. Dennett, *Darwin's Dangerous Idea: Evolution and the Meanings of Life* (New York: Simon and Schuster, 1995), 343.

52. Daniel C. Dennett, *Consciousness Explained* (New York: Little, Brown and Co., 1991), 207.

53. Deacon, *The Symbolic Species,* 221.

54. James Gardner, "Genes Beget Memes and Memes Beget Genes: Modeling a New Catalytic Closure," *Complexity* 4, no. 5 (May/June 1999): 22–23. Gardner cites the Crick test in his paper.

55. Deacon, *The Symbolic Species,* 322. Emphasis added.

56. It is, of course, necessary to recall the important differences between the Hegelian system and complex adaptive systems. Unlike the closed structure of Hegel's archeo-teleological system, complex adaptive systems are open and have no prescribed design. Order emerges spontaneously in patterns that bear abiding traces of the aleatory processes through which they form.

57. Jacques Derrida, *Archive Fever,* 19. The most obvious example of the prosthesis of the inside is nanotechnology. See K. Eric Drexler, *Engines of Creation* (Garden City, N.Y.: Anchor, 1986), and *Nanosystems: Molecular Machinery, Manufacturing, and Computation* (New York: Wiley, 1992); and K. Eric Drexler and Chris Peterson, with Gayle Pergamit, *Unbounding the Future: The Nanotechnology Revolution* (New York: William Morrow, 1991).

58. Crumey, *Pfitz,* 123.

CODA

1. Edward Wyatt, "Investors See Room for Profit in the Demand for Education," *New York Times,* November 4, 1999.

2. Michael Moe, Kathleen Bailey, and Rhoda Lau, *The Book of Knowledge: Investing in the Growing Education and Training Industry* (New York: Merrill Lynch, 1999), 2. More recently, Bank of America Securities has issued a 293-page report on investment opportunities in on-line education.

3. David Collis, "'When Industries Change,' Revisited: New Scenarios for Higher Education," *Forum Futures 99,* ed. Maureen E. Devlin, 3. Collis insists that this response is both shortsighted and dangerous: "The first threat posed by entrants serving new market segments is that of a 'disruptive technology.' Over time the entrants' new business model, or technology, will offer programs at such attractive price/feature combinations that they will begin to attract traditional students. Obviously those existing institutions most at threat here are the smaller, non-research state schools, which today come as close as any college or university to serving such students. Higher end residential and research universities that offer a liberal arts, tutorial based education will clearly be the last to face competitive pres-

sures from this source. But as the example of the US auto industry, and the UK motor-bike industry, and the US steel industry, and many other industries illustrate, the period of protection for those at the higher end might be only twenty years at the very most, and may be substantially shorter."

4. James Perley and Denise Marie Tanguary, "Accrediting On-Line Institution Diminishes Higher Education," *Chronicle of Higher Education,* October 29, 1999, B4.

5. Immanuel Kant, *The Conflict of the Faculties,* trans. Mary J. Gregor (Lincoln: University of Nebraska Press, 1992), 23.

6. Ibid., 25.

7. Ibid., 27, 29.

8. Kant never offers a satisfactory explanation for the reason he labels the theological, legal, and medical faculties high and the philosophical faculty low. His only comment is the concluding sentence of his introduction: "The reason why this faculty, despite its great prerogative (freedom), is called the lower faculty lies in human nature; for a man to give commands, even though he is someone else's humble servant, is considered more distinguished than a free man who has no one under his command" (29).

9. Clement Greenberg, "The Avant-Garde and Kitsch," in *Perceptions and Judgments, 1939–1944,* ed. John O'Brian (Chicago: University of Chicago Press, 1986), 1:11–12.

10. Pierre Bourdieu, *The Rules of Art: Genesis and Structure of the Literary Field,* trans. Susan Emanuel (Stanford: Stanford University Press, 1996), 142. In developing my account of the social conditions surrounding the emergence of the distinction between high and low art, I have been guided by Bourdieu's analysis.

11. Quoted by Bourdieu, *The Rules of Art,* 82.

12. Ibid., 129.

13. Charles E. McClelland, *State, Society, and University in Germany, 1700–1914* (Cambridge: Cambridge University Press, 1990), 114.

14. For a thorough investigation of the notion of *Bildung,* see W. H. Bruford, *The German Tradition of Self-Cultivation: "Bildung" from Humboldt to Thomas Mann* (Cambridge: Cambridge University Press, 1975). The emphasis on self-development should not obscure the political dimension of Bildung. As the tradition of Bildung develops, it comes to serve a political function for the nation-state by encouraging the inward appropriation of national cultural values. The most sophisticated philosophical analysis of this process is developed by Hegel in chapter 6 of *Phenomenology of Spirit.*

15. McClelland, *State, Society, and University in Germany,* 118–19.

16. See Mark C. Taylor, *Journeys to Selfhood: Hegel and Kierkegaard.*

17. John Henry Newman, *The Idea of the University Defined and Illustrated* (Oxford: Oxford University Press, 1976), I.v.4; I.vii.1. For an informed account of Newman's understanding of the university in relation to the contemporary situation, see Jaroslav Pelikan, *The Idea of the University: A Reexamination* (New Haven: Yale University Press, 1992). The association of liberal learning with inutility can actually be traced back to Aristotle's *Politics:* "It is therefore not difficult to see that the young must be taught those useful arts that are indispensably necessary; but, those pursuits that are liberal *(eleutheron)* having been distinguished from those that are illiberal, it is clear that they should not be taught all the useful arts, and that they must participate in such that will not make the participant a

philistine *(banauson)*" (Aristotle, *Politics,* ed. Gerald Else [Cambridge, Mass.: Harvard University Press, 1963], VIII, 1337b.)

18. This is not to suggest a sharp opposition between the British and the continental traditions on this issue. In post-Kantian romanticism, literature gradually emerges as the privileged site for critical reflection. For an important exploration of this point, see Philippe Lacoue-Labarthe and Jean-Luc Nancy, *The Literary Absolute: The Theory of Literature in German Romanticism,* trans. Philip Barnard and Cheryl Lester (Albany: State University of New York Press, 1988).

19. Frederick Rudolph, *The American College and University: A History* (New York: Knopf, 1962), 24. Though published almost thirty years ago, Rudolph's study remains the most useful account of the history of American colleges and universities. Rudolph's Williams colleague Francis Oakley effectively complements his study in his incisive analysis of historical roots of controversies that have dominated debates about higher education for the past several decades. See Francis Oakley, *Community of Learning: The American College and the Liberal Arts Tradition* (New York: Oxford University Press, 1992).

20. Rudolph, *The American College and University,* 236, 65.

21. Ibid., 219, 135.

22. Ibid., 331.

23. Cary Nelson and Stephen Watt, *Academic Keywords: A Devil's Dictionary for Higher Education* (New York: Routledge, 1999), 9.

24. Walter W. Powell and Jason Owen-Smith, "Universities as Creators and Retailers of Intellectual Property: Life-Sciences Research and Commercial Development," *To Profit or Not to Profit: The Commercial Transformation of the Nonprofit Sector,* ed. Burton A. Weisbrod (New York: Cambridge University Press, 1998), 171.

25. Jacques Derrida, "The Principle of Reason: The University in the Eyes of Its Pupils," *Diacritics* 13 (fall 1983): 3–20, and "Mochlos: or, The Conflict of the Faculties," *Logomachia: The Conflict of the Faculties,* ed. Richard Rand (Lincoln: University of Nebraska Press, 1992), 1–34. Studies that derive from these essays include Richard Rand, ed., *Logomachia: The Conflict of the Faculties*; Peggy Kamuf, *The Division of Literature; or, The University in Deconstruction* (Chicago: University of Chicago Press, 1997); and Bill Readings, *The University in Ruins* (Cambridge, Mass.: Harvard University Press, 1996).

26. Derrida, "The Principle of Reason," 12–13.

27. Heidegger, "The Question Concerning Technology," 4–5.

28. Ibid., 21. Heidegger's primary aim in this influential essay is to propose an alternative account of technology in which it is not the result of human self-assertion but of being's self-disclosure, which can be apprehended only in the state of passive acceptance.

29. Derrida, "The Principle of Reason," 14.

30. Ibid., 16, 18.

31. Readings, *The University in Ruins,* 175. Derrida's most explicit analysis of the "anti-economic logic" of waste can be found in his comparative analysis of Hegel and Bataille in "From Restricted to General Economy: A Hegelianism without Reserve," *Writing and Difference,* 251–77.

32. For an extensive elaboration of this point, see Kamuf, *The Division of Literature; or, The University in Deconstruction.*

33. Derrida, "The Principle of Reason," 19.

34. See, inter alia, Geoffrey Hartman, *Criticism in the Wilderness: The Study of Literature Today* (New Haven: Yale University Press, 1980).

35. Derrida, "The Principle of Reason," 17.

36. It should now be clear why Derrida cites Kierkegaard in the closing paragraph of "The Principle of Reason." Derrida's criticism of the university is, in effect, a secularized version of Kierkegaard's attack on the church. Derrida writes: "Then the time of reflection is also another time, it is heterogeneous with what it reflects and perhaps gives time for what calls for and is called thought. It is the chance for an event about which one does not know whether or not, presenting itself *within* the university, it belongs to the history of the university. It may also be brief and paradoxical, it may tear up time, like the instant invoked by Kierkegaard, one of those thinkers who are foreign, even hostile to the university, who gives us more to think about, with respect to the essence of the university, than academic reflections themselves" (19–20).

37. Ibid., 16, 19.

38. Ibid., 14, 13.

39. See, for example, a recent issue of the bulletin of the American Association of University Professors devoted to the defense of tenure: *Academe,* May–June 2000.

40. It is important to note that modernist and postmodernist artists have been quick to appropriate various technologies for artistic purposes and, during the last half of the twentieth century, have become quite adept at marketing their work. In movements as diverse as Russian Constructivism, de Stijl, and the Bauhaus, artists use mechanical technologies in creating their work. In the effort to produce art that is socially effective, many of the leading modernists enthusiastically embrace utilitarianism. Commenting on the program at Malevich's Vitebksk art school, Ilia Chashnik goes so far as to claim: "The constructions of Suprematism are blueprints for the building and assembling of forms of utilitarian organism. Consequently, any Suprematist project is Suprematism extended into functionality. The Department of Architecture and Technology is the builder of new forms of utilitarian Suprematism" (R. L. Rutsky, *High Techne: Art and Technology from the Machine Aesthetic to the Posthuman* [Minneapolis: University of Minnesota Press, 1999], 86). This point of view is expressed most concisely in the modernist adage: "Form is function." In the last several decades, artists have turned from mechanical to electronic technology to create works using not only video but also digital and virtual reality technology. With the emergence of New York's gallery system after the Second World War, artists had to develop strategies for marketing their work. No one saw this better than Andy Warhol. With characteristic insight and irony, he writes: "Business art is the step that comes after Art. I started as a commercial artist, and I want to finish as a business artist. After I did the thing called 'art' or whatever it's called, I went into business art. I wanted to be an Art Businessman or a Business Artist. Being good in business is the most fascinating kind of art. During the hippie era people put down the idea of business—they'd say, 'Money is bad,' and 'Working is bad,' but making money is art and working is art and good business is the best art" (Andy Warhol, *The Philosophy of Andy Warhol: From A to B and Back Again* [New York: Harcourt Brace Jovanovich, 1975], 92).

There are lessons for academics in these artistic developments. New technologies can be effectively appropriated for educational ends in enterprises that generate profits, which can be used to support educational institutions and programs. For further discussions of these artistic tendencies, see Mark C. Taylor, *Hiding, About Religion: Economies of Faith in Virtual Culture,* and *Disfiguring: Art, Architecture, Religion.*

41. For an account of the 1999 Sun Valley conference, see Ken Auletta, "What I Did at Summer Camp: A Reporter Gets inside Herb Allen's CEO Retreat," *The New Yorker,* July 26, 1999, 46–51. Auletta has published an essay on Allen in *The Highwaymen: Warriors of the Information Superhighway* (New York: Random House, 1997).

42. Andy Grove, *Only the Paranoid Survive: How to Exploit the Crisis Points That Challenge Every Company* (New York: Doubleday, 1996). Grove repeatedly uses images and metaphors of turbulence to describe the strategic inflection point. He stresses that in this complex moment, equilibrium breaks down to create unpredictable situations.

43. The advisory board includes John W. Chandler, president emeritus, Williams College; James O. Freedman, president emeritus, Dartmouth College; Theodore M. Hesburg, president emeritus, the University of Notre Dame; and Frank H. T. Rhodes, president emeritus, Cornell University.

44. Dan Carnevale and Jeffrey Young, "Who Owns On-Line Courses: Colleges and Universities Start to Sort It Out," *The Chronicle of Higher Education,* December 17, 1999.

45. The distance separating the parties in these discussions is underscored by a comment by a faculty member quoted in the *Chronicle* article: "'One of the dangers I see in modeling contracts and agreements of other industries is the insensitivity to the kind of language we tend to use in higher education,' says Rodney J. Petersen, director of policy and planning at the University of Maryland at College Park. 'The word "employee" is not a popular term for a university professor,' he says. Mr. Peterson faces the issue himself this year in preparing to teach an on-line course at Maryland about intellectual-property rights for faculty members. 'In the contract, I'm called a "producer,"' says Mr. Peterson. 'I would really like to think that I'm a scholar or an educator.'" Whether or not they like it, faculty members *undeniably* are employees in the "education industry."

46. Ibid.

47. New software also makes it possible for students to offer continuous feedback, which professors can use to modify courses while they are being given.

48. As colleges and universities attempt to stay competitive by responding rapidly to student interests and demands, it will become more difficult for them to commit to lifelong employment for any faculty members. Tenure will give way to term contracts, which allow institutions more flexibility. With the decline of tenure, the governance of colleges and universities will change. Faculty power will decrease and the power of administrators and boards of trustees will increase.

49. Cary Nelson and Stephen Watt, *Academic Keywords,* 116. Obviously for Nelson, Watt, and their colleagues, the purported disadvantages of distributed learning for faculty outweigh the admitted advantages for students.

bibliography

Abrams, Meyer H. *Natural Supernaturalism: Tradition and Revolution in Romantic Literature.* New York: W. W. Norton and Co., 1971.

Adleman, Leonard M. "Computing with DNA." *Scientific American,* August 1998, 54–61.

Anderson, Philip W., Kenneth J. Arrow, and David Pines, eds. *The Economy as an Evolving Complex System.* Santa Fe Institute Studies, vol. 5. Reading, Mass.: Addison-Wesley, 1988.

Appignanesi, Lisa, ed. *Postmodernism: ICA Documents.* London: Free Association Books, 1989.

Aristotle. *Poetics.* Edited by Gerald Else. Cambridge: Harvard University Press, 1963.

Arrighi, Giovanni, and Beverly Silver. *Chaos and Governance in the Modern World System.* Minneapolis: University of Minnesota Press, 1999.

Atlan, Henri. *Enlightenment to Enlightenment: Intercritique of Science and Myth.* Translated by Lenn Schramm. Albany, N.Y.: SUNY Press, 1993.

———. *Entre le cristal et la fumée: Essai sur l'organisation du vivant.* Paris: Éditions du Seuil, 1979.

———. *L'organisation biologique et la théorie de l'information.* Paris: Hermann, 1972.

Augaitis, Diana, and Dan Lander, eds. *Radio Rethink: Art, Sound, and Transmission.* Alberta, Canada: Walter Phillips Gallery, 1994.

Auge, Marc. *Non-places: Introduction to an Anthropology of Supermodernity.* Translated by John Howe. New York: Verson, 1995.

Auletta, Ken. *The Highwaymen: Warriors of the Information Superhighway.* New York: Random House, 1997.

———. "What I Did at Summer Camp: A Reporter Gets inside Herb Allen's CEO Retreat." *The New Yorker,* July 26, 1999, 46–51.

Axelrod, Robert. *The Evolution of Cooperation.* New York: Basic Books, 1984.

Axelrod, Robert, and Michael Cohen. *Harnessing Complexity: Organizational Implications of a Scientific Frontier.* New York: Free Press, 1999.

Axtell, James. *The Educational Writings of John Locke.* New Haven: Yale University Press, 1998.

Bak, Per. *How Nature Works: The Science of Self-Organized Criticality.* New York: Springer-Verlag, 1996.

Balkin, J. M. *Cultural Software: A Theory of Ideology.* New Haven: Yale University Press, 1998.

Ballard, J. G. *The Atrocity Exhibition.* Hong Kong: RE/SEARCH Publications, 1990.

Barthes, Roland. *The Rustle of Language.* Translated by Richard Howard. New York: Hill and Wang, 1986.

Bateson, Gregory. *Mind and Nature: A Necessary Unity.* New York: E. P. Dutton, 1979.

————. *Steps to an Ecology of Mind.* New York: Ballentine Books, 1972.

Bateson, Mary Catherine. *Our Own Metaphor: A Personal Account of a Conference on the Effects of Conscious Purpose on Human Adaptation.* Washington, D.C.: Smithsonian Institution Press, 1972.

Baty, S. Paige. *e-mail trouble: love and addiction @ the matrix.* Austin: University of Texas Press, 1999.

Baudrillard, Jean. *The Ecstasy of Communication.* Translated by Bernard and Caroline Schutze. New York: Semiotext(e), 1987.

————. *The Evil Demon of Images.* Sydney: Power Institute of Fine Arts, 1987.

————. *The Illusion of the End.* Translated by Chris Turner. Stanford: Stanford University Press, 1994.

————. *Seduction.* Translated by Brian Singer. New York: St. Martin's Press, 1990.

————. *Simulacra and Simulation.* Translated by Sheila Faria Glaser. Ann Arbor: University of Michigan Press, 1997.

————. *Simulations.* Translated by Paul Foss, Paul Patton, and Philip Beitchman. New York: Semiotext(e), 1983.

————. *Symbolic Exchange and Death.* Translated by Iain Hamilton Grant. London: Sage Publications, 1993.

Bauman, Zygmunt. *Intimations of Postmodernity.* London: Routledge, 1993.

Beaudoin, Tom. *Virtual Faith: The Irreverent Spiritual Quest of Generation X.* San Francisco: Jossey-Bass, 1998.

Beckmann, John, ed. *The Virtual Dimension: Architecture, Representation, and Crash Culture.* Princeton: Princeton University Press, 1998.

Begley, Sharon. "Aping Language." *Newsweek,* January 19, 1998, 56–58.

————. "Talking from Hand to Mouth." *Newsweek,* March 15, 1999, 56–58.

Belew, Richard K., and Melanie Mitchell, eds. *Adaptive Individuals in Evolving Populations.* Santa Fe Institute Studies, vol. 26. Reading, Mass.: Addison-Wesley, 1996.

Bell, Daniel. *The Cultural Contradictions of Capitalism.* New York: Basic Books, 1996.

Bell, George I., and Thomas G. Marr, eds. *Computers and DNA.* Santa Fe Institute Studies, vol. 7. Reading, Mass.: Addison-Wesley, 1990.

Bender, Gretchen, and Timothy Druckrey, eds. *Culture on the Brink: Ideologies of Technology.* Seattle: Bay Press, 1994.

Ben-Jacob, Eshel, and Herbert Levine. "The Artistry of Microorganisms." *Scientific American,* October 1998, 82–87.

Bennett, Charles. "Demons, Engines, and the Second Law." *Scientific American,* November 1987, 108–16.

Berger, Peter, and Thomas Luckmann. *The Social Construction of Reality: A Treatise in the Sociology of Knowledge.* New York: Doubleday, 1966.

Bersani, Leo. *The Culture of Redemption.* Cambridge, Mass.: Harvard University Press, 1990.

"Birds do it, bees do it . . ." *The Economist,* August 30, 1997, 59–61.

Blackmore, Susan. *The Meme Machine.* New York: Oxford University Press, 1999.

Blake, Peter. *God's Own Junkyard: The Planned Deterioration of America's Landscape.* New York: Holt, Rinehart and Winston, 1964.

Blakeslee, Sandra. "Computer 'Life Form' Mutates in an Evolution Experiment." *New York Times,* November 25, 1997.

———. "Some Biologists Ask 'Are Genes Everything?'" *New York Times,* September 2, 1997.

Blaser, Walter. *Mies van der Rohe.* New York: Praeger Publishers, 1972.

Bloom, Howard. *Global Brain: The Evolution of Mass Mind from the Big Bang to the Twenty-first Century.* New York: John Wiley & Sons, Inc., 2000.

Boden, A. Margaret. *The Creative Mind: Myths and Mechanisms.* New York: Basic Books, 1990.

Boenau, Bruce A., and Katsuyuki Niiro, eds. *Post-industrial Society.* New York: University Press of America, 1983.

Bohm, David, and B. J. Hiley. *The Undivided Universe: An Ontological Interpretation of Quantum Theory.* New York: Routledge, 1993.

Bolter, David Jay, and Richard Grusin. *Remediation: Understanding New Media.* Cambridge, Mass.: MIT Press, 1999.

Bonabeau, Eric, Marco Dorigo, and Guy Theraulaz. *Swarm Intelligence: From Natural to Artificial Systems.* New York: Oxford University Press, 1999.

Bonner, John Tyler. *The Evolution of Culture in Animals.* Princeton: Princeton University Press, 1980.

Boole, George. *An Investigation of the Laws of Thought.* New York: Dover Publications, 1954.

Boorstin, Daniel J. *The Image: A Guide to Pseudo-events in America.* New York: Vintage Books, 1992.

Borgman, Christine. *From Guttenberg to the Global Information Infrastructure: Access to Information in the Networked World.* Cambridge, Mass.: MIT Press, 2000.

Borgmann, Albert. *Holding On to Reality: The Nature of Information at the Turn of the Millennium.* Chicago: University of Chicago Press, 1999.

Bossomaier, Terry, and David Green. *Patterns in the Sand: Computers, Complexity, and Everyday Life.* Reading, Mass.: Perseus Books, 1998.

Boudon, Raymond. *The Uses of Structuralism.* Translated by Michalina Vaughan. London: Heinmann, 1971.

Bourdieu, Pierre. "The Market of Symbolic Goods." *Poietics* 14 (1985): 13–44.

———. *On Television.* Translated by Priscilla Parkhurst Ferguson. New York: The New Press, 1996.

———. *Outline of a Theory of Practice.* Translated by Richard Nice. Cambridge: Cambridge University Press, 1992.

———. *The Rules of Art: Genesis and Structure of the Literary Field.* Translated by Susan Emanuel. Stanford: Stanford University Press, 1996.

Bova, Ben. *Cyberbooks.* New York: Tom Doherty Associates, 1989.

Brann, Eva. *Paradoxes of Education in a Republic.* Chicago: University of Chicago Press, 1989.

Briggs, John. *Fractals: The Patterns of Chaos.* New York: Simon and Schuster, 1992.

Bringsjord, Selmer. "Chess Is Too Easy." *Technology Review,* March/April, 1998, 23–28.

Brodie, Richard. *The Virus of the Mind: The New Science of the Meme.* Seattle: Integral Press, 1996.

Broeckmann, Andreas, Joke Brouwer, Bart Lootsma, Arjen Mulder, and Lars Spuybroek, eds. *The Art of the Accident.* Translated by Frances Brettell, Ronald Fritz, and Leo Reijnen. Rotterdam: NAI Publishers/V2_Organisation, 1998.

Brown, Shona, and Kathleen Eisenhardt. *Competiting on the Edge: Strategy as Structured Chaos.* Boston: Harvard Business School Press, 1998.

Browne, Malcolm W. "Neuron Talks to Chip, and Chip to Nerve Cell." *New York Times,* August 22, 1995.

Bruford, W. H. *The German Tradition of Self-Cultivation: "Bildung" from Humboldt to Thomas Mann.* Cambridge: Cambridge University Press, 1975.

Bukatman, Scott. *Terminal Identity: The Virtual Subject in Postmodern Science Fiction.* Durham: Duke University Press, 1993.

Burke, James. *The Knowledge Web: From Electronic Agents to Stonehenge and Back—and Other Journeys through Knowledge.* New York: Simon and Schuster, 1999.

Burroughs, William. *The Adding Machine.* New York: Arcade Publishing, 1986.

———. *The Job.* New York: Penguin Books, 1974.

———. *Naked Lunch.* New York: Grove Weidenfield, 1990.

———. *The Soft Machine.* New York: Grove Press, 1996.

Butler, Judith. *Bodies That Matter: On the Discursive Limits of "Sex."* New York: Routledge, 1993.

———. *Gender Trouble: Feminism and the Subversion of Identity.* New York: Routledge, 1990.

Byrne, David. *Complexity Theory and the Social Sciences: An Introduction.* New York: Routledge, 1998.

Cadigan, Pat. *Synners.* New York: Bantam Books, 1991.

Caillois, Roger. *Man, Play, and Games.* Translated by Meyer Barash. New York: Thames and Hudson, 1962.

Calvin, William, H. *How Brains Think: Evolving Intelligence, Then and Now.* New York: Basic Books, 1996.

Calvino, Italo. *Cosmicomics.* Translated by William Weaver. New York: Harcourt Brace Jovanovich, 1968.

———. *If on a winter's night a traveler.* Translated by William Weaver. New York: Harcourt Brace Jovanovich, 1981.

———. *Invisible Cities.* Translated by William Weaver. New York: Harcourt Brace Jovanovich, 1974.

———. *T Zero.* Translated by William Weaver. New York: Harcourt Brace Jovanovich, 1969.

———. *The Uses of Literature.* Translated by Patrick Creagh. New York: Harcourt Brace Jovanovich, 1982.

Campbell, Jeremy. *Grammatical Man: Information, Entropy, Language, and Life.* New York: Simon and Schuster, 1982.

Canguilhem, Georges. *The Normal and the Pathological.* Cambridge, Mass.: Zone Books, 1991.

———. *A Vital Rationalist.* Translated by Arthur Goldhammer. Edited by François Delaporte. Cambridge, Mass.: Zone Books, 1994.

Capra, Fritjof. *The Turning Point: Science, Society, and the Rising Culture.* New York: Simon and Schuster, 1982.

Carnevale, Dan, and Jeffrey Young. "Who Owns On-Line Courses: Colleges and Universities Start to Sort It Out." *Chronicle of Higher Education,* December 17, 1999.

Carroll, Lewis. *Through the Looking Glass.* London: Macmillan, 1871.

Casdagli, Martin, and Stephen Eubank, eds. *Nonlinear Modeling and Forecasting.* Santa Fe Institute Studies, vol. 12. Reading, Mass.: Addison-Wesley, 1992.

Cassidy, John. "The Force of an Idea." *The New Yorker,* January 12, 1998, 32–37.

————. "Moniconomics 101." *The New Yorker,* September 21, 1998, 73–77.

————. "The New World Disorder." *The New Yorker,* October 26–November 2, 1998, 198–207.

Castells, Manuel. *End of the Millennium.* Cambridge, Mass.: Blackwell, 1998.

————. *The Power of Identity.* Cambridge, Mass.: Blackwell, 1997.

————. *The Rise of Network Society.* Cambridge, Mass.: Blackwell, 1996.

Casti, John. *The Cambridge Quintet: A Work of Scientific Speculation.* Reading, Mass.: Addison-Wesley, 1998.

————. *Complexification: Explaining a Paradoxical World through the Science of Surprise.* New York: HarperCollins, 1994.

————. *Would-Be Worlds: How Simulation Is Changing the Frontiers of Science.* New York: John Wiley, 1997.

Chaitin, Gregory. *Information, Randomness, and Incompleteness.* Singapore: World Scientific Co., 1990.

Change, Yahlin. "Roll Over, Beethoven." *Newsweek,* July 29, 1996, 71.

Changeux, Jean-Pierre. *Neuronal Man: The Biology of Mind.* Translated by Laurence Garey. Princeton: Princeton University Press, 1997.

Chomsky, Noam. *Cartesian Linguistics.* New York: Harper and Row, 1966.

————. *Current Issues in Linguistic Theory.* The Hague: Mouton, 1964.

————. *Syntactic Structure.* The Hague: Mouton, 1957.

Churchland, Patricia S., and Terence J. Sejnowski. *The Computational Brain.* Cambridge, Mass.: MIT Press, 1992.

Chytry, Josef. *The Aesthetic State: A Quest in Modern German Thought.* Berkeley and Los Angeles: University of California Press, 1989.

Cilliers, Paul. *Complexity and Postmodernism: Understanding Complex Systems.* New York: Routledge, 1998.

Cladis, P. E., and P. Palffy-Muhoray, eds. *Spatio-Temporal Patterns in Nonequilibrium Complex Systems.* Santa Fe Institute Studies, vol. 21. Reading, Mass.: Addison-Wesley, 1995.

Clippinger, John Henry. *The Biology of Business: Decoding the Natural Laws of Enterprise.* San Francisco: Jossey-Bass Publishers, 1999.

Cobb, Jennifer. *Cybergrace: The Search for God in the Digital World.* New York: Crown Publishers, 1998.

Cohen, Jack. "Thinking about Thinking." *Scientific American,* July 1998, 113–14.

Collis, David. "'When Industries Change,' Revisited: New Scenarios for Higher Education." *Forum Futures 99.* Edited by Maureen E. Devlin.

Conley, Verena Andermatt. *Rethinking Technologies.* Minneapolis: University of Minnesota Press, 1993.

Cook, David, and Arthur Kroker. *The Postmodern Scene: Excremental Culture and Hyper-aesthetics.* New York: St. Martin's Press, 1986.

Coupland, Douglas. *Generation X: Tales for an Accelerated Culture.* New York: St. Martin's Press, 1991.

———. *Microserfs.* New York: HarperCollins, 1995.

Coveney, Peter, and Roger Highfield. *Frontiers of Complexity: The Search for Order in a Chaotic World.* New York: Fawcett Columbine, 1995.

Coyle, Diane. *The Weightless World: Strategies for Managing the Digital Economy.* Cambridge, Mass.: MIT Press, 1998.

Cowan, George A., David Pines, and David Meltzer, eds. *Complexity: Metaphors, Models, and Reality.* Santa Fe Institute Studies, vol. 19. Reading, Mass.: Addison-Wesley, 1994.

Cowley, Geoffrey, and Anne Underwood. "Memory." *Newsweek,* June 15, 1998, 49–54.

Crary, Jonathan. *Techniques of the Observer: On Vision and Modernity in the Nineteenth Century.* Cambridge, Mass.: MIT Press, 1996.

Crary, Jonathan, and Sanford Kwinter, eds. *Incorporations.* New York: Zone, 1992.

Crichton, Michael. *Jurassic Park.* New York: Ballantine Books, 1990.

Crumey, Andrew. *Pfitz.* New York: Picador, 1995.

Csanyi, Vilmos. *Evolutionary Systems and Society: A General Theory of Life, Mind, and Culture.* Durham: Duke University Press, 1989.

D'Amato, Barbara. *Killer.app.* New York: Tom Doherty Associates, 1996.

Darwin, Charles. *On the Origin of Species by Means of Natural Selection.* London: Murray, 1859.

Davidson, Martin. *The Consumerist Manifesto: Advertising in Postmodern Times.* New York: Routledge, 1992.

Davis, Erik. *Techgnosis: Myth, Magic, and Mysticism in the Age of Information.* New York: Harmony Books, 1998.

Dawkins, Richard. *The Extended Phenotype: The Long Reach of the Gene.* New York: Oxford University Press, 1982.

———. *River out of Eden: A Darwinian View of Life.* New York: Basic Books, 1995.

———. *The Selfish Gene.* New York: Oxford University Press, 1989.

———. *Unweaving the Rainbow: Science, Delusion, and the Appetite for Wonder.* New York: Houghton Mifflin, 1998.

Dawson, John W., Jr. "Gödel and the Limits of Logic." *Scientific American,* June 1999, 76–81.

Deacon, Terrence. *The Symbolic Species: The Co-evolution of Language and the Brain.* New York: Norton, 1997.

Debord, Guy. *Society of the Spectacle.* Detroit: Black and Red, 1983.

de Certeau, Michel. *Culture in the Plural.* Translated by Luce Giard. Minneapolis: University of Minnesota Press, 1997.

———. *Heterologies: Discourse on the Other.* Translated by Brian Massumi. Minneapolis: University of Minnesota Press, 1986.

———. *The Practice of Everyday Life.* Translated by Steven Rendall. Berkeley and Los Angeles: University of California Press, 1988.

———. *The Writing of History.* Translated by Tom Conley. New York: Columbia University Press, 1988.

De Landa, Manuel. *A Thousand Years of Nonlinear History.* Cambridge, Mass.: Zone, 1997.

———. *War in the Age of Intelligent Machines.* New York: Swerve Editions, 1991.

Deleuze, Gilles. *Cinema I: The Movement-Image.* Translated by Hugh Tomlinson and Barbara Habberjam. Minneapolis: University of Minnesota Press, 1986.

———. *Difference and Repetition.* Translated by Paul Patton. New York: Columbia University Press, 1994.

———. *The Fold: Leibniz and the Baroque.* Translated by Tom Conley. Minneapolis: University of Minnesota Press, 1993.

———. *Foucault.* Translated by Seán Hand. Minneapolis: University of Minnesota Press, 1988.

———. *Kant's Critical Philosophy: The Doctrine of the Faculties.* Translated by H. Tomlinson and B. Habberjam. Minneapolis: University of Minnesota Press, 1984.

———. *Logique du sens.* Paris: Les Éditions de Minuit, 1969.

———. *Negotiations, 1972–1990.* Translated by Martin Joughin. New York: Columbia University Press, 1995.

Deleuze, Gilles, and Felix Guattari. *Anti-Oedipus: Capitalism and Schizophrenia.* Translated by Robert Hurley, Mark Seem, and Helen Lane. New York: Viking Press, 1977.

———. *A Thousand Plateaus: Capitalism and Schizophrenia.* Translated by Brian Massumi. Minneapolis: University of Minnesota Press, 1987.

———. *What Is Philosophy?* Translated by Hugh Tomlinson and Graham Burchell. New York: Columbia University Press, 1994.

Delillo, Don. *End Zone.* New York: Houghton Mifflin, 1972.

———. *White Noise.* New York: Penguin, 1986.

Denby, David. "In Darwin's Wake." *The New Yorker,* July 21, 1997, 50–62.

Dennett, Daniel C. *Consciousness Explained.* New York: Little, Brown and Co., 1991.

———. "'Darwinian Fundamentalism': An Exchange." *New York Review of Books,* August 14, 1997, 64–65.

———. *Darwin's Dangerous Idea: Evolution and the Meanings of Life.* New York: Simon and Schuster, 1995.

Depew, David, and Bruce Weber. *Darwinism Evolving: Systems Dynamics and the Genealogy of Natural Selection.* Cambridge, Mass.: MIT Press, 1995.

Derrida, Jacques. *Archive Fever: A Freudian Impression.* Translated by Eric Prenowitz. Chicago: University of Chicago Press, 1996.

———. *Dissemination.* Translated by Barbara Johnson. Chicago: University of Chicago Press, 1981.

———. *Edmund Husserl's Origin of Geometry.* Translated by John Leavey. Stonybrook, N.Y.: Nicholas Hays, 1978.

———. *Glas.* Paris: Galilée, 1979.

———. *Margins of Philosophy.* Translated by Alan Bass. Chicago: University of Chicago Press, 1982.

———. *Memoirs of the Blind: The Self-Portrait and Other Ruins.* Translated by P. A. Brault and M. Nass. Chicago: University of Chicago Press, 1993.

———. *Of Grammatology.* Translated by Gayatri C. Spivak. Baltimore: Johns Hopkins University Press, 1976.

———. *Of Spirit: Heidegger and the Question.* Translated by Geoffrey Bennington and Rachel Bowlby. Chicago: University of Chicago Press, 1989.

―――. *Positions.* Translated by Alan Bass. Chicago: University of Chicago Press, 1981.

―――. *The Post Card: From Socrates to Freud and Beyond.* Translated by Alan Bass. Chicago: University of Chicago Press, 1987.

―――. "The Principle of Reason: The University in the Eyes of Its Pupils." *Diacritics* 13, no. 3 (Fall 1983): 3–20.

―――. *Specters of Marx: The State of the Debt, the Work of Mourning, and the New International.* Translated by Peggy Kamuf. New York: Routledge, 1994.

―――. *Writing and Difference.* Translated by Alan Bass. Chicago: University of Chicago Press, 1978.

Dery, Mark. *Escape Velocity: Cyberculture at the End of the Century.* New York: Grove Press, 1996.

Descartes, René. *Discourse on the Method and Meditations on First Philosophy.* Edited by David Weissman. New Haven: Yale University Press, 1996.

―――. *Discourse on the Method of Rightly Conducting One's Reason and of Seeking Truth in the Sciences.* Translated by Donald A. Cress. Indianapolis: Hackett Publishing Company, 1988.

―――. *The Geometry of René Descartes.* Translated by David Eugene Smith and Marcia L. Latham. New York: Dover Publications, 1954.

Descombes, Vincent. *Modern French Philosophy.* Translated by L. Scott-Fox and J. M. Harding. Cambridge: Cambridge University Press, 1982.

Deutsch, Karl W. *The Nerves of Government: Models of Political Communication and Control.* New York: Free Press, 1963.

"Distinctively American: The Residential Liberal Arts Colleges." *Daedalus* 128, no. 1 (1999).

Ditto, William L., and Louis M. Pecora. "Mastering Chaos." *Scientific American,* August 1993, 78–84.

Docker, John. *Postmodernism and Popular Culture: A Cultural History.* Cambridge: Cambridge University Press, 1994.

Donato, Eugenio, and Richard Macksey, eds. *The Structuralist Controversy: The Languages of Criticism and the Sciences of Man.* Baltimore: Johns Hopkins University Press, 1977.

Doolen, Gary D., ed. *Lattice Gas Methods for Partial Differential Equations.* Santa Fe Institute Studies, vol. 4. Reading, Mass.: Addison-Wesley, 1990.

Dosse, François. *History of Structuralism.* Vol. 1, *The Rising Sign, 1945–1966.* Minneapolis: University of Minnesota Press, 1997.

―――. *History of Structuralism.* Vol. 2, *The Sign Sets, 1967–Present.* Minneapolis: University of Minnesota Press, 1997.

Dreifus, Claudia. "A Mathematician at Play in the Fields of Space-Time." *New York Times,* January 19, 1999.

Drexler, K. Eric. *Engines of Creation.* Garden City, N.Y.: Anchor, 1986.

―――. *Nanosystems: Molecular Machinery, Manufacturing, and Computation.* New York: Wiley, 1992.

Drexler, K. Eric, and Chris Peterson, with Gayle Pergamit. *Unbounding the Future: The Nanotechnology Revolution.* New York: Quill, 1991.

Dreyfus, Hubert. *What Computers Can't Do: The Limits of Artificial Intelligence.* New York: Harper and Row, 1972.

Dyson, George. *Darwin among the Machines: The Evolution of Global Intelligence.* New York: Addison-Wesley, 1997.

Eco, Umberto. *A Theory of Semiotics.* Bloomington: Indiana University Press, 1976.

———. *Travels in Hyperreality.* Translated by William Weaver. San Diego: Harcourt Brace Jovanovich, 1986.

Eisenstein, Sergi M. *Selected Works.* Vol. 2, *Towards a Theory of Montage E.* Edited by M. Glenny and R. Taylor. London: British Film Institute, 1991.

Edelman, Gerald. *Neural Darwinism: The Theory of Group Selection.* New York: Basic Books, 1987.

Epstein, Joshua M. *Nonlinear Dynamics, Mathematical Biology, and Social Science.* Santa Fe Institute Studies, Lecture Notes, vol. 4. Reading, Mass.: Addison-Wesley Publishing, 1997.

Everdell, William R. *The First Moderns: Profiles in the Origins of Twentieth-Century Thought.* Chicago: University of Chicago Press, 1997.

Ewen, Stuart. *All Consuming Images: The Politics of Style in Contemporary Culture.* New York: Basic Books, 1988.

Featherstone, Mike. *Consumer Culture and Postmodernism.* London: Sage Publications, 1993.

Feher, Michel, Ramona Naddaf, and Nadia Tazi, eds. *Fragments for a History of the Human Body.* 3 vols. New York: Zone, 1989.

Feldman, Tony. *An Introduction to Multimedia.* New York: Routledge, 1997.

Ferris, Timothy. "Frauds! Fakes! Phonies!" *New York Times Book Review,* January 10, 1999, 7.

Feyerabend, Paul. *Farewell to Reason.* New York: Verso, 1987.

Fixmer, Rob. "The Soul of the Next New Machine: Humans." *New York Times,* November 6, 1992.

Forrest, Stephanie. *Emergent Computation: Self-Organizing, Collective, and Cooperative Phenomena in Natural and Artificial Computing Networks.* Cambridge, Mass.: MIT Press, 1991.

Foster, Hal, ed. *The Anti-aesthetic: Essays on Postmodern Culture.* Port Townsend, Wash.: Bay Press, 1983.

Foucault, Michel. *"The Archaeology of Knowledge" and "The Discourse on Language."* Translated by A. M. Sheridan Smith. New York: Harper and Row, 1972.

———. *The Birth of the Clinic: An Archaeology of Medical Perception.* Translated by A. M. Sheridan Smith. New York: Vintage Books, 1975.

———. *The Birth of the Prison.* Translated by Alan Sheridan. New York: Vintage Books, 1979.

———. *The Care of the Self.* Vol. 3 of *The History of Sexuality.* Translated by Robert Hurley. New York: Vintage Books, 1988.

———. *Language, Counter-Memory, Practice.* Edited by Donald Bouchard. Translated by Donald Bouchard and Sherry Simon. Ithaca: Cornell University Press, 1977.

———. *Madness and Civilization: A History of Insanity in the Age of Reason.* Translated by Richard Howard. New York: Vintage Books, 1965.

———. *The Order of Things: An Archaeology of the Human Sciences.* New York: Vintage, 1971.

———. *Power/Knowledge.* Translated by Colin Gordon, Leo Marshall, John Mephan, and Kate Soper. New York: Pantheon Books, 1980.

―――. *This Is Not a Pipe.* Translated by James Harkness. Berkeley: University of California Press, 1983.

―――. *The Use of Pleasure.* Vol. 2 of *The History of Sexuality.* Translated by Robert Hurley. New York: Vintage Books, 1990.

Friedberg, Anne. *Window Shopping: Cinema and the Postmodern.* Berkeley and Los Angeles: University of California Press, 1993.

Friedman, Daniel, and John Rust, eds. *The Double Auction Market.* Santa Fe Institute Studies, vol. 14. Reading, Mass.: Addison-Wesley, 1991.

Friedman, Thomas L. *The Lexus and the Olive Tree: Understanding Globalization.* New York: Farrar, Strauss, and Giroux, 1999.

Fukuyama, Francis. *The End of History and the Last Man.* New York: Free Press, 1992.

―――. *The Great Disruption: Human Nature and the Reconstitution of the Social Order.* Free Press, 1999.

Galison, Peter. *Image and Logic: A Material Culture of Microphysics.* Chicago: University of Chicago Press, 1997.

Garber, Marjorie, Jann Matlock, and Rebecca L. Walkowitz, eds. *Media Spectacles.* New York: Routledge, 1993.

Gardner, James. "Genes Beget Memes and Memes Beget Genes: Modeling a New Catalytic Closure." *Complexity* 4, no. 5 (May/June, 1999): 29–37.

Gatlin, Lila. *Information Theory and the Living System.* New York: Columbia University Press, 1972.

Gell-Mann, Murray. *The Quark and the Jaguar: Adventures in the Simple and the Complex.* New York: W. H. Freeman, 1994.

Gerbel, Karl, and Peter Weibel, eds. *Intelligent Environment,* vol. 1. Vienna: PVS Verleger, 1994.

Gershenfeld, Neil, and Isaac L. Chuang. "Quantum Computing with Molecules." *Scientific American,* June 1998, 66–71.

Gibson, William. *Idoru.* New York: Berkley Books, 1996.

―――. *Mona Lisa Overdrive.* New York: Bantam Books, 1989.

―――. *Neuromancer.* New York: Ace Books, 1984.

―――. *Virtual Light.* New York: Bantam Books, 1993.

Giere, Ronald N. *Science without Laws.* Chicago: University of Chicago Press, 1999.

Gilder, George. *Life after Television: The Coming Transformation of Media and American Life.* New York: Norton, 1994.

Giovannini, Joseph. "Art into Architecture." *Guggenheim* 21.

Gladwell, Malcom. *The Tipping Point: How Little Things Can Make a Big Difference.* New York: Little, Brown and Co., 2000.

Glass, William. *On Being Blue: A Philosophical Inquiry.* Boston: David R. Godine, 1991.

Gleick, James. *Chaos: Making a New Science.* New York: Viking, 1987.

―――. *Faster: The Acceleration of Just About Everything.* New York: Pantheon Books, 1999.

Goertzel, Ben. *Chaotic Logic: Language, Thought, and Reality from the Perspective of Complex Systems Science.* New York: Plenum Press, 1994.

Golding, Sue, ed. *The Eight Technologies of Otherness.* New York: Routledge, 1997.

Goodwin, Brian. *How the Leopard Changed Its Spots: The Evolution of Complexity.* New York: Scribner's, 1994.

Goodwin, Brian, Atuhiro Sibatani, and Gerry Webster, *Dynamic Structures in Biology.* Edinburgh: Edinburgh University Press, 1989.

Gould, Elizabeth, Alison Reeves, Michael Graziano, and Charles Gross. "Neurogenesis in the Neocortex of Adult Primates." *Science* 286 (October 15, 1999): 548–52.

Gould, Stephen Jay. "Evolution: The Pleasures of Pluralism." *New York Review of Books,* June 16, 1997, 47–52.

———. *Full House: The Spread of Excellence from Plato to Darwin.* New York: Harmony Books, 1996.

———. *Hen's Teeth and Horse's Toes.* New York: W. W. Norton, 1983.

———. *Wonderful Life: The Burgess Shale and the Nature of History.* New York: W. W. Norton, 1989.

Gray, John. *False Dawn: The Delusions of Global Capitalism.* New York: New Press, 1999.

Greenberg, Clement. "The Avant-Garde and Kitsch." In *Perceptions and Judgments, 1939–1944,* edited by John O'Brian. Chicago: University of Chicago Press, 1986.

———. "Modernist Painting." *Modernism with a Vengeance, 1957–69,* edited by John O'Brian. Chicago: University of Chicago Press, 1993.

Greenberg, Jan, and Sandra Jordan. *Chuck Close, Up Close.* New York: DK Publishing, 1998.

Gross, Paul R., and Norman Levitt. *Higher Superstition: The Academic Left and Its Quarrels with Science.* Baltimore: Johns Hopkins University Press, 1994.

Grove, Andy. *Only the Paranoid Survive: How to Exploit the Crisis Points That Challenge Every Company.* New York: Doubleday, 1996.

Guattari, Felix. *Chaosmosis: An Ethic-Aesthetic Paradigm.* Translated by Paul Bains and Julian Pefanis. Bloomington: Indiana University Press, 1995.

Guggenheim Bilbao Museoa. Bilbao: FMGB Guggenheim Bilbao Museoa, 1997.

Guilbaut, Serge, ed. *Reconstructing Modernism: Art in New York, Paris, and Montreal, 1945–1964.* Cambridge, Mass.: MIT Press, 1992.

Gumerman, George J., and Murray Gell-Mann, eds. *Understanding Complexity in the Prehistoric Southwest.* Santa Fe Institute Studies, vol. 16. Reading, Mass.: Addison-Wesley, 1994.

Habermas, Jürgen. *Knowledge and Human Interests.* Translated by Jeremy Shapiro. Boston: Beacon Press, 1971.

Hacking, Ian. *The Social Construction of What?* Cambridge, Mass.: Harvard University Press, 1999.

Haken, Hermann. *The Science of Structure Synergetics.* Translated by Fred Bradley. New York: Van Nostrand Reinhold Co., 1984.

Hall, Stephen S. "Our Memories, Our Selves." *New York Times Magazine,* February 15, 1998, 26–33, 49, 56–57.

Hannerz, Ulf. *Cultural Complexity: Studies in the Social Organization of Meaning.* New York: Columbia University Press, 1992.

Haraway, Donna, *Modest_Witness@Second_Millennium.FemaleMan_Meets_OncoMouse: Feminism and Technoscience.* New York: Routledge, 1997.

Hardison, O. B., Jr., *Disappearing through the Skylight: Culture and Technology in the Twentieth Century.* New York: Viking, 1989.

Harland, Richard. *Beyond Superstructuralism: The Syntagmatic Side of Language.* New York: Routledge, 1993.

———. *Superstructuralism: The Philosophy of Structuralism and Post-structuralism.* New York: Routledge, 1991.

Harris, Craig, ed. *Art and Invention: The Xerox Parc Artist-in-Residence Program.* Cambridge, Mass.: MIT Press, 1995.

Hart, Kevin. "Flying Home." *Wicked Heat.* Sydney: Paper Bark Press, 1999.

Hartman, Geoffrey. *Criticism in the Wilderness: The Study of Literature Today.* New Haven: Yale University Press, 1980.

Harvey, David. *The Condition of Postmodernity: An Inquiry into the Origins of Cultural Change.* Cambridge, Mass.: Blackwell, 1990.

Hawkins, John A., and Murray Gell-Mann, eds. *The Evolution of Human Languages.* Santa Fe Institute Studies, vol. 11. Reading, Mass.: Addison-Wesley, 1992.

Hayashi, Alden M. "Pinker and the Brain." *Scientific American,* July 1999, 32–34.

Hayles, N. Katherine. "Virtual Bodies and Flickering Signifiers." *October* 66 (fall 1993): 91.

———. *How We Became Posthuman: Virtual Bodies in Cybernetics, Literature, and Informatics.* Chicago: University of Chicago Press, 1999.

———, ed. *Chaos and Order: Complex Dynamics in Literature and Science.* Chicago: University of Chicago Press, 1991.

Heeger, David J., and James R. Bergen. "Pyramid-Based Texture Analysis/Synthesis." Proceedings of SIGGRAPH 95 (Los Angles, August 6–11, 1995). In *Computer Graphics: Proceedings,* Annual Conference Series, 1995, ACM SIGGRAPH: 229–38.

Hegel, G. W. F. *Lectures on the History of Philosophy.* Translated by E. S. Haldane. 3 vols. New York: Humanities Press, 1968.

———. *Lectures on the Philosophy of Religion.* Translated by E. B. Speirs and J. B. Sanderson. 3 vols. New York: Humanities Press, 1968.

———. *The Logic of Hegel.* Translated by W. Wallace. New York: Oxford University Press, 1968.

———. *Phenomenology of Spirit.* Translated by A. V. Miller. New York: Oxford University Press, 1977.

———. *Philosophy of Mind.* Translated by W. Wallace and A. V. Miller. New York: Oxford University Press, 1971.

———. *Philosophy of Nature.* Translated by A. V. Miller. New York: Humanities Press, 1970.

———. *Philosophy of Subjective Spirit.* Translated by M. J. Petry. Boston: D. Reidel Publishing Co., 1978.

———. *Science of Logic.* Translated by A. V. Miller. New York: Humanities Press, 1969.

Heidegger, Martin. *Hegel's Concept of Experience.* Translated by Kenley Dove. New York: Harper and Row, 1970.

———. *Identity and Difference.* Translated by Joan Stambaugh. New York: Harper and Row, 1969.

———. *Kant and the Problem of Metaphysics.* Translated by James Churchill. Bloomington: Indiana University Press, 1962.

———. *Poetry, Language, Thought.* Translated by Albert Hofstadter. New York: Harper Colophon, 1975.

———. *"The Question Concerning Technology" and Other Essays.* Translated by William Lovitt. New York: Harper and Row, 1977.

Heim, Michael. *Virtual Realism.* New York: Oxford, 1998.

Henrich, Dieter. *Hegel im Kontext.* Frankfurt: Suhrkamp Verlag, 1971.

Hess, Deborah M. *Complexity in Maurice Blanchot's Fiction: Relations between Science and Literature.* New York: Peter Lang Publishing, 1999.

Hilts, Philip J. "Listening to the Conversation of Neurons." *New York Times,* May 27, 1997.

Hine, Thomas. *Populuxe.* New York: Knopf, 1986.

Hobart, Michael, and Zachary Schiffman. *Information Ages: Literacy, Numeracy, and the Computer Revolution.* Baltimore: Johns Hopkins University Press, 1998.

Hofstadter, Douglas. *Gödel, Escher, Bach: An Eternal Golden Braid.* New York: Vintage, 1980.

Hohenberg, Paul, and Lynn Lees. *The Making of Urban Europe, 1000–1950.* Cambridge, Mass.: Harvard University Press, 1985.

Holland, John H. *Emergence: From Chaos to Order.* Reading, Mass.: Addison-Wesley, 1998.

———. *Hidden Order: How Adaptation Builds Complexity.* New York: Addison-Wesley, 1995.

Holston, James. *The Modernist City: An Anthropological Critique of Brasilia.* Chicago: University of Chicago Press, 1989.

Hume, David. *A Treatise on Human Nature.* New York: Prometheus, 1991.

Hyman, Anthony. *Charles Babbage: Pioneer of the Computer.* Princeton: Princeton University Press, 1982.

Hyppolite, Jean. *Genesis and Structure of Hegel's "Phenomenology of Spirit."* Translated by Samuel Cherniak and John Heckman. Evanston, Ill.: Northwestern University Press, 1974.

———. *Logic and Existence.* Translated by Leonard Lawor and Amit Sen. Albany, N.Y.: SUNY Press, 1997.

Idhe, Don. *Sense and Significance.* Pittsburgh: Duquesne University Press, 1973.

Ingber, Donald E. "The Architecture of Life." *Scientific American,* January 1998, 48–57.

"The Internet: Fulfilling the Promise." *Scientific American* 276, no. 3 (March 1997): 49–83.

Jacob, François. *The Logic of Life: A History of Heredity.* Translated by Betty Spillman. Princeton: Princeton University Press: 1973.

———. *The Possible and the Actual.* Seattle: University of Washington Press, 1982.

Jakobson, Roman. *Fundamentals of Language.* The Hague: Mouton, 1956.

Jameson, Fredric. *The Political Unconscious: Narrative as a Socially Symbolic Act.* Ithaca: Cornell University Press, 1985.

———. *Postmodernism, or, The Cultural Logic of Late Capitalism.* Durham: Duke University Press, 1991.

———. *The Prison-House of Language: A Critical Account of Structuralism and Russian Formalism.* Princeton: Princeton University Press, 1972.

Jantsch, Erich. *The Self-Organizing Universe: Scientific and Human Implications of the Emerging Paradigm of Evolution.* New York: Pergamon Press, 1980.

Jay, Martin. *Downcast Eyes: The Denigration of Vision in Twentieth-Century French Thought.* Berkeley and Los Angeles: University of California Press, 1993.

Jervis, Robert. *System Effects: Complexity in Political and Social Life.* Princeton: Princeton University Press, 1997.

Jibu, Mari, and Kunio Yasue. *Quantum Brain Dynamics and Consciousness: An Introduction.* Philadelphia: John Benjamins, 1995.

Johnson, George. *Fire in the Mind: Science, Faith and the Search for Order.* New York: Random House, 1995.

———. "From Grains of Sand: A World of Order." *New York Times,* September 8, 1996.

———. "Mindless Creatures, Acting 'Mindfully.'" *New York Times,* March 23, 1999.

———. "Of Mice and Elephants: A Matter of Scale." *New York Times,* January 12, 1999.

———. "Separating the Unsolvable and the Merely Difficult." *New York Times,* July 13, 1999.

———. *Strange Beauty: Murray Gell-Mann and the Revolution in Twentieth-Century Physics.* New York: Knopf, 1999.

———. "Useful Invention or Absolute Truth: What Is Math?" *New York Times,* February 10, 1998.

———. "Yes, There Is Such a Thing as Mind over Matter." *New York Times,* February 25, 1996.

Johnson, Mark. *The Body in the Mind: The Bodily Basis of Meaning, Imagination, and Reason.* Chicago: University of Chicago Press, 1987.

Johnson, Philip, and Henry-Russell Hitchcock. *The International Style: Architecture since 1922.* New York: W. W. Norton and Co., 1932.

Johnson, Steven. *Interface Culture: How New Technology Transforms the Way We Create and Communicate.* New York: Harper Edge, 1997.

Johnston, John. *Information Multiplicity: American Fiction in the Age of Media Saturation.* Baltimore: Johns Hopkins University Press, 1998.

Jones, Caroline, and Peter Galison, eds. *Picturing Science Producing Art.* New York: Routledge, 1998.

Jonscher, Charles. *The Evolution of Wired Life: From the Alphabet to the Soul-Catcher Chip—How Information Technologies Change Our World.* New York: John Wiley, 1999.

Julesz, Bela, and Ilona Kovács, eds. *Maturational Windows and Adult Cortical Plasticity.* Santa Fe Institute Studies, vol. 23. Reading, Mass.: Addison-Wesley, 1995.

Kahn, Douglas, and Gregory Whitehead, eds. *Wireless Imagination: Sound, Radio, and the Avant-Garde.* Cambridge, Mass.: MIT Press, 1992.

Kamuf, Peggy. *The Division of Literature; or, The University in Deconstruction.* Chicago: University of Chicago Press, 1997.

Kant, Immanuel. *The Conflict of the Faculties.* Translated by Mary J. Gregor. Lincoln: University of Nebraska Press, 1992.

———. *Critique of Judgment.* Translated by James Meredith. New York: Oxford University Press, 1973.

———. *Critique of Practical Reason.* Translated by Lewis White Beck. Indianapolis: Bobbs-Merrill Company, 1956.

———. *Critique of Pure Reason.* Translated by Norman Kemp Smith. New York: St. Martin's Press, 1965.

Kauffman, Stuart. *At Home in the Universe: The Search for the Laws of Self-Organization and Complexity.* New York: Oxford University Press, 1995.

———. *The Origins of Order: Self-Organization and Selection in Evolution.* New York: Oxford University Press, 1993.

———. "The Sciences of Complexity and 'Origins of Order.'" In *Principles of Organization in Organisms,* edited by Jay Mittenthal and Arthur Baskin, 303–19. Santa Fe Institute Studies, vol. 13. Reading, Mass.: Addison-Wesley, 1992.

Keating, Ann B., and Joseph Hargitai. *The Wired Professor: A Guide to Incorporating the World Wide Web in College Instruction.* New York: New York University Press, 1999.

Kellener, Douglas. *Media Culture: Cultural Studies, Identity, and Politics between the Modern and the Postmodern.* New York: Routledge, 1995.

Kellert, Stephen H. *In the Wake of Chaos: Unpredictable Order in Dynamical Systems.* Chicago: University of Chicago Press, 1993.

Kelly, Kevin. *New Rules for the New Economy: Ten Radical Strategies for a Connected World.* New York: Viking, 1998.

Kelso, J. A. Scott. *Dynamic Patterns: The Self-Organization of Brain and Behavior.* Cambridge, Mass.: MIT Press, 1995.

Kierkegaard, Søren. *Concluding Unscientific Postscript.* Translated by David F. Swenson and Walter Lowrie. Princeton: Princeton University Press, 1941.

———. *Either-Or.* 2 vols. Vol. 1 translated by David F. and Lillian Marvin Swenson. Vol. 2 translated by Walter Lowrie. Princeton: Princeton University Press, 1971.

Kittler, Friedrich A. *Discourse Networks, 1800/1900.* Translated by Michael Metteer. Stanford: Stanford University Press, 1990.

Klotz, Heinrich. *Contemporary Art: The Collection of the ZKM/Center for Art and Media, Karlsruhe.* Munich: Prestel-Verlag, 1997.

Knabb, Ken, ed. *Situationist International Anthology.* Translated by Ken Knabb. Berkeley: Bureau of Public Secrets, 1989.

Kojève, Alexandre. *Introduction to the Reading of Hegel.* Translated by James H. Nicholas. Edited by A. Bloom. New York: Basic Books, 1969.

Kolata, Gina. "Biology's Big Project Turns into Challenge for Computer Experts." *New York Times,* June 11, 1996.

———. "Novel Kind of Computing: Calculation with DNA." *New York Times,* November 22, 1994.

Kramer, Gregory, ed. *Auditory Display.* Santa Fe Institute Studies, vol. 18. Reading, Mass.: Addison-Wesley, 1994.

Krauss, Rosalind. "Grids." *The Originality of the Avant-Garde and Other Modernist Myths.* Cambridge, Mass.: MIT Press, 1986.

———. *The Optical Unconscious.* Cambridge, Mass.: MIT Press, 1993.

Krauss, Rosalind, and Yve-Alain Bois. *Formless: A User's Guide.* Cambridge, Mass.: MIT Press, 1997.

Kristof, Nicholas D. "Robokitty." *New York Times Magazine,* August 1, 1999, 42–45.

Kroker, Arthur. *Spasm: Virtual Reality, Android Music, and Electric Flesh.* New York: St. Martin's Press, 1993.

Kroker, Arthur, and Michael A. Weinstein. *Data Trash: The Theory of the Virtual Class.* New York: St. Martin's Press, 1994.

Kroker, Arthur, and Marilouise Kroker, eds. *Body Invaders: Panic Sex in America.* New York: St. Martin's Press, 1987.

———. *Digital Delirium.* New York: St. Martin's Press, 1997.

Kruger, Barbara. *Remote Control: Power, Culture, and the World of Appearances.* Cambridge, Mass.: MIT Press, 1994.

Krugman, Paul. *Accidental Theorist: And Other Dispatches from the Dismal Science.* New York: Norton, 1998.

Kurzweil, Edith. *The Age of Structuralism: Lévi-Strauss to Foucault.* New York: Columbia University Press, 1980.

Kurzweil, Ray. *The Age of Spiritual Machines: When Computers Exceed Human Intelligence.* New York: Viking, 1999.

Lacoue-Labarthe, Philippe, and Jean-Luc Nancy. *Absolue littéraire: Théorie de la littérature du romantisme allemand.* Paris: Éditions du Seuil, 1978.

———. *The Literary Absolute: The Theory of Literature in German Romanticism.* Translated by Philip Barnard and Cheryl Lester. Albany, N.Y.: SUNY Press, 1988.

Lakoff, George. *Women, Fire, and Dangerous Things: What Categories Reveal about the Mind.* Chicago: University of Chicago Press, 1990.

Lakoff, George, and Mark Johnson. *Philosophy in the Flesh: The Embodied Mind and Its Challenge to Western Thought.* New York: Basic Books, 1999.

Langton, Christopher G., ed. *Artificial Life.* Santa Fe Institute Studies, vol. 6. Reading, Mass.: Addison-Wesley, 1989.

———. *Artificial Life III.* Santa Fe Institute Studies, vol. 17. Reading, Mass.: Addison-Wesley, 1994.

Laszlo, Ervin. *Introduction to Systems Philosophy: Toward a New Paradigm of Contemporary Thought.* New York: Gordon and Breach, 1972.

———. *The Systems View of the World: The Natural Philosophy of the New Developments in the Sciences.* New York: George Braziller, 1972.

Leary, Timothy. *Chaos and CyberCulture.* Berkeley: Ronin, 1994.

Le Corbusier. *The City of To-morrow and Its Planning.* Translated by Frederick Etchells. Cambridge, Mass.: MIT Press, 1986.

———. *Towards a New Architecture.* Translated by Frederick Etchells. New York: Dover Publications, n.d.

Leibniz, Gottfried Wilhelm, Freiherr von. *"Discourse on Metaphysics" and Related Writings.* Translated by R. N. D. Martin and Stuart Brown. New York: Manchester University Press, 1988.

———. *"The Monadology" and Other Philosophical Writings.* Translated by Robert Latta. Oxford: Clarendon Press, 1898.

———. *New Essays Concerning Human Understanding.* Translated by Alfred Gideon Langley. New York and London: Macmillan, 1896.

Lem, Stanislaw. *The Cyberiad.* Translated by Michael Kandel. New York: Harcourt Brace Jovanovich, 1985.

Lenoir, Timothy. *The Strategy of Life: Teleology and Mechanics in Nineteenth Century German Biology.* Boston: D. Reidel Publishing, 1982.

Levinson, Paul. *The Soft Edge: A Natural History and Future of the Information Revolution.* New York: Routledge, 1997.

Lévi-Strauss, Claude. *The Elementary Structures of Kinship.* Translated by J. H. Bell, J. R. von Sturman, and R. Needham. Boston: Beacon Press, 1969.

————. *The Raw and the Cooked: Introduction to the Sciences of Mythology.* Translated by J. D. Weightman. New York: Harper and Row, 1969.

————. *The Savage Mind.* Chicago: University of Chicago Press, 1970.

————. *Structural Anthropology.* Translated by Claire Jacobson and Brooke Grundfest. New York: Basic Books, 1963.

————. *Tristes Tropiques.* Translated by John and Doreen Weightman. New York: Atheneum, 1974.

Levy, Steven. *Artificial Life: A Report from the Frontier Where Computers Meet Biology.* New York: Random House, 1993.

Lewin, Roger. *Complexity: Life at the Edge of Chaos.* New York: Macmillan, 1992.

Lewis, Michael. "How the Eggheads Cracked." *New York Times Magazine,* January 24, 1999, 24–31, 42, 67, 69, 71, 77.

Lewontin, Richard C. *Biology as Ideology: The Doctrine of DNA.* New York: Harper, 1993.

————. *The Genetic Basis of Evolutionary Change.* New York: Columbia University Press, 1974.

Lewontin, Richard C., Steven Rose, and Leon Kamin. *Not in Our Genes: Biology, Ideology, and Human Nature.* New York: Pantheon Books, 1984.

Lilienfeld, Robert. *The Rise of Systems Theory: An Ideological Analysis.* New York: Wiley, 1978.

Livingston, Paisley, ed. *Disorder and Order.* Saratoga, Calif.: Anma Libri & Co., 1984.

Lodder, Christina. *Russian Constructivism.* New Haven: Yale University Press, 1983.

Lovelock, J. E. *Gaia: A New Look at Life on Earth.* New York: Oxford University Press, 1987.

Lowe, Donald M. *The Body in Late-Capitalist USA.* Durham: Duke University Press, 1995.

Luhmann, Niklas. "Deconstruction as Second-Order Observing." *New Literary History* 24 (1993): 763–82.

————. *Essays on Self-Reference.* New York: Columbia University Press, 1990.

————. *Observations on Modernity.* Translated by William Whobrey. Stanford: Stanford University Press, 1998.

————. *Social Systems.* Translated by John Bednarz, Jr., with Dirk Baeker. Stanford: Stanford University Press, 1995.

Lukach, Joan. *Hilla Rebay: In Search of the Spirit in Art.* New York: George Braziller, 1983.

Lumley, Ted. "Complexity and the 'Learning Organization,'" *Complexity* 2, no. 5 (May/June 1997): 14–22.

Lumsden, Charles J., and Edward O. Wilson. *Genes, Mind, and Culture: The Coevolutionary Process.* Cambridge, Mass.: Harvard University Press, 1981.

Lundenfeld, Peter, ed. *The Digital Dialectic: New Essays on New Media.* Cambridge, Mass.: MIT Press, 1995.

Lynch, Aaron. *Thought Contagion: How Belief Spreads through Society.* New York: Basic Books, 1999.

Lyotard, Jean-François. *The Postmodern Condition: A Report on Knowledge.* Translated by G. Bennington and B. Massumi. Minneapolis: University of Minnesota Press, 1984.

MacKay, Donald M. *Information, Mechanism, and Meaning.* Cambridge, Mass.: MIT Press, 1969.

Mandelbrot, Benoit. "A Multifractal Walk Down Wall Street." *Scientific American,* February 1999, 70–73.

———. *The Fractal Geometry of Nature.* New York: W. H. Freeman, 1983.

Marcus, George E., ed. *Connected: Engagements with Media.* Chicago: University of Chicago Press, 1996.

Margulis, Lynn. *Symbiosis in Cell Evolution.* San Francisco: Freeman, 1981.

Margulis, Lynn, and Dorion Sagan. *What Is Life?* New York: Simon and Schuster, 1995.

Martin, Emily. *Flexible Bodies: Tracking Immunity in American Culture from the Days of Polio to the Age of Aids.* Boston: Beacon Press, 1994.

Marx, Karl. *Grundrisse.* Translated by Martin Nicolaus. London: Penguin, 1973.

Maturana, Humberto R., and Francisco J. Varela. *Autopoiesis and Cognition: The Realization of the Living.* Boston: Reidel, 1980.

Mazlish, Bruce. *The Fourth Discontinuity: The Co-evolution of Humans and Machines.* New Haven: Yale University Press, 1993.

McCaffery, Larry, ed. *Storming the Reality Studio: A Casebook of Cyberpunk and Postmodern Science Fiction.* Durham: Duke University Press, 1992.

McCaughey, Robert A. *Scholars and Teachers: The Faculties of Select Liberal Arts Colleges and Their Place in American Higher Learning.* New York: Conceptual Litho Reproductions, 1994.

McClelland, Charles E. *State, Society, and University in Germany, 1700–1914.* Cambridge: Cambridge University Press, 1990.

McClelland, James L., David E. Rumelhart, and the PDP Research Group. *Parallel Distributed Processing: Explorations in the Microstructure of Cognition.* Vol. 1, *Foundations.* Cambridge, Mass.: MIT Press, 1987.

———. *Parallel Distributed Processing: Explorations in the Microstructure of Cognition.* Vol. 2, *Psychological and Biological Models.* Cambridge, Mass.: MIT Press, 1987.

Merleau-Ponty, Maurice. *The Prose of the World.* Translated by J. O'Neill. Evanston, Ill.: Northwestern University Press, 1973.

———. *Sense and Non-sense.* Translated by H. L. Dreyfus and P. A. Dreyfus. Evanston, Ill.: Northwestern University Press, 1964.

———. *Signs.* Translated by R. C. McCleary. Evanston, Ill.: Northwestern University Press, 1964.

———. *The Visible and the Invisible.* Translated by A. Lingis. Evanston, Ill.: Northwestern University Press, 1968.

Merrell, Floyd. *Simplicity and Complexity: Pondering Literature, Science, and Painting.* Ann Arbor: University of Michigan Press, 1998.

Messaris, Paul. *Visual "Literacy": Image, Mind, and Reality.* Boulder, Colo.: Westview Press, 1994.

Michod, Richard. *Darwinian Dynamics: Evolutionary Transitions in Fitness and Individuality.* Princeton: Princeton University Press, 1999.

Miller, J. Hillis. "The Critic as Host." In *Deconstruction and Criticism,* 217–53. New York: Seabury Press, 1979.

Miller, James Grier. *Living Systems.* New York: McGraw-Hill, 1978.

Minsky, Marvin. *The Society of Mind.* New York: Simon and Schuster, 1986.

Mirzoeff, Nicholas. *An Introduction to Visual Culture.* New York: Routledge, 1999.

Mitchell, W. J. T. *The Last Dinosaur Book.* Chicago: University of Chicago Press, 1998.

Mitchell, William J. *City of Bits: Space, Place, and the Infobahn.* Cambridge, Mass.: MIT Press, 1995.

———. *E-topia: Urban Life, Jim—but Not as We Know It.* Cambridge, Mass.: MIT Press, 1999.

———. *The Reconfigured Eye: Visual Truth in the Post-photographic Era.* Cambridge, Mass.: MIT Press, 1992.

Mittenthal, Jay, and Arthur Baskin, eds. *The Principles of Organization in Organisms.* Santa Fe Institute Studies, vol. 13. Reading, Mass.: Addison-Wesley, 1992.

Moe, Michael, Kathleen Bailey, and Rhoda Lau. *The Book of Knowledge: Investing in the Growing Education and Training Industry.* New York: Merrill Lynch, 1999.

Monod, Jacques. *Chance and Necessity.* New York: Knopf, 1971.

Monsom, Ingrid. *Saying Something: Jazz Improvisation and Interaction.* Chicago: University of Chicago Press, 1996.

Moore, Geoffrey. *Living on the Fault Line: Managing for Shareholder Value in the Age of the Internet.* New York: HarperBusiness, 2000.

Moravec, Hans. *Robot: Mere Machine to Transcendent Mind.* New York: Oxford University Press, 1999.

Morowitz, Harold J., and Jerome L. Singer, eds. *The Mind, the Brain, and Complex Adaptive Systems.* Santa Fe Institute Studies, vol. 22. Reading, Mass.: Addison-Wesley, 1995.

Morse, Margaret. *Virtualities: Television, Media Art, and Cyberculture.* Bloomington: Indiana University Press, 1998.

Negativland. *Fair Use: The Story of the Letter U and the Numeral 2.* Concord, Calif.: Seeland, 1995.

Nelkin, Dorothy, and M. Susan Lindee. *The DNA Mystique: The Gene as a Cultural Icon.* New York: W. H. Freeman and Company, 1995.

Nelson, Cary, and Stephen Watt. *Academic Keywords: A Devil's Dictionary for Higher Education.* New York: Routledge, 1999.

Newman, John Henry. *The Idea of the University Defined and Illustrated.* Oxford: Oxford University Press, 1976.

Nicolis, Gregoier, and Ilya Prigogine. *Exploring Complexity: An Introduction.* New York: W. H. Freeman, 1989.

———. *Self-Organization in Nonequilibrium Systems: From Dissipative Structures to Order through Fluctuations.* New York: John Wiley, 1977.

Nietzsche, Friedrich. *Beyond Good and Evil.* Translated by W. Kaufmann. New York: Random House, 1966.

———. *The Will to Power.* Translated by Walter Kaufmann and R. J. Hollingdale. New York: Vintage Books, 1968.

Nightingale, Virginia. *Studying Audience: The Shock of the Real.* New York: Routledge, 1996.

Nijhout, H. E., Lynn Nadel, and Daniel L. Stein, eds. *Pattern Formation in the Physical and Biological Sciences.* Santa Fe Institute Studies, Lecture Notes, vol. 5. Reading, Mass.: Addison-Wesley, 1997.

Nørretranders, Tor. *The User Illusion: Cutting Consciousness Down to Size.* Translated by Jonathan Sydenham. New York: Viking, 1998.

Oakley, Francis. *Community of Learning: The American College and the Liberal Arts Tradition.* New York: Oxford University Press, 1992.

Ohmann, Richard. *Selling Culture: Magazines, Markets, and Class at the Turn of the Century.* New York: Verso, 1996.

"The Origins of Technology." *Scientific American.* Special issue, 1997.

Ormerod, Paul. *Butterfly Economics: A New General Theory of Social and Economic Behavior.* New York: Pantheon Books, 2000.

Overbye, Dennis. "The Cosmos According to Darwin." *New York Times Magazine,* July 13, 1997, 24, 26–27.

Pagels, Heinz. *The Dreams of Reason: The Computer and the Rise of the Sciences of Complexity.* New York: Simon and Schuster, 1988.

Papineau, David. "Don't Know Much Biology." *New York Times Book Review,* January 18, 1998, 7.

Parsons, Talcott. *The Evolution of New Societies.* Edited by Jackson Toby. Englewood Cliffs: Prentice-Hall, 1977.

Pasachoff, Naomi. *Alexander Graham Bell: Making Connections.* New York: Oxford University Press, 1996.

Pattee, Howard. *Hierarchy Theory: The Challenge of Complex Systems.* New York: George Braziller, 1973.

Paul, Robert. "German Academic Science and the Mandarin Ethos, 1850–1880." *British Journal of the History of Science* 17 (1984): 17–29.

Paulson, William R. *The Noise of Culture: Literary Texts in a World of Information.* Ithaca: Cornell University Press, 1988.

Pelikan, Jaroslav. *The Idea of the University: A Reexamination.* New Haven: Yale University Press, 1992.

Perelson, Alan S., ed. *Theoretical Immunology.* Santa Fe Institute Studies, vols. 2–3. Reading, Mass.: Addison-Wesley, 1988.

Perelson, Alan S., and Frederik Wiegel. "Some Design Principles for Immune System Recognition." *Complexity* 4, no. 5 (May/June 1999): 29–37.

Perley, James, and Denise Marie Tanguary. "Accrediting On-Line Institution Diminishes Higher Education." *Chronicle of Higher Education,* October 29, 1999, B4.

Perloff, Marjorie. *The Futurist Moment: Avant-Garde, Avant Guerre, and the Language of Rupture.* Chicago: University of Chicago Press, 1986.

Pfohl, Stephen. *Death at the Parasite Café: Social Science (Fictions) and the Postmodern.* New York: St. Martin's Press, 1992.

Phillips, D. C. *Holistic Thought in Social Science.* Stanford: Stanford University Press, 1976.

Pines, David, ed. *Emerging Syntheses in Science.* Santa Fe Institute Studies, vol. 1. Reading, Mass.: Addison-Wesley, 1988.

Pinker, Steven. *How the Mind Works.* New York: Norton, 1997.

———. *The Language Instinct.* New York: William Morrow, 1994.

Pinkerton, James P. *What Comes Next: The End of Big Government—and the New Paradigm Ahead.* New York: Hyperion, 1995.

Pollan, Michael. "Playing God in the Garden." *New York Times Magazine,* October 25, 1998, 44–51, 62–62, 82, 92–93.

Popper, Karl. *The Open Society and Its Enemies.* Vol. 2, *The High Tide of Prophecy: Hegel, Marx, and the Aftermath.* London: George Routledge and Sons, 1945.

Posner, Michael I., and Marcus E. Raichle. *Images of Mind.* New York: Scientific American Library, 1994.

Poster, Mark. *The Mode of Information: Poststructuralism and Social Context.* Chicago: University of Chicago Press, 1990.

———. *The Second Media Age.* Cambridge: Polity Press, 1995.

Powell, Walter W., and Jason Owen-Smith. "Universities as Creators and Retailers of Intellectual Property: Life-Sciences Research and Commercial Development." In *To Profit or Not to Profit: The Commercial Transformation of the Nonprofit Sector,* edited by Burton A. Weisbrod, 171–93. New York: Cambridge University Press, 1998.

Powers, Richard. *Gain.* New York: Farrar, Straus, and Giroux, 1998.

———. *Galeta 2.2.* New York: Farrar, Straus, and Giroux, 1995.

———. *The Gold Bug Variations.* New York: Harper and Row, 1991.

Prigogine, Ilya. *From Being to Becoming: Time and Complexity in the Physical Sciences.* San Francisco: W. H. Freeman, 1980.

———. *The End of Certainty: Time, Chaos, and the New Laws of Nature.* New York: Free Press, 1997.

Prigogine, Ilya, and Isabelle Stengers. *Order out of Chaos: Man's New Dialogue with Nature.* New York: Bantam Books, 1984.

Propp, Vladimir. *Morphology of the Folktale.* Bloomington: Indiana Research Center in Anthropology, 1958.

Rabinow, Paul. *Making PCR: A Story of Biotechnology.* Chicago: University of Chicago Press, 1996.

Rand, Richard, ed. *Logomachia: The Conflict of the Faculties.* Lincoln: University of Nebraska Press, 1992.

Readings, Bill. *The University in Ruins.* Cambridge, Mass.: Harvard University Press, 1996.

Regalado, Antonio. "The Next Genome Project." *Technology Review,* May/June 1998, 51–53.

Resnick, Mitchel. *Turtles, Termites, and Traffic Jams: Explorations in Massively Parallel Microworlds.* Cambridge, Mass.: MIT Press, 1994.

Riffaterre, Michael. *Semiotics of Poetry.* Bloomington: Indiana University Press, 1978.

Roe, Ann, and George Gaylord Simpson, eds. *Behavior and Evolution.* New Haven: Yale University Press, 1969.

Romanyshyn, Robert D. *Technology as Symptom and Dream.* New York: Routledge, 1989.

Ronell, Avital. *The Telephone Book: Technology, Schizophrenia, Electric Speech.* Lincoln: University of Nebraska Press, 1989.

Rorty, Richard. *Philosophy and the Mirror of Nature.* Princeton: Princeton University Press, 1980.

Rosenau, Pauline Marie. *Post-modernism and the Social Sciences: Insights, Inroads, and Intrusions.* Princeton: Princeton University Press, 1992.

Rosenheim, Shawn James. *The Cryptographic Imagination: Secret Writing from Edgar Poe to the Internet.* Baltimore: Johns Hopkins University Press, 1997.

Ross, Andrew, ed. *Science Wars.* Durham: Duke University Press, 1996.

———. *Strange Weather: Culture, Science, and Technology in the Age of Limits.* London: Verso, 1991.

Ross, Kristin. *Fast Cars, Clean Bodies: Decolonization and the Reordering of French Culture.* Cambridge, Mass.: MIT Press, 1995.

Roszak, Theodore. *The Cult of Information: A Neo-Luddite Treatise on High Tech, Artificial Intelligence, and the True Art of Thinking.* New York: Pantheon Books, 1994.

Rothblatt, Sheldon. *The Modern University and Its Discontents: The Fate of Newman's Legacies in Britain and America.* New York: Cambridge University Press, 1997.

Rudolph, Frederick. *The American College and University: A History.* New York: Knopf, 1962.

Rundle, John B., Donald L. Turcotte, and William Klein, eds.. *Reduction and Predictability of Natural Disasters.* Santa Fe Institute Studies, vol. 25. Reading, Mass.: Addison-Wesley, 1996.

Ruthen, Russell. "Adapting to Complexity." *Scientific American,* January 1993, 138.

Rutsky, R. L. *High Techne: Art and Technology from the Machine Aesthetic to the Posthuman.* Minneapolis: University of Minnesota Press, 1999.

Ryan, Marie-Laure, ed. *Cyberspace Textuality: Computer Technology and Literary Theory.* Bloomington: Indiana University Press, 1999.

Safranski, Rudiger. *Martin Heidegger: Between Good and Evil.* Translated by Ewald Osers. Cambridge, Mass.: Harvard University Press, 1998.

Sahlins, Marshall. *Culture and Practical Reason.* Chicago: University of Chicago Press, 1976.

Samuelson, Robert J. "The Way the World Works." *Newsweek,* January 12, 1998, 52.

Sappington, Rodney, and Tyler Stallings, eds. *Uncontrollable Bodies: Testimonies of Identity and Culture.* Seattle: Bay Press, 1994.

Sassen, Saskia. *Globalization and Its Discontents: Essays on the New Mobility of People and Money.* New York: Free Press, 1998.

———. *Losing Control? Sovereignty in an Age of Globalization.* New York: Columbia University Press, 1996.

Sassower, Raphael. *Cultural Collisions: Postmodern Technoscience.* New York: Routledge, 1995.

———. *Technoscientific Angst: Ethics and Responsibility.* Minneapolis: University of Minnesota Press, 1997.

Schiller, Friedrich. *On the Aesthetic Education of Man: In a Series of Letters.* Translated by Reginald Snell. New York: Frederick Ungar, 1965.

Schiller, Paul. *Irrational Exuberance.* Princeton: Princeton University Press, 2000.

Schneider, Cynthia, and Brian Wallis, eds. *Global Television.* New York: Wedge Press, 1988.

Schrage, Michael. *Shared Minds: The New Technologies of Collaboration.* New York: Random House, 1990.

Schrödinger, Ernst. *What Is Life?* Cambridge: Cambridge University Press, 1967.

Schwarz, Hans-Peter. *Media-Art-History: Media Museum, ZKM/Center for Art and Media Karlsruhe.* Munich: Prestel-Verlag, 1997.

Searle, John R. "Consciousness and the Philosophers." *New York Review of Books,* March 6, 1997, 43–50.

———. "The Mystery of Consciousness: Part II." *New York Review of Books,* November 16, 1995, 54–61.

Sebeok, Thomas. *A Sign Is Just a Sign.* Bloomington: Indiana University Press, 1991.

Sennett, Richard. *The Fall of Public Man.* New York: W. W. Norton, 1992.

Sercarz, Eli. *The Semiotics of Cellular Communication in the Immune System.* New York: Springer Verlag, 1988.

Serres, Michel. *Detachment.* Translated by Genevieve James and Raymond Ferderman. Athens: Ohio University Press, 1989.

———. *Genesis.* Translated by Genevieve James and James Nielson. Ann Arbor: University of Michigan Press, 1995.

———. *Hermes: Literature, Science, Philosophy.* Edited by Josué V. Harari and David F. Bell. Baltimore: Johns Hopkins University Press, 1982.

———. *The Natural Contract.* Translated by Elizabeth MacArthur and William Paulson. Ann Arbor: University of Michigan Press, 1992.

———. *The Parasite.* Translated by Lawrence R. Schehr. Baltimore: Johns Hopkins University Press, 1982.

———. *Rome: The Book of Foundations.* Translated by Felicia McCarren. Stanford: Stanford University Press, 1991.

Serres, Michel, with Bruno Latour. *Conversations on Science, Culture, and Time.* Translated by Roxanne Lapidus. Ann Arbor: University of Michigan Press, 1995.

Shannon, Claude, and Warren Weaver. *The Mathematical Theory of Communication.* Urbana: University of Illinois Press, 1949.

Shapin, Steven. *The Scientific Revolution.* Chicago: University of Chicago Press, 1998.

Shih, Lawrence Kai, and David Greene. "Simulating Acquisition on the Economic Web." *Complexity* 3, no 4 (March/April 1998): 41–45.

Shubik, Martin. "Game Theory, Complexity, and Simplicity Part III: Critique and Prospective." *Complexity* 3, no. 5 (May–June 1998): 34–46.

Shulman, Helene. *Living at the Edge of Chaos: Complex Systems in Culture and Psyche.* Einsiedeln: Daimon Verlag, 1997.

Sigmund, Karl. *Games of Life: Explorations in Ecology, Evolution, and Behaviour.* New York: Oxford University Press, 1993.

Silverstone, Roger, and Eric Hirsch, eds. *Consuming Technologies: Media and Information in Domestic Spaces.* New York: Routledge, 1992.

Simpson, David. *The Academic Postmodern and the Rule of Literature: A Report on Half-Knowledge.* Chicago: University of Chicago Press, 1995.

Smith, John Maynard. "Genes, Memes, and Minds." *New York Review of Books,* November 30, 1995, 46–48.

Smith, Terry. *Making the Modern: Industry, Art, and Design in America.* Chicago: University of Chicago Press, 1993.

Smolin, Lee. *Life in the Cosmos.* New York: Oxford University Press, 1997.

Sokal, Alan, and Jean Bricmont. *Intellectual Impostures: Postmodern Intellectuals' Abuse of Science.* New York: Picador, 1998.

Space, Time, Place. South to the Future, vol. 1, issue 3. San Francisco: South to the Future, 1997.

Stafford, Barbara Maria. *Visual Analogy: Consciousness and the Art of Connecting.* Cambridge, Mass.: MIT Press, 1999.

Stein, Wilfred D., and Francisco J. Varela, eds. *Thinking about Biology: An Invitation to Current Theoretical Biology.* Santa Fe Institute Studies in the Complexity of Biology, Lecture Notes, vol. 3. Reading, Mass.: Addison-Wesley, 1993.

Stephens, Mitchell. *The Rise of the Image and the Fall of the World.* New York: Oxford University Press, 1998.

Stephenson, Neal. *Cryptonomicon.* New York: Avon Books, 1999.

Sterling, Bruce. *Islands in the Net.* New York: Ace Books, 1989.

Stewart, Thomas. *Intellectual Capital: The New Wealth of Organizations.* New York: Doubleday, 1997.

Stix, Gary. "Domesticating Cyberspace." *Scientific American,* August 1993, 100–110.

Stone, Allucquère Rosanne. *The War of Desire and Technology at the Close of the Mechanical Age.* Cambridge, Mass.: MIT Press, 1995.

Stoppard, Tom. *Arcadia.* Boston: Farber and Farber, 1993.

Storr, Robert, ed. *Chuck Close.* New York: Museum of Modern Art, 1998.

Sturrock, John. *Structuralism.* New York: Paladin, 1986.

Sulloway, Frank J. "Darwinian Virtues." *New York Review of Books,* April 9, 1998, 34–39.

Tainter, Joseph A., and Bonnie B. Tainter, eds. *Evolving Complexity and Environmental Risk in the Prehistoric Southwest.* Santa Fe Institute Studies, vol. 24. Reading, Mass.: Addison-Wesley, 1996.

Talbott, Stephen L. *The Future Does Not Compute: Transcending the Machines in Our Midst.* Sebastopol, Calif.: O'Reilly and Associates, 1995.

Taylor, Frederick Winslow. *The Principles of Scientific Management.* New York: Harper, 1929.

Taylor, Mark C. *About Religion: Economies of Faith in Virtual Culture.* Chicago: University of Chicago Press, 1999.

———. *Disfiguring: Art, Architecture, Religion.* Chicago: University of Chicago Press, 1992.

———. *Hiding.* Chicago: University of Chicago Press, 1997.

———. *Journeys to Selfhood: Hegel and Kierkegaard.* Berkeley: University of California Press, 1980; New York: Fordham University Press, 2000.

———. *The Picture in Question: Mark Tansey and the Ends of Representation.* Chicago: University of Chicago Press, 1999.

Taylor, Mark C., and José Márquez. *The Réal: Las Vegas, Nevada.* Williamstown, Mass.: Williams College Museum of Art and Massachusettes Museum of Contemporary Art, 1977.

Taylor, Mark C., and Esa Saarinen. *Imagologies: Media Philosophy.* New York: Routledge, 1994.

Taylor, Todd, and Irene Ward. *Literacy Theory in the Age of the Internet.* New York: Columbia University Press, 1998.

Teilhard de Chardin, Pierre. *The Phenomenon of Man.* New York: Harper Brothers, 1959.

Theraulaz, Guy, Eric Bonabeau, and Jean-Louis Deneubourg. "The Origin of Nest Complexity in Social Insects." *Complexity* 3, no. 6 (July/August 1998): 15–25.

Thom, René. *Structural Stability and Morphogenesis: An Outline of a General Theory of Models.* Translated by D. H. Fowler. Reading, Mass.: Addison-Wesley, 1989.

Thompson, D'Arcy W. *On Growth and Form.* Cambridge: Cambridge University Press, 1917.

Thompson, John N. *The Coevolutionary Process.* Chicago: University of Chicago Press, 1994.

Thurow, Lester. *Building Wealth: The New Rules for Individuals, Companies, and Nations in a Knowledge-Based Economy.* New York: HarperCollins, 1999.

Toffler, Alvin. *The Culture Consumers: Art and Affluence in America.* Baltimore, Md.: Penguin Books, 1965.

————. *The Third Wave.* New York: Bantam, 1990.

Toy, Maggie, Iona Spens, Iona Baird, Rachel Bean, Stephen Watt, and Sara Parkin, eds. "Architects in Cyberspace." *Architectural Design,* no. 118 (1995).

Traub, Joseph, and A. G. Werschulz. *Complexity and Information.* New York: Cambridge University Press, 1998.

Tschumi, Bernard. *Architecture and Disjunction.* Cambridge, Mass.: MIT Press, 1994.

————. Special issue. *Architecture and Urbanism.* March 1994.

Turner, J. Scott. *The Extended Organism: The Physiology of Animal-Built Structures.* Cambridge, Mass.: Harvard University Press, 2000.

Twitchell, James. *ADCULTusa: The Triumph of Advertising in American Culture.* New York: Columbia University Press, 1996.

Ulmer, Gregory L. *Applied Grammatology: Post(e)-Pedagogy from Jacques Derrida to Joseph Beuys.* Baltimore: Johns Hopkins University Press, 1987.

————. *Heuretics: The Logic of Invention.* Baltimore: Johns Hopkins University Press, 1994.

Van Bruggen, Coosje. *Frank O. Gehry: Guggenheim Museum Bilbao.* New York: Guggenheim Museum Publications, 1997.

van de Wetering, Janwillem. *The Maine Massacre.* New York: Ballantine Books, 1979.

Varela, Francisco J. *Principles of Biological Autonomy.* New York: North Holland, 1979.

Varnedoe, Kirk, and Adam Gopnik. *High and Low: Modern Art and Popular Culture.* New York: Museum of Modern Art, 1990.

Venturi, Robert. *Complexity and Contradiction in Architecture.* New York: Museum of Modern Art, 1966.

————. *Iconography and Electronics upon a Generic Architecture: A View from the Drafting Room.* Cambridge, Mass.: MIT Press, 1996.

Venturi, Robert, Denise Scott Brown, and Steven Izenour. *Learning from Las Vegas: The Forgotten Symbolism of Architectural Form.* Cambridge, Mass.: MIT Press, 1988.

Virilio, Paul. *The Aesthetic of Disappearance.* Translated by Philip Beitchman. New York: Semiotext(e), 1991.

————. *The Lost Dimension.* Translated by Daniel Moshenberg. New York: Semiotext(e), 1991.

————. *Speed and Politics: An Essay on Dromology.* Translated by Mark Polizzotti. New York: Semiotext(e), 1986.

————. *The Vision Machine.* Translated by Julie Rose. Bloomington: Indiana University Press, 1994.

Virilio, Paul, and Sylvère Virilio. *Pure War.* Translated by Mark Polizzotti. New York: Semiotext(e), 1983.

Virtual Reality. National Gallery of Australia Exhibition, December 10, 1994–February 5, 1995. Canberra: National Gallery of Australia Publications Department, 1994.

von Bertalanffy, Ludwig. *General System Theory: Foundations, Development, Applications.* New York: George Braziller, 1968.

Von Laue, Theodore H. *The World Revolution of Westernization: The Twentieth Century in Global Perspective.* New York: Oxford University Press, 1987.

von Neumann, John. *Theory of Self-Reproducing Automata.* Champaign-Urbana: University of Illinois Press, 1966.

Waddington, C. H., ed. *Organizers and Genes.* Cambridge: Cambridge University Press, 1940.

———. *Principles of Embryology.* New York: Macmillan, 1956.

———. *Towards a Theoretical Biology.* 3 vols. Chicago: Aldine Publishing Company, 1968–70.

Wade, Nicholas. "From Ants to Ethics: A Biologist Dreams of Unity of Knowledge." *New York Times,* May 12, 1998.

———. "The Struggle to Decipher Human Genes." *New York Times,* March 10, 1998.

Wallerstein, Immanuel. *Geopolitics and Geoculture: Essays on the Changing World-System.* Cambridge: Cambridge University Press, 1991.

Wallis, Brian, ed. *Art after Modernism: Rethinking Representation.* New York: New Museum of Contemporary Art, 1984.

Warhol, Andy. *America.* New York: Harper and Row, 1985.

———. *The Andy Warhol Diaries.* Edited by Pat Hackett. New York: Warner Books, 1989.

———. *The Philosophy of Andy Warhol: From A to B and Back Again.* New York: Harcourt Brace Jovanovich, 1977.

———. *A Retrospective.* Edited by Kynaston McShine. New York: Museum of Modern Art, 1989.

———. "What Is Pop Art? Answers from Eight Painters." *Artnews* 62 (November 1963).

Wark, McKenzie. *Virtual Geography: Living with Global Media Events.* Bloomington: Indiana University Press, 1994.

Watson, James D. *The Double Helix: A Personal Account of the Discovery of the Structure of DNA.* New York: New American Library, 1968.

Weber, Samuel. *Mass Mediarus: Form, Technics, Media.* Stanford: Stanford University Press, 1996.

Weigend, Andreas S., and Neil A. Gershenfeld, eds. *Time Series Prediction.* Santa Fe Institute Studies, vol. 15. Reading, Mass.: Addison-Wesley, 1994.

Wells, H. G. *World Brain.* London: Methuen & Co., 1938.

Wheatley, Margaret. *Leadership and the New Science: Learning about Organization from an Orderly Universe.* San Francisco: Berrett-Koehler Publishers, 1992.

Wheeler, David L. "Evolutionary Economics." *Chronicle of Higher Education,* July 5, 1996, A8, A12.

Whitehead, Alfred North. *Process and Reality: An Essay in Cosmology.* New York: Free Press, 1969,

Whiting, Cecile. *A Taste for Pop: Pop Art, Gender, and Consumer Culture.* Cambridge: Cambridge University Press, 1997.

Wiener, Norbert. *Cybernetics; or, Control and Communication in the Animal and Machine.* New York: John Wiley, 1948.

———. *The Human Use of Human Beings.* Boston: Houghton Mifflin, 1950.

Wilden, Anthony. *The Rules Are No Game: The Strategy of Communication.* New York: Routledge, 1987.

———. *System and Structure.* New York: Routledge, 1980.

Wills, David. *Prosthesis.* Stanford: Stanford University Press, 1995.

Wilson, Alexander. *The Culture of Nature: North American Landscape from Disney to the Exxon Valdez.* Cambridge, Mass.: Blackwell, 1992.

Wilson, Edward O. *Consilience: The Unity of Knowledge.* New York: Knopf, 1998.

———. *Sociobiology: The New Synthesis.* Cambridge, Mass.: Harvard University Press, 1975.

Wolpert, David H., ed. *The Mathematics of Generalization.* Santa Fe Institute Studies, vol. 20. Reading, Mass.: Addison-Wesley, 1995.

Wood, John, ed. *The Virtual Embodied: Presence/Practice/Technology.* New York: Routledge, 1998.

Woodcock, Alexander, and Monte Davis. *Catastrophe Theory.* New York: Avon Books, 1978.

Wright, Robert. "Can Machines Think?" *Time,* March 25, 1996, 50–58.

Wyatt, Edward. "Investors See Room for Profit in the Demand for Education." *New York Times,* November 4, 1999.

Yates, F. Eugene. *Self-Organizing Systems: The Emergence of Order.* New York: Plenum Press, 1988.

Young, Jeffrey R. "Using Computer Models to Study the Complexities of Human Society." *Chronicle of Higher Education,* July 24, 1998, A17, A19–20.

Zha, Jianying. *China Pop: How Soap Operas, Tabloids, and Bestsellers Are Transforming a Culture.* New York: New Press, 1995.

Zimmerman, Michael E. *Heidegger's Confrontation with Modernity: Technology, Politics, Art.* Bloomington: Indiana University Press, 1990.

Zizek, Slavoj. *The Plague of Fantasies.* New York: Verso, 1997.

Zurek, Wojciech H., ed. *Complexity, Entropy and the Physics of Information.* Santa Fe Institute Studies, vol. 8. Reading, Mass.: Addison-Wesley, 1990.

index

abstract expressionism, 131

academic freedom, 242, 255, 265

Academic Keywords (Nelson and Watt), 250

adaptation: of ant colony, 163; in Darwinian evolution, 178, 179, 180, 181; in swarm behavior, 156. *See also* complex adaptive systems

aleatory aggression, 136, 137

aleatory phenomena: in complex adaptive systems, 169, 291n. 56; creative appropriation of, 137; in emergent systems, 155; Hegel's system and, 93; of postmodern architecture, 37; in quantum physics, 116; undecidability and, 97. *See also* chance

algorithmic information theory, 138–39, 166–67, 171, 184

allelomimesis, 153

Allen, Herbert A., 10, 259–60

Altarity (Taylor), 12

alterity, 48

ant colonies, 152, 153, 162–65. *See also* Aunt Hillary; swarms

Ant Fugue (Escher), 164

Apollinaire, Guillaume, 74

Arcadia (Stoppard), 285n. 2

The Archaeology of Knowledge (Foucault), 55

architecture, 14; grids in, 20–23, 25–28, 32–34, 35, 41; network culture and, 14, 20, 40, 46; networks in, 20–23, 41; religion and, 7. *See also* Gehry, Frank; Le Corbusier; Mies van der Rohe; Venturi, Robert

Archive Fever (Derrida), 63–64

Aristotle: Cuvier's morphology and, 175; Darwinian evolution and, 177, 179; on education, 292–93n. 17

art: avant-garde, 7–8, 31, 32–34, 245, 254; as business, 9, 45, 243–45, 294–95n. 40; Derrida's ambivalence toward, 7; high vs. low, 243–45; Kant on, 31, 74, 84, 131, 243, 253, 254; modernism in, 31–32, 74; religion and, 7, 9, 13, 245; self-reflexivity in, 74, 75, 125–26; technology appropriated by, 8–9, 294–95n. 40; Wilson on, 60. *See also* Close, Chuck; Magritte, René; Tansey, Mark; Warhol, Andy

artificial life, 145–46, 152–53, 285n. 3

assembly line, 28–29, 30; Lévi-Strauss's structures and, 52; modern university and, 240–41, 257

At Home in the Universe (Kauffman), 24–25, 272n. 5

Atlan, Henri, 16, 135–37, 139, 167, 169

AT&T Building (Johnson), 38, 40

attractors, 147, 185, 186, 191, 283n. 36

Augustine, 200–202, 207, 231

Aunt Hillary, 163–64, 169, 181, 194, 205, 220, 226

Autobiography (Darwin), 178

autocatalytic sets, 88, 186, 187

automobile, 4, 28, 29, 36, 44

Autopoiesis and Cognition (Maturana and Varela), 89–90

autopoietic systems, 15, 88, 89–93; constructivism and, 277n. 38; symmetry-breaking and, 150; undecidability and, 97

avant-garde, 7–8, 31, 32–34, 245, 254

"The Avant-Garde and Kitsch" (Greenberg), 244

Babbage, Charles, 103–4
Bach, Johann Sebastian, 157–59, 161
Bak, Per, 148, 149, 168, 190, 259, 283n. 43
Barthes, Roland, 131
Bataille, Henry, 280n. 33
Bateson, Gregory, 105, 204
Baudrillard, Jean, 15, 55, 65–72; Derrida and, 65–66, 94, 101; Foucault and, 65–66, 75; Hofstadter's recursive images and, 78; paralogic and, 101; Tansey's study of, 12; on terrorism of the code, 69, 82; Warhol and, 8. *See also* signs; simulacra; simulation
Bauhaus, 8, 33–34, 294n. 40
Belousov-Zhabotinsky reaction, 283n. 38
Benjamin, Walter, 131
Bennett, Charles, 118, 280–81n. 44
Berlin, University of, 17, 240, 245–46, 255
Berlin Wall, 5, 14, 20, 49, 235
Beyond the Pleasure Principle (Freud), 63
bifurcation point, 150; at instant of decision, 62; of morphological types, 193
Big Self-Portrait (Close), 127, 128, 129, 131
Bildung, 246, 247, 292n. 14
binary oppositions: of assembly line, 29; of Baudrillard's arguments, 71–72; of capitalism, 30; Close's paintings and, 129; of Cold War, 23, 49–50, 52; of form vs. matter, 105–6, 223–24, 230; in higher education, 242–43, 269; of information and noise, 100–101; of Le Corbusier's argument, 26–27, 29, 273n. 10; in Lévi-Strauss's structures, 52; of madness and reason, 62; of mind/body dualism, 29, 106, 223–24, 226; of screening, 200; transformed by network culture, 72, 230; undecidability and, 94, 96–97; in Venturi's critique of modernism, 36, 273n. 21; of Western metaphysics, 104–5; Zeno of Elea and, 160. *See also* dialectical logic
biology. *See* life
bits, 109, 129
Boltzmann, Ludwig, 114–15
Bonaventura, 200
Boole, Charles, 185
Boolean networks, 147–48, 185–86, 190–91
Born, Max, 116
Bourdieu, Pierre, 244–45
brain: coadaptation with culture, 218, 226, 227–30; complex dynamics of, 204–5, 226–27; new

cells in adult, 228; in vat, 222, 226. *See also* mind
Bricomont, Jean, 59
Brown, Denise Scott, 34
Burroughs, William, 288–89n. 26
business: art and, 9, 45, 243–45, 294–95n. 40; complexity theory and, 272n. 5; education and, 17, 234–39, 260–61, 291–92n. 3; Kant's vision of university and, 242, 247, 258. *See also* capitalism; economy
Byron, Lord, 104

calculation, mechanical: Hegel on, 81–82; Heidegger on, 82–83, 84. *See also* computers
Calvino, Italo, 99, 111, 114, 151, 214
Camus, Albert, 51
canon: musical, 158–59; Western, 48
capitalism: architecture and, 34; art and, 8–9; Baudrillard on, 68; Cold War and, 5, 23, 49–50; Darwinism and, 16, 179–80; in Moravec's utopia, 223; new forms of, 4, 20; rationality and, 31; social criticism of 1960s and, 50–51; as total way of life, 29–30. *See also* business; economy; industrial revolution; industrial society
Carnap, Rudolf, 95
Cartesian dualism, 223. *See also* dualism
Cartesian rationalism: Derrida on, 62; grids and, 30
Castells, Manuel, 272n. 2
Casti, John, 16, 24, 139, 149
catastrophe theory, 13–14, 40, 53
catastrophists, 174
Catia software, 14, 41
causality: in assembly line, 29; in autopoietic systems, 93; chance and, 149; in chaotic systems, 13, 24; in complex systems, 143, 218; in Darwinism, 182; in hive, 284n. 60; in Lévi-Strauss's system, 52; in Newtonian physics, 79, 115; in nonlinear systems, 165; in organism vs. mechanism, 85; self-organized criticality and, 148–49; in teleonomic processes, 169. *See also* determinism
cellular automata, 136, 143–46, 147; Close's paintings and, 16; swarms modeled by, 152, 153–54
Center for Technology in the Arts and Humanities, 258
central dogma of molecular biology, 228–29

and, 97; Venturi on, 37. *See also* far-from-equi-
librium systems
Escape Velocity (Dery), 105
Escher, M. C., 159, 161
essentialism: constructivists' critique of, 58, 60;
    Darwin's suspicion of, 176, 177, 179; genetic,
    225
Estes, Richard, 127
Euclidean geometry, 14, 26, 40, 41
evolution: of ant colony, 163; of complex adaptive
    systems, 143, 156, 167, 171–72, 184, 206; toward
    complexity, 137, 175–76, 177, 179, 193, 287n.
    56; constraints acting in, 224–25; cultural,
    216–19, 220, 224–25, 228, 229, 289–90n. 30; of
    far-from-equilibrium systems, 150; Kurzweil's
    technological fantasy of, 221, 222; Lamarck's
    theories, 86, 175–76, 177, 179, 193, 229; of lan-
    guages, 215; morphology and, 176, 182–84, 192,
    193; self-organization in, 172, 184, 186, 190–91,
    192–93; social and political, 194, 287n. 60;
    structuralism and, 54, 183–84; as teleonomic
    process, 169, 193, 194; as theological issue,
    172–74, 175, 176, 177, 180, 194. *See also* coevo-
    lution; Darwin, Charles; natural selection
evolutionary psychology, 289n. 30
exformation, 203

Fable of the Bees (Mandeville), 179
The Fair Captive (Magritte), 76
far-from-equilibrium systems, 14, 16, 143, 171; be-
    tween academic disciplines, 265; cultural, 57,
    214; dissipative structures in, 120, 121, 184;
    Gehry's architecture and, 46; vs. Newtonian
    systems, 80; symmetry-breaking in, 150; think-
    ing as, 198
fashion, 70–71
feedback and feed-forward loops: in autopoietic
    systems, 93; in complex adaptive systems, 165,
    168, 206, 207; in connectionist networks,
    154–55; cybernetics and, 15, 107; in far-from-
    equilibrium systems, 143; in gene regulation,
    185; information revolution and, 106; in me-
    chanical systems, 107; of nature and culture,
    225
Fermat's Last Theorem, 160–61, 284–85n. 2
Fichte, Johann Gottlieb, 240
fitness, 179, 190–91, 197, 207
Flaubert, Gustave, 245

Force of Habit (Magritte), 74
"Force of Law" (Derrida), 274n. 21
Ford, Henry, 28, 29, 30
"form is function," 294n. 40
form vs. matter, 105–6, 223–24, 230. *See also* Pla-
    tonic forms
Foucault, Michel, 15, 55–58, 72; Baudrillard and,
    65–66, 75; on "death of the author," 131; Der-
    rida's critique of, 61–63, 64; Jacob's influence
    on, 277n. 30; on Magritte, 73, 74–75; scien-
    tists' attacks on, 58–59, 60; Tansey's study
    of, 12
Fourier, Jean-Joseph, 112, 113
fractal geometry, 40–41; autopoietic systems
    and, 92; myths and, 212; of recursive images,
    77
The Fractal Geometry of Nature (Mandelbrot),
    40–41
Freud, Sigmund, 63–64
"Freud and the Scene of Writing" (Derrida), 63
Friedman, Thomas L., 23
fugue, 158, 159, 161, 194
"Fumées" (Apollinaire), 74

Game of Life, 144–45, 147, 152
game theory, 141
Gardner, James, 229
Gehry, Frank, 14, 20, 22, 40–45, 47, 269; biologi-
    cal morphology and, 193; Close compared to,
    132; Condé Nast building, 44–46, 257; emer-
    gence and, 147
Gell-Mann, Murray: on complex adaptive systems,
    166–68, 169, 171, 189–90, 205–7, 282n. 17; on
    cultural evolution, 216; on ecological commu-
    nities, 190; on environment of genotype,
    189–90; Kauffman and, 184, 287n. 56; on the-
    ories, 206, 211, 288n. 18
GEN. *See* Global Education Network
gender: capitalism and, 30; Le Corbusier's mod-
    ernism and, 273n. 10
General System Theory (von Bertalanffy), 140
genes: as complex adaptive system, 188; cultural
    evolution and, 216–17, 225, 228, 289–90n. 30;
    environment and, 189–90; evolution and,
    180–81, 182; holism and, 187; memes and, 217,
    218, 228–29; networks of, 25, 185, 188
Genes, Mind, and Culture (Lumsden and Wilson),
    216

hypertext: author's experiments with, 10, 11; future curriculum and, 234, 260, 262; techno-fantasies and, 219, 220

The Idea of the University (Newman), 247
identity: of autopoietic system, 92; parasitic nature of, 196. *See also* self
If on a winter's night a traveler (Calvino), 111, 151
images: Baudrillard on, 68–69, 71; economic value of, 9; as memes, 218; screening of, 200; self-reflexive, 77–78, 88, 126; viral nature of, 288–89n. 26; and words in Surrealist art, 73, 74, 75. *See also* signs
Imagologies: Media Culture (Taylor), 10
industrial revolution: artists and, 7, 84, 243–44; calculating machines and, 104; Newtonian mechanics and, 79; nineteenth century reaction against, 84; thermodynamics and, 135
industrial society: alienation in, 80–81; Bauhaus and, 33–34; Gehry's architecture and, 41; Le Corbusier's argument and, 27, 28, 29; Lévi-Strauss's structures and, 52; Mies' architecture and, 32, 34; scientific management, 28–29; transition to network culture, 4, 14, 20, 71–72, 100; Venturi's critique of modernism and, 36, 40. *See also* capitalism
information: algorithmic complexity and, 138–39, 166–67, 184; in ant colony, 163; Bateson's definition of, 105; Baudrillard's signs and, 69; bifurcation point and, 150; in complex adaptive systems, 166–67, 168; Derrida on technology of, 253; economic value of, 9; entropy and, 111–12, 114, 118, 120–21, 280–81n. 44, 281n. 50; etymology of, 105; excess of, 100, 203; expanded notion of, 4–5; in genome, 229; identity of the self and, 196; language and, 215–16; life and, 135, 145; material conditions of, 105–6; memes and, 217–19, 228–29; in network culture, 100, 230–31; noise in interplay with, 121–23, 134, 135–37, 163, 203; preserved by isomorphism, 159; probability and, 109–10, 118–19; Santa Fe conference on, 99–100; schemata and, 206; screening of, 207–11, 212, 228, 231; selective destruction of, 203–4; in speech vs. writing, 104–5. *See also* noise
Information, Randomness, and Incompleteness (Chaitin), 139
information economy, 44, 66

information revolution, 20, 100, 103–6
information technology. *See* communication technology; technology
information theory, 15; algorithmic, 138–39, 166–67, 171, 184; Atlan's use of, 136, 137; paralogic and, 102; self-organizing systems and, 88; Shannon's contribution, 107–9, 111–12, 120–21, 136, 281n. 50; systems theory and, 141; thermodynamics and, 135
Intellectual Impostures (Sokal and Bricomont), 59
intellectual property rights, 263–64, 295n. 45
Internet: education and, 235, 236–37, 239, 257; globalization and, 23. *See also* World Wide Web
In the Wake of Chaos (Kellert), 13
irreversible processes, 113–14, 116–17; in data processing, 118, 280–81n. 44
isomorphism, 159; of complex adaptive networks, 165; of mind and brain, 206, 226
Izenour, Steven, 34

Jacob, François, 85–86, 87, 90, 185, 193, 277n. 30
Jantsch, Erich, 150
Jefferson, Thomas, 30, 249
Jewish tradition, Derrida and, 7
Johnson, Philip, 34, 38, 40
Johnston, John, 288–89n. 26

Kant, Immanuel: apriori categories of, 89, 200, 204, 278n. 52; on art, 31, 74, 84, 131, 243, 253, 254; on organisms, 84–85, 87–88, 90, 173, 182, 186–87
Kant's vision of university, 17, 240–43, 245, 292n. 8; Derrida and, 251, 254, 255; vs. Newman's vision, 247–48; resistance to e-Ed and, 258; University of Berlin and, 240, 246, 255
Kauffman, Stuart, 16, 24–25, 184–88, 190–92, 272n. 5; on autopoietic systems, 88; on Boolean networks, 147–48, 185–86, 190–91; on coevolution, 287n. 50; collaboration with Goodwin, 191–92; on emergence, 146, 147–48; on evolution, 171–72, 182, 184, 185, 190, 191–92; on phase transitions, 25, 146, 149, 191; philosophical beliefs, 193–94
Kellert, Stephen, 13, 283n. 36
Kelly, Kevin, 284n. 60
Kelso, J. A. Scott, 151, 204, 205, 226
Keynes, John Maynard, 179

Maturana, Humberto, 15, 88–91, 277n. 38

Maxwell, James Clerk, 114, 115, 117

Maxwell's demon, 117, 118, 280–81n. 44

McClelland, Charles, 246

McCulloch, Warren, 143

meaning: Baudrillard on, 66–68; context dependence of, 204–5; creation of, 214; Foucault on, 75; in information theory, 108–9; from randomness, 136; from screened information, 203, 204, 210–11, 212; self-reference and, 75–76. *See also* reference

media: Baudrillard on, 68–69, 71; new, 20, 66, 258. *See also* multimedia

memes, 217–19, 226, 228–29

Memoirs of the Blind (Derrida), 7

memory: Augustine on, 201, 202; in complex adaptive systems, 167, 168, 204; of computer, 103; consciousness and, 207; Derrida on, 63–64, 275–76n. 47; of Maxwell's demon, 118, 280–81n. 44; Plato on, 200–201; writing process and, 197, 198

Mendel, Gregor, 180

Menger, Kurt, 95

Merrill Lynch, 235–36

Mies van der Rohe, 14, 20, 22, 32–34, 38; Baudrillard and, 275n. 43; Gehry compared to, 42, 257; Lévi-Strauss and, 51; Venturi's critique of, 35, 36, 273n. 19

Milken, Michael, 233, 235

Miller, Arthur, 264

Miller, J. Hillis, 97–98

Millonas, Mark M., 152–53, 154, 155

mind: coadaptation with brain, 226; coevolution with brain, 229–30; complex dynamics of, 204–5, 206; cultural evolution of, 217; disembodied, 105–6, 221–24; distributed quality of, 230–31; dualism and, 223–24, 226, 230; emerging from brain, 163; language and, 215; mechanistic philosophy and, 141; memes and, 228; self and, 231

modernism: Gehry's alternative to, 40, 41; Greenberg on, 74; grids in, 25–26, 27–28, 31–32, 35; industrial capitalism and, 31; Tansey's study of, 12; technology and, 4, 294n. 40; Venturi's critique of, 14, 35–40, 273n. 21. *See also* Le Corbusier; Mies van der Rohe

Modern Times (Chaplin), 28

moment of complexity, 3–4, 5; critical emergency

at, 47; at edge of chaos, 23–24, 25; enlightenment at, 98; Gehry's architecture and, 46; global culture and, 194; from grids to networks at, 20, 23, 35; higher education in, 233–34, 257, 270; the I as, 232; information revolution and, 100, 106; living organisms and, 137; moment of decision and, 149, 151; self-organizing systems and, 24, 78. *See also* complexity

Mondrian, Piet, 31

Monod, Jacques, 177, 185, 194

Moravec, Hans, 105, 221, 222–24

Morely, Michael, 128

morphology: of Cuvier, 174–75; emergence of new types, 193; Goodwin on, 182–84, 192; Lamarck and, 176; Waddington and, 191

multimedia, 10–11, 65, 257, 258, 262

music: Hofstadter on, 157–59, 161; Lévi-Strauss on, 274n. 8

Musical Offering (Bach), 158, 159

myth, 210–11, 212–14; as meme, 218; Wagner's analysis of, 274n. 8

Nasdaq, 45, 257, 260

natural selection: in brain evolution, 218, 230; Darwin on, 174, 177, 178, 179, 180, 181; Gould on, 181; for greater complexity, 287n. 56; Kauffman on, 185, 190, 191–92, 193, 194, 287n. 50. *See also* evolution; selection

Natural Theology (Paley), 172–73

negentropy, 101, 119–20, 121, 281n. 50; in living systems, 119, 123, 137, 184–85. *See also* entropy

Nelson, Cary, 250, 255, 265–66

Nelson, Ted, 219

neo-Darwinism, 172, 182, 184, 188

network culture, 5; architecture and, 14, 20, 40, 46; complex systems and, 155; Derrida's critique of, 65; emergence of, 14, 20, 25, 49, 72, 100; experiments in new media and, 11; globalization and, 23, 194, 202, 214; grids as obsolete in, 37; Hegel's system and, 48–49; higher education and, 233–34, 256, 257 (*see also* Global Education Network); hope for creative change in, 270; information in, 100, 230–31; recursive images and, 78; religion and, 6, 9; screens in, 200, 214; structuralism and, 55; subjectivity in, 200, 202, 231; Times Square and, 44

networks: autopoietic, 89–90, 91; Boolean, 147–48,

and, 138; Kauffman's vision and, 193–94; in network culture, 6, 9; of Newton, 138; Platonic forms and, 200; post-structuralism and, 6–7; rational defenses of, 172–74; as universal cultural dimension, 6, 7. *See also* evolution, as theological issue; God

reproduction vs. production, 4, 72, 100; Baudrillard on, 66, 68, 70, 71. *See also* mass production

reversible systems, 79, 80, 113, 116, 134–35

Richter, Gerhard, 131

River out of Eden (Dawkins), 217–18

Robot (Moravec), 105

robotic intelligence, 223

Rock, Michael, 11

Roosevelt, Theodore, 248

Roszak, Theodore, 105

Rudolph, Frederick, 248

Russell, Bertrand, 95, 96

Russian Constructivism, 8, 294n. 40

Saarinen, Esa, 9

Santa Fe Institute, 13, 99, 146, 152; Kauffman and, 146, 171–72; Tansey influenced by, 12

Sartre, Jean-Paul, 51

Saussure, Ferdinand de, 67

The Savage Mind (Lévi-Strauss), 51

schemata: of complex adaptive systems, 166–68, 169, 189–90, 205–7; in cultural change, 214; for society, 216; in thinking process, 198

Schiffman, Zachary, 104, 105

Schiller, Friedrich, 80

Schleiermacher, Friedrich, 240

Schumpeter, Joseph, 28

Science of Logic (Hegel), 81, 82, 92, 283n. 40

scientific theories, 206, 211, 225, 288n. 18

screening: in brain development, 227–28; defined, 199–200, 288n. 3; of experience, 207–11; Gehry's architecture and, 41; in higher education, 260, 267, 269; of information, 207–11, 212, 228, 231; knowledge and, 118, 200, 203, 204, 208, 210; in learning, 228; of reality, 78; subjectivity and, 200, 205; undecidability and, 98, 200; walls and, 20, 199, 214, 231, 235; in writing process, 196, 197

Seagram Building (Mies), 22, 32, 34, 35, 38, 40

second law of thermodynamics, 93, 113; Maxwell's demon and, 117, 118, 280–81n. 44; organism and, 135. *See also* entropy

selection: in brain development, 227–28; in evolution of languages, 215; among schemata, 207. *See also* natural selection

self, 196, 202, 231. *See also* subjectivity

self-assembly, 76

self-consciousness: of Augustine, 200–202; Descartes on, 115; Hegel on, 278n. 51; knowledge and, 200–202, 208, 209, 210, 212. *See also* consciousness

selfish gene theory, 180–81, 229, 286n. 30

self-organized criticality, 148–49, 165, 283n. 43; in coadapting systems, 190; memory and, 168

self-organizing systems, 24, 87–88, 142, 143, 156; autopoietic, 88, 90; Bak's theory of, 148–49, 165, 168, 190, 283n. 43; Boolean networks as, 185–86; brain as, 204, 226; of chemical reactions, 187, 283n. 38; Close's paintings as, 131; coadaptive, 190–91; conditions for, 205; in evolution, 172, 184, 186, 190–91, 192–93; Kant on, 85, 87–88; languages as, 215; of network culture, 78; noise and, 136–37; organisms as, 182; origin of life in, 184, 185, 186, 187, 277n. 35; scientific inquiry as, 92; thinking as, 198, 205. *See also* complex adaptive systems; emergent phenomena

Self-Portrait (Close), 125–26, 127

self-reference: in autopoietic systems, 91, 92; Gödel's theorem and, 95–96; in Magritte's art, 74, 75–76; of modernist art, 31. *See also* reference

self-reflexivity: in art, 74, 75, 125–26; of autopoietic systems, 90, 92; consciousness and, 89, 210; cybernetics and, 88, 277n. 36; of information revolution, 106; in literature, 151; observer of system and, 88, 89, 115–16; of open systems, 93–94; of post-structuralists, 15; of strange loops, 75, 78, 200, 209–10; undecidability and, 97, 98. *See also* recursive systems

Sellars, Susan, 11

Serra, Richard, 127

Serres, Michel, 15, 101–3, 119, 121–22, 135; ant colony and, 163; Close's paintings and, 134

Seurat, Georges, 129

Shannon, Claude, 15, 107–9, 111–12; Atlan influenced by, 136; Wiener and, 120–21, 281n. 50

signs: Baudrillard on, 66–67, 70, 71, 101; Close's self-portrait as, 126; Derrida on, 105; Foucault on, 75; in Lévy's utopia, 220–21; modernist architecture and, 36, 38; self-reflexive, 75, 78. *See also* images; simulacra; simulation

Simon, Herbert A., 141

simplicity: complexity and, 54, 137–38, 142, 199, 221; utopian visions of, 221, 223

simulacra: Baudrillard on, 69, 71, 78; Derrida on, 94, 96; Foucault on, 75; network culture and, 12, 72. *See also* signs; simulation

simulation: defined, 67; in fashion, 70; impasse of, 71; theory of, 15, 55, 72. *See also* Baudrillard, Jean; signs; simulacra

singularities, in catastrophe theory, 13, 53, 272n. 12

skepticism, 58, 59

Smith, Adam, 16, 178, 179, 180, 287n. 50

Smolin, Lee, 84

Snow, C. P., 58

social constructivism. *See* constructivism

Social Systems (Luhmann), 91

social systems, autopoietic, 90–91. *See also* cultural evolution

sociobiology, 216–17, 289n. 30

Sokal, Alan, 59

Solomonoff, Ray, 138

Specters of Marx (Derrida), 101, 275–76n. 47

Spinoza, Baruch, 134

stability, longing for, 3

stable systems: autopoiesis and, 92–93; of Cold War, 5, 23, 49; complexity and, 78, 97, 146; Hegel's dialectic and, 92–93; of Lévi-Strauss, 52, 53, 54; Newtonian, 78–80, 112; part/whole relationship in, 81; quantum theory and, 116; swarms, 155–56

state cycles, 185, 186

statistical mechanics, 100, 114–15, 172

Stoppard, Tom, 19, 40, 285n. 2

Storr, Robert, 125, 131

strange loops: of autocatalytic sets, 186, 187; in Close's art, 125–26; of complex adaptive systems, 169; in Gehry's architecture, 46; in Gödel's theorem, 96; Hofstadter on, 75–77, 186; in network culture, 72, 78; paralogic and, 98, 101, 102; self-consciousness and, 200, 210; in writing process, 198

structuralism, 51–55; Baudrillard and, 55, 67, 68, 70, 71; complexity and, 54; Derrida's critique of, 55, 61–63, 97; in developmental biology, 183, 184; Foucault's critique of, 55–58; as Hegelianism, 61, 62; organisms and, 86;

Tansey's study of, 12; undecidability and, 97. *See also* Lévi-Strauss, Claude; post-structuralism

Structural Stability and Morphogenesis (Thom), 13, 53

subjectivity, 207–10; in network culture, 200, 202, 231; nodular, 16, 231, 232; screening and, 200, 205; techno-fantasies and, 220. *See also* self

superstructure and infrastructure, 66, 68

Suprematism, 294n. 40

Surrealism, 73, 74

swarms, 152–154, 155–156, 284n. 52, 60. *See also* ant colonies

The Symbolic Species (Deacon), 229–30

symbols, 204, 205, 210–12, 214; of language, 215; as memes, 218

symmetry-breaking, 150

systems: organism as, 134–35; social, 91. *See also* autopoietic systems; closed systems; complex adaptive systems; far-from-equilibrium systems; open systems; self-organizing systems

systems theory, 140–41

Szilard, Leo, 280–81n. 44

Tanguary, Denise Marie, 239

Tansey, Mark, 12–13

Tappan, Henry, 248

Taylor, Frederick Winslow, 28, 30

Techgnosis (Davis), 105

technological unconscious, 16, 203, 231

technology: art transformed by, 8–9, 294–95n. 40; author's experiments with, 9–12, 257–58 (*see also* Global Education Network); Baudrillard on, 66, 68, 69–70; complexity associated with, 138; cultural change and, 4, 5, 19–20, 218–19; Derrida on, 63, 64–65, 231, 251–53; education and, 233–34 (*see also* e-Ed); globalization and, 23, 194, 202; Hegel on logic of machine, 81–82; Heidegger on, 82–84, 251, 252, 253, 254, 277n. 22; information revolution and, 20, 106–7; utopian fantasies and, 219–24. *See also* communication technology

technoscience, Derrida on, 64, 251–53

Teilhard de Chardin, Pierre, 290n. 38

teleology: complex adaptive systems and, 168–69, 193; existence of God and, 172, 173; Hegel on, 87; Kant on, 85, 87, 173, 186–87, 243, 246, 251

teleonomic processes, 169, 193, 194